Interactive Student Edition

Reveal **MATH**® Accelerated

Volume 1

Mc
Graw
Hill

Cover: (tl)shuige/Moment/Getty Images, (tr)Kletr/Shutterstock.com, (bl)kuritafsheen/RooM/Getty Images, (br) Elles Rijsdijk/EyeEm/Getty Images

my.mheducation.com

Send all inquiries to:
McGraw-Hill Education
STEM Learning Solutions Center
8787 Orion Place
Columbus, OH 43240

ISBN: 978-0-07-667377-3
MHID: 0-07-667377-4

Reveal Math Accelerated
Interactive Student Edition, Volume 1

Printed in the United States of America.

9 10 LMN 28 27 26 25 24 23 22

Contents in Brief

Reveal Math® Accelerated
Makes Math Meaningful...

Interactive Student Edition

Example 2 Find Volume of Cones

A cone-shaped paper cup is filled with water. The height of the cup is 9 centimeters and the diameter is 8 centimeters.

What is the volume of the paper cup? Round to the nearest tenth.

$V = \frac{1}{3}\pi r^2 h$ Volume of a cone

$V = \frac{1}{3}\pi$ ☐   Replace r and h.

$V = $ ☐ π Multiply.

$V = $ ☐   Use a calculator. Round to the nearest tenth.

So, the volume of the paper cup is about 150.8 cubic centimeters.

Check

Find the volume of a cone with a radius of 1.5 inches and a height of 9 inches. Round to the nearest tenth.

Go Online You can complete an Extra Example online.

Pause and Reflect

What part(s) of finding the volume of cones did you feel most confident? Why?

Think About It!
What dimensions of the cone are needed to solve the problem?

Talk About It!
Suppose a smaller conical paper cup has a diameter of 5 centimeters, and a height of 8 centimeters. Compare and contrast the volumes of the cones using 3.14 for π versus using the π button on a calculator, and rounding the volume to the nearest tenth.

Apply Popcorn

A family-owned movie theater offers popcorn in the sizes shown. Their cost for the popcorn is $0.09 per cubic inch. If each container is filled to the top, what is the difference between the costs of the popcorn in the two containers?

1 What is the task?
Make sure you understand exactly what question to answer or problem to solve. You may want to read the problem three times. Discuss these questions with a partner.

First Time Describe the context of the problem, in your own words.
Second Time What mathematics do you see in the problem?
Third Time What are you wondering about?

2 How can you approach the task? What strategies can you use?

3 What is your solution?
Use your strategy to solve the problem.

4 How can you show your solution is reasonable?
Write About It! Write an argument that can be used to defend your solution.

Talk About It!
What do you notice about the relationship between the cost of a cylindrical container and the cost of a conical container?

818 Module 12 · Area, Surface Area, and Volume

Student Digital Center

Apply
Popcorn

A family-owned movie theater offers popcorn in the sizes shown. Their cost for the popcorn is $0.09 per cubic inch. If each container is filled to the top, what is the difference between the costs of the popcorn in the two containers?

Learning on the Go!

The flexible approach of *Reveal Math Accelerated* can work for you using digital only or digital and your *Interactive Student Edition* together.

...to Reveal YOUR Full Potential!

Reveal Math® Accelerated Brings Math to Life in Every Lesson

Reveal Math Accelerated is a blended print and digital program that supports access on the go. You'll find the *Interactive Student Edition* aligns to the Student Digital Center, so you can record your digital observations in class and reference your notes later, or access just the digital center, or a combination of both! The Student Digital Center provides access to the interactive lessons, interactive content, animations, videos, and technology-enhanced practice questions.

Write down your username and password here

Username: _____

Password: _____

Go Online!
my.mheducation.com

Web Sketchpad® Powered by The Geometer's Sketchpad®- Dynamic, exploratory, visual activities embedded at point of use within the lesson.

Animations and Videos – Learn by seeing mathematics in action.

Interactive Tools – Get involved in the content by dragging and dropping, selecting, and completing tables.

Personal Tutors – See and hear a teacher explain how to solve problems.

eTools – Math tools are available to help you solve problems and develop concepts.

Module 1
Proportional Relationships

e Essential Question
What does it mean for two quantities to be in a proportional relationship?

Module 2
Solve Percent Problems

e Essential Question
How can percent describe the change of a quantity?

TABLE OF CONTENTS

Module 3
Operations with Integers and Rational Numbers

e Essential Question
How are operations with rational numbers related to operations with integers?

Module 4
Exponents and Scientific Notation

e **Essential Question**
Why are exponents useful when working with very large or very small numbers?

Module 5
Real Numbers

e Essential Question
Why do we classify numbers?

Module 6
Algebraic Expressions

e Essential Question
Why is it beneficial to rewrite expressions in different forms?

Module 7
Equations and Inequalities

ⓔ Essential Question
How can equations be used to solve everyday problems?

Module 8

Linear Relationships and Slope

e Essential Question

How are linear relationships related to proportional relationships?

Copyright © McGraw-Hill Education

TABLE OF CONTENTS

Module 9
Probability

ⓔ Essential Question

How can probability be used to predict future events?

Copyright © McGraw-Hill Education

Module 10
Sampling and Statistics

e Essential Question
How can you use a sample to gain information about a population?

TABLE OF CONTENTS

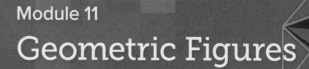

Module 11
Geometric Figures

e Essential Question
How does geometry help to describe objects?

Module 12

Area, Surface Area, and Volume

e Essential Question
How can we measure objects to solve problems?

Module 13

Transformations, Congruence, and Similarity

℮ Essential Question

What does it mean to perform a transformation on a figure?

Module 1

Proportional Relationships

@ Essential Question

What does it mean for two quantities to be in a proportional relationship?

What Will You Learn?

Place a checkmark (✓) in each row that corresponds with how much you already know about each topic **before** starting this module.

KEY — I don't know. — I've heard of it. — I know it!	Before			After		
	⬛	◈	★	⬛	◈	★
computing unit rates involving ratios of fractions						
determining whether a relationship is proportional by looking at a table of values						
determining whether a relationship is proportional by looking at a graph						
finding and interpreting the constant of proportionality						
interpreting the points (0, 0) and (1, r) on the graph of a proportional relationship						
representing proportional relationships with equations						
solving problems involving proportional relationships						

📖 Foldables Cut out the Foldable and tape it to the Module Review at the end of the module. You can use the Foldable throughout the module as you learn about proportional relationships.

What Vocabulary Will You Learn?

Check the box next to each vocabulary term that you may already know.

☐ constant of proportionality   ☐ proportional

☐ nonproportional   ☐ proportional relationship

☐ proportion   ☐ unit rate

Are You Ready?

Complete the Quick Review to see if you are ready to start this module.
Then complete the Quick Check.

Quick Review

Example 1
Write ratios.

Write the ratio of wins to losses.

Mavericks	
Wins	10
Losses	12
Ties	8

wins : losses
   10 : 12

The ratio of wins to losses is 10 : 12.

Example 2
Determine if ratios are equivalent.

Determine whether the ratios 250 miles in 4 hours and 500 miles in 8 hours are equivalent.

250 miles : 4 hours → $\dfrac{250}{4} = \dfrac{125}{2}$ or $62\dfrac{1}{2}$

$\div 2$ $\div 2$

500 miles : 8 hours → $\dfrac{500}{8} = \dfrac{125}{2}$ or $62\dfrac{1}{2}$

$\div 4$ $\div 4$

The ratios are equivalent because the ratio $\dfrac{125}{2}$ is maintained.

Quick Check

1. Refer to the table in Example 1. Write the ratio of wins to total games.

2. Determine whether the ratios 20 nails for every 5 shingles and 12 nails for every 3 shingles are equivalent.

How Did You Do?

Which exercises did you answer correctly in the Quick Check?
Shade those exercise numbers at the right.

 ① ②

Unit Rates Involving Ratios of Fractions

I Can... find unit rates when one or both quantities are fractions.

What Vocabulary Will You Learn?

unit rate

Explore Find Unit Rates with Fractions

Online Activity You will use bar diagrams to explore how to find a unit rate when one or both quantities of a given rate are fractions.

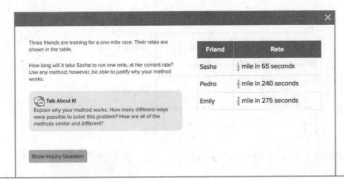

Three friends are training for a one-mile race. Their rates are shown in the table.

How long will it take Sasha to run one mile, at her current rate? Use any method; however, be able to justify why your method works.

Talk About It!
Explain why your method works. How many different ways were possible to solve this problem? How are all of the methods similar and different?

Show Inquiry Question

Friend	Rate
Sasha	$\frac{1}{4}$ mile in 65 seconds
Pedro	$\frac{2}{3}$ mile in 240 seconds
Emily	$\frac{5}{6}$ mile in 275 seconds

Learn Unit Rates Involving Ratios of Fractions

A recipe to make a diluted cleaning solution calls for 4 gallons of water mixed with $\frac{1}{3}$ cup of cleaner. The ratio of gallons of water to cups of cleaner is 4 to $\frac{1}{3}$ or $4 : \frac{1}{3}$ You have 1 cup of cleaner and you want to use all of it. How many gallons of water do you need to mix with 1 cup of cleaner to maintain the ratio $4 : \frac{1}{3}$?

To find the *unit rate*, the number of gallons of water needed to mix with 1 cup of cleaner, you can use various strategies.

Study Tip

You learned about ratios, rates, and unit rates in a previous grade.

A *ratio* is a comparison of two quantities, in which for every *a* units of one quantity, there are *b* units of another quantity.

A *rate* is a ratio that compares two quantities with unlike units.

A *unit rate* compares the first quantity per every 1 unit of the second quantity.

Talk About It!

Explain to a partner why the rate $4 : \frac{1}{3}$ is not a unit rate.

Study Tip

The expression $\frac{4}{\frac{1}{3}}$ is called a *complex fraction*. A complex fraction is a fraction in which the numerator or denominator, or both, are also fractions.

Talk About It!

How is finding a unit rate when one of the quantities is a fraction similar to finding a unit rate when both quantities are whole numbers? How is it different?

The double number line shows that, for every $\frac{1}{3}$ cup of cleaner, 4 gallons of water are needed. So, 12 gallons of water are needed to mix with 1 cup of cleaner.

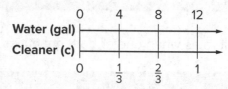

The bar diagram also shows that 12 gallons of water are needed to mix with 1 cup of cleaner. For every $\frac{1}{3}$ cup of cleaner, 4 gallons of water are needed.

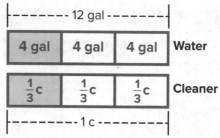

By creating a ratio table, you can scale forward by multiplying both $\frac{1}{3}$ and 4 by 3. The ratio table confirms that 12 gallons of water are needed to mix with 1 cup of cleaner.

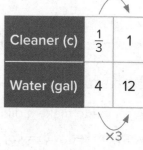

You can also use division when finding a unit rate. Recall that a ratio or rate can be written in fraction form. The ratio 4 : 1 can be written in fraction form as $\frac{4}{1}$.

The ratio $4 : \frac{1}{3}$ can be written in fraction form as $\frac{4}{\frac{1}{3}}$.

Because a fraction bar indicates division, you can divide the numerator by the denominator to find the unit rate.

$\frac{4}{\frac{1}{3}} = 4 \div \frac{1}{3}$ The fraction bar indicates division.

$= \frac{4}{1} \div \frac{1}{3}$ Write 4 as $\frac{4}{1}$.

$= \frac{4}{1} \cdot \frac{3}{1}$ Multiply by the multiplicative inverse of $\frac{1}{3}$.

$= \frac{12}{1}$, or 12   Multiply the fractions.

So, $\frac{4}{\frac{1}{3}} = 12$.

Using any of these strategies, the unit rate is 12 gallons of water for every 1 cup of cleaner.

🌐 Example 1 Find Unit Rates

Tia is painting one side of her shed. She paints 36 square feet in 45 minutes.

At this rate, how many square feet can she paint each hour?

You know that 45 minutes is $\frac{3}{4}$ of an hour. So, Tia's rate is 36 square feet per $\frac{3}{4}$ hour. You need to find the unit rate, the number of square feet she can paint per 1 hour.

Method 1 Use a bar diagram.

Draw two bars to model the ratio $36 : \frac{3}{4}$. Divide each bar into four sections, because $\frac{3}{4}$ is a multiple of $\frac{1}{4}$, and there are 4 sections of $\frac{1}{4}$ hour in 1 hour.

To find the unit rate, first find the value of each section in the bar representing square feet. Because three sections have a value of 36 square feet, each section has a value of $36 \div 3$, or 12 square feet. Because $4(12) = 48$, the unit rate is 48 square feet per hour.

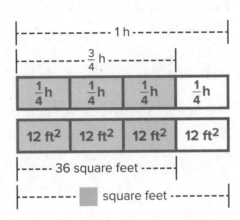

Method 2 Use a double number line.

The top number line represents the number of hours. The bottom number line represents the number of square feet. Mark and label the ratio $36 : \frac{3}{4}$.

Mark and label four equal increments of $\frac{1}{4}$ on the top number line. Mark the same number of equal increments on the bottom number line.

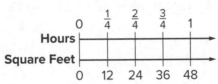

Each increment on the bottom number line represents 12 square feet. Because $4(12) = 48$, the unit rate is 48 square feet per hour.

(continued on next page)

💬 **Think About It!**

Why is 45 minutes equal to $\frac{3}{4}$ of an hour?

📲 **Talk About It!**

Use mathematical reasoning to explain why Tia can paint more than 36 square feet per hour.

Talk About It!

Why do you need to scale backward first before scaling forward?

Method 3 Use a ratio table.

The ratio table shows the number of square feet painted in $\frac{3}{4}$ hour. Scale backward to find the number of square feet painted in $\frac{1}{4}$ hour.

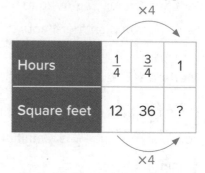

Scale forward to find the number of square feet Tia can paint in 1 hour. This is the unit rate.

Because $\frac{1}{4}(4) = 1$, multiply 12(4). So, Tia can paint 12(4), or 48 square feet, in one hour.

Method 4 Use division.

The rate 36 square feet in $\frac{3}{4}$ hour can be written as $\frac{36}{\frac{3}{4}}$.

$$\frac{36}{\frac{3}{4}} = 36 \div \frac{3}{4} \qquad \text{Write the complex fraction as a division problem.}$$

$$= \frac{36}{1} \div \frac{3}{4} \qquad \text{Write 36 as } \frac{36}{1}.$$

$$= \frac{36}{1} \cdot \frac{4}{3} \qquad \text{Multiply by the reciprocal of } \frac{3}{4}, \text{ which is } \frac{4}{3}.$$

$$= \frac{144}{3} \text{ or } 48 \qquad \text{Multiply. The unit rate is 48 ft}^2 \text{ per hour.}$$

So, Tia can paint 48 square feet in one hour.

Talk About It!

Compare the four methods. What operation(s) did you use with each of the methods?

Check

Doug entered a canoe race. He paddled 5 miles in $\frac{2}{3}$ hour. What is his average speed in miles per hour? Use any strategy.

Show your work here

Go Online You can complete an Extra Example online.

🌐 **Example 2** Find Unit Rates

Josiah can jog $\frac{5}{6}$ mile in 15 minutes.

Find his average speed in miles per hour.

You know that 15 minutes is $\frac{1}{4}$ hour. So, Josiah's rate is $\frac{5}{6}$ mile per $\frac{1}{4}$ hour. You need to find the unit rate, the number of miles he can jog per 1 hour.

Method 1 Use a bar diagram.

Draw two bars to model the ratio $\frac{5}{6} : \frac{1}{4}$. Divide each bar into 4 sections, because there are 4 sections of $\frac{1}{4}$ in 1 hour.

To find the unit rate, first find the value of each section in the bar representing miles. Each section has a value of $\frac{5}{6}$ mile. Because $4\left(\frac{5}{6}\right) = \frac{20}{6}$ or $3\frac{1}{3}$, the unit rate is $3\frac{1}{3}$ miles per hour.

Method 2 Use a double number line.

The top number line represents the number of hours. The bottom number line represents the number of miles. Mark and label the ratio $\frac{5}{6} : \frac{1}{4}$.

Mark and label four equal increments of $\frac{1}{4}$ on the top number line. Mark the same number of equal increments on the bottom number line.

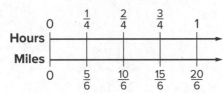

Each increment on the bottom number line represents $\frac{5}{6}$ mile.

Because $4\left(\frac{5}{6}\right) = \frac{20}{6}$, or $3\frac{1}{3}$, the unit rate is $3\frac{1}{3}$ miles per hour.

(continued on next page)

Copyright © McGraw-Hill Education

💭 **Think About It!**
What do you notice about both quantities of the rate?

Method 3 Use a ratio table.

The ratio table shows the number of miles jogged in $\frac{1}{4}$ hour. Scale forward to find the number of miles Josiah can jog in 1 hour. This is the unit rate.

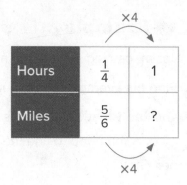

Because $\frac{1}{4}(4) = 1$, multiply $\frac{5}{6}(4)$. So, Josiah can jog $\frac{5}{6}(4)$, or $3\frac{1}{3}$ miles in one hour.

Method 4 Use division.

The rate $\frac{5}{6}$ mile per $\frac{1}{4}$ hour can be written as $\frac{\frac{5}{6}}{\frac{1}{4}}$.

$\dfrac{\frac{5}{6}}{\frac{1}{4}} = \frac{5}{6} \div \frac{1}{4}$ Write the complex fraction as a division expression.

$= \frac{5}{6} \cdot \frac{4}{1}$ Rewrite division as multiplication.

$= \frac{20}{6}$ Multiply.

$= 3\frac{1}{3}$ Simplify. The unit rate is $3\frac{1}{3}$ miles per hour.

With each representation, because there are 4 quarter-hours in one hour, his average speed is found by multiplying by 4. So, Josiah can jog $3\frac{1}{3}$ miles per hour.

Check

A garden hose was left on in a yard and spilled $\frac{7}{8}$ gallon every $\frac{2}{3}$ minute. Find the average number of gallons spilled per minute. Use any strategy.

 Go Online You can complete an Extra Example online.

Talk About It!

How could you word this problem using minutes instead of hours? Does it change the outcome of Josiah's average speed?

🌎 Apply Kayaking

Carolina and her friends are training separately for a kayaking competition. The average distance and time traveled by each is shown in the table. If the distance kayaked in the competition is 3 miles, predict who will win based on the rates shown. Predict how long it will take the winner to complete the race, if their rate remains constant.

Person	Carolina	Leslie	Bryan	Javier
Average Distance (mi)	$\frac{7}{8}$	1	$\frac{3}{4}$	1
Average Time (h)	$\frac{1}{2}$	$\frac{1}{3}$	$\frac{3}{4}$	$\frac{2}{3}$

1 What is the task?

Make sure you understand exactly what question to answer or problem to solve. You may want to read the problem three times. Discuss these questions with a partner.

First Time Describe the context of the problem, in your own words.
Second Time What mathematics do you see in the problem?
Third Time What are you wondering about?

2 How can you approach the task? What strategies can you use?

3 What is your solution?

Use your strategy to solve the problem.

4 How can you show your solution is reasonable?

✏️ **Write About It!** Write an argument that can be used to defend your solution.

Copyright © McGraw-Hill Education

💬 **Talk About It!**

Which method is more advantageous to use when solving this problem?

Check

A walk-a-thon was held at a local middle school. The table gives the average distances and times for four walkers for certain periods of the walk-a-thon. If the total distance of the route was 2 miles, who completed the route first? How long did it take her to complete the route if she walked at a constant rate?

Person	Distance (mi)	Time (min)
Lakeisha	$\frac{7}{8}$	18
Baydan	$\frac{9}{10}$	22
Madison	$\frac{4}{5}$	17

🧭 **Go Online** You can complete an Extra Example online.

Pause and Reflect

Compare the process for finding unit rates involving fractions with what you know about dividing fractions. How are they similar? How are they different?

Practice

🔵 **Go Online** You can complete your homework online.

Solve each problem. Use any strategy, such as a bar diagram, double number line, ratio table, or division.

1. A truck driver drove 48 miles in 45 minutes. At this rate, how many miles can the truck driver drive in one hour? (Example 1)

2. Russell runs $\frac{9}{10}$ mile in 5 minutes. At this rate, how many miles can he run in one minute? (Example 1)

3. A small airplane flew 104 miles in 50 minutes. At this rate, how many miles can it fly in one hour? (50 minutes = $\frac{5}{6}$ hour) (Example 1)

4. DeAndre downloaded 8 apps onto his tablet in 12 seconds. At this rate, how many apps could he download in one minute? (12 seconds = $\frac{1}{5}$ minute) (Example 1)

5. In Lixue's garden, the green pepper plants grew 5 inches in $\frac{3}{4}$ month. At this rate, how many feet can they grow in one month? (Let 5 inches = $\frac{5}{12}$ foot) (Example 2)

6. Thunder from a bolt of lightning travels $\frac{1}{10}$ mile in $\frac{1}{2}$ second. At this rate, how many miles can it travel in one second? (Example 2)

Test Practice

7. The average sneeze can travel $\frac{3}{100}$ mile in 3 seconds. At this rate, how far can it travel in one minute? (3 seconds = $\frac{1}{20}$ minute) (Example 2)

8. **Multiselect** Anita is making headbands for her softball team. She needs a total of $\frac{3}{4}$ yard of fabric. Select all types of fabric that cost less than $8 per yard. (Example 2)

☐ cotton

☐ flannel

☐ fleece

☐ terry cloth

Fabric	Total Cost for $\frac{3}{4}$ Yard ($)
Cotton	5.54
Flannel	2.62
Fleece	4.27
Terry Cloth	6.52

Apply

9. During the first seconds after takeoff, a rocket traveled 208 kilometers in 50 minutes at a constant rate. Suppose a penny is dropped from a skyscraper and could travel 153 kilometers in $\frac{1}{2}$ hour at a constant rate. Which of these objects has a faster unit rate per hour? How much faster?

10. To prepare for a downhill skiing competition, Roman completed three training sessions. The table shows his average time and distance for each session. Did Roman's rate, in miles per hour, increase from session to session? Write an argument that can be used to defend your solution.

Session	Time (h)	Distance (mi)
1	$\frac{2}{125}$	$\frac{9}{10}$
2	$\frac{3}{200}$	$\frac{11}{12}$
3	$\frac{3}{250}$	$\frac{17}{20}$

11. (MP) **Reason Abstractly** Explain why a student who runs $\frac{3}{4}$ mile in 6 minutes is faster than a student who runs $\frac{1}{2}$ mile in 5 minutes.

12. Compare and contrast the rates $\frac{4}{5}$ mile in 8 minutes and 4 minutes to travel $\frac{2}{5}$ mile.

13. (MP) **Find the Error** Carli made 9 greeting cards in $\frac{3}{4}$ hour. She determined her unit rate to be $\frac{1}{12}$ card per hour. Find her error and correct it.

14. (MP) **Be Precise** A standard shower drain can drain water at the rate of 480 gallons in $\frac{2}{3}$ hour. Create three different rates, using the same or different units, that are all equivalent to this rate. Be precise in the units you choose. Then find the unit rate, in gallons per minute.

Understand Proportional Relationships

I Can... use models and ratio reasoning to understand how a proportional relationship can exist between quantities.

Learn Proportional Relationships

When baking, the ratio(s) of ingredients is important to maintain. The table shows a common recipe for pizza dough. Too much or too little of any one ingredient will not create a good dough.

Ingredient	Amount
Flour	3 c
Salt	$\frac{1}{2}$ tsp
Yeast	2 tsp
Water	1 c
Olive Oil	4 tsp

In this recipe, the ratio of cups of flour to cups of water is 3 : 1. Suppose you wanted to make two batches of dough. Is the ratio of flour to water the same? What if you wanted to make three batches?

The bar diagrams show the relationship between flour and water for different batches of dough. In each batch, 3 equal-size sections represent cups of flour, and 1 section of the same size represents cups of water.

One Batch

For one batch, the ratio is 3 cups of flour to 1 cup of water.

Two Batches

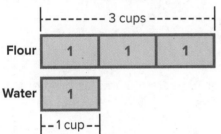

For two batches, the ratio is 6 cups of flour to 2 cups of water. For every 3 cups of flour, you need 1 cup of water. The ratio 3 : 1 is maintained.

Three Batches

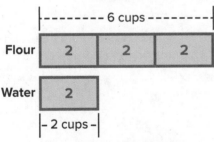

For three batches, the ratio is 9 cups of flour to 3 cups of water. For every 3 cups of flour, you need 1 cup of water. The ratio 3 : 1 is maintained.

(continued on next page)

What Vocabulary Will You Learn?
proportional relationship

💬 Talk About It!
Would this ratio be maintained if you wanted to make half a batch of dough? Explain.

The ratio of cups of flour to cups of water is maintained regardless of how many batches of pizza dough you make. In each batch, there are 3 cups of flour for every 1 cup of water.

Two quantities are in a **proportional relationship** if the two quantities vary and have a constant ratio between them. For example, if a recipe calls for 2 cups of flour for every 1 cup of sugar, the ingredients are in a proportional relationship because, while the number of cups of flour or sugar can vary, the ratio of cups of flour to sugar is constant, 2 : 1.

Some relationships are not proportional relationships. In these cases, a ratio is not maintained. For example, suppose that Pedro is 14 years old and his little brother is 7 years old. The ratio between their current ages is 14 : 7. Pedro is currently twice as old as his brother. Will he always be twice as old?

The bar diagram represents this relationship. Because Pedro is currently twice as old as his brother, the bar diagram representing Pedro's age has twice as many sections as the bar diagram representing his brother's age.

Current Age

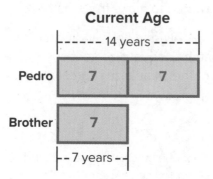

The ratio between Pedro's age and his brother's age is 14 : 7. This ratio is equivalent to 2 : 1 because 14 is twice as great as 7.

In five years, Pedro will be 14 + 5, or 19 years old and his brother will be 7 + 5, or 12 years old. Add a section to each bar diagram to represent the additional 5 years.

Age in Five Years

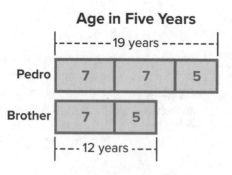

The ratio between Pedro's age and his brother's age in 5 years is 19 : 12.

💬 **Talk About It!**

Will there ever be an age, other than 14 and 7, where Pedro is twice as old as his brother? Explain.

The bar diagram representing their ages in 5 years shows that the ratio 19 : 12 is not equivalent to 14 : 7 or 2 : 1. In 5 years, Pedro will not be twice as old as his brother because 19 ≠ 2(12). Because Pedro will not always be twice as old as his brother and the ratio 14 : 7 or 2 : 1 is not maintained, the relationship is not proportional.

Example 1 Identify Proportional Relationships

The recipe for a homemade glass cleaner indicates to use a ratio of 1 part vinegar to 4 parts water. Elyse used 3 tablespoons of vinegar and 12 tablespoons of water to make the cleaner.

Is the relationship between the vinegar and water in the recipe and the vinegar and water in Elyse's cleaning solution a proportional relationship? Explain.

To determine if the relationship is proportional, Elyse must maintain the ratio of vinegar to water. Draw a bar diagram to represent the ratio of ingredients in the recipe.

Think About It!

What is the relationship between the parts of water and the parts of vinegar in the recipe?

For every 1 part of vinegar, there are 4 parts of water. The units representing the part do not matter. The units can be cups, quarts, gallons, etc.

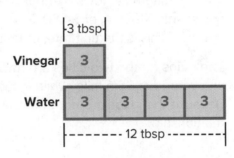

Draw a bar diagram to represent the ratio of ingredients in Elyse's glass cleaner.

Elyse used 3 tablespoons of vinegar, which is 1 part. She used 12 tablespoons of water. Because there are four parts of 3 in 12, she used 4 parts of water.

The ratio between vinegar and water was maintained when Elyse used 3 tablespoons of vinegar and 12 tablespoons of water. Because the ratio was maintained, this represents a proportional relationship.

Talk About It!

Would this ratio be maintained if she used 1 cup of vinegar and 4 cups of water? Explain.

Check

Refer to the recipe for homemade glass cleaner in Example 1. Marcus mixed 1.5 cups of vinegar and 6 cups of water to make his cleaner. Is the relationship between the vinegar and water in the recipe and the vinegar and water in Marcus' cleaning solution a proportional relationship? Explain.

Go Online You can complete an Extra Example online.

🌐 Example 2 Identify Proportional Relationships

A 5-mile taxi ride in one city costs Ayana $25. A 4-mile taxi ride in a different city costs $18. Assume the cost in each city is the same amount per mile.

Is the relationship between the number of miles and the total cost between the two cities a proportional relationship? Explain.

Draw a diagram to represent the relationship between miles and cost for each city.

In the first city, it costs $25 to travel 5 miles.

The bar diagram shows 5 equal-size sections. Each section represents $25 ÷ 5, or $5. So, in the first city, it costs $5 per mile. This is the unit rate, or unit cost.

In the second city, it costs $18 to travel 4 miles.

The bar diagram shows 4 equal-size sections. Each section represents $18 ÷ 4, or $4.50. So, in the second city, it costs $4.50 per mile. This is the unit rate, or unit cost.

The ratios for the two cities are not equivalent. In the first city, it costs an average of $5 a mile, while it costs $4.50 a mile in the second city. Because the ratios are not equivalent, this is not a proportional relationship.

Check

One type of yarn costs $4 for 100 yards. Another type of yarn costs $5 for 150 yards. Is the relationship between the number of yards and the cost a proportional relationship between the two types of yarn? Explain.

Show your work here

🌀 **Go Online** You can complete an Extra Example online.

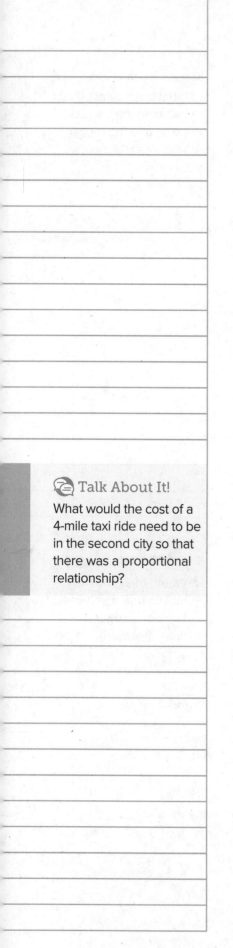

💬 Talk About It!

What would the cost of a 4-mile taxi ride need to be in the second city so that there was a proportional relationship?

 Apply Construction

Christine is building a deck in her backyard. In order to place the posts, she will make concrete using a mixture. The mix requires 1 part water to 2 parts cement to 3 parts sand. The relationship between water, cement, and sand is proportional by weight. If she has 25 pounds of cement and will use it all, how many pounds of sand will she need? One gallon of water weighs about 8.34 pounds. How many gallons of water will she need? Round to the nearest tenth.

1 What is the task?

Make sure you understand exactly what question to answer or problem to solve. You may want to read the problem three times. Discuss these questions with a partner.

First Time Describe the context of the problem, in your own words.
Second Time What mathematics do you see in the problem?
Third Time What are you wondering about?

2 How can you approach the task? What strategies can you use?

3 What is your solution?

Use your strategy to solve the problem.

4 How can you show your solution is reasonable?

Write About It! Write an argument that can be used to defend your solution.

Talk About It!
Compare and contrast your method for solving this problem with a classmate's method.

Check

A basic slime recipe calls for 1 part borax, 24 parts white glue, and 48 parts water. The relationship between borax, white glue, and water is proportional by volume. If Catalina has 3 tablespoons of borax and will use it all, how many cups of white glue will she need? (*Hint*: 1 cup equals 16 tablespoons)

Go Online You can complete an Extra Example online.

Pause and Reflect

Have you ever wondered when you might use the concepts you learn in math class? What are some everyday scenarios in which you might use what you learned today?

Practice

🔵 **Go Online** You can complete your homework online.

Determine if each situation represents a proportional relationship. Explain your reasoning. (Examples 1 and 2)

1. A salad dressing calls for 3 parts oil and 1 part vinegar. Manuela uses 2 tablespoons of vinegar and 6 tablespoons of oil to make her salad dressing.

2. A specific shade of orange paint calls for 2 parts yellow and 3 parts red. Catie uses 3 cups of yellow paint and 4 cups of red paint to make orange paint.

3. A saltwater solution for an aquarium calls for 35 parts salt to 1000 parts water. Tareq used 7 tablespoons of salt and 200 tablespoons of water.

4. A conveyor belt moves at a constant rate of 12 feet in 3 seconds. A second conveyor belt moves 16 feet in 4 seconds.

5. A tectonic plate in Earth's crust moves at a constant rate of 4 centimeters per year. In a different part of the world, another tectonic plate moves at a constant rate of 30 centimeters in ten years.

6. A strand of hair grows at a constant rate of $\frac{1}{2}$ inch per month. A different strand of hair grows at a constant rate of 4 inches per year.

Test Practice

7. **Multiselect** One blend of garden soil is 1 part minerals, 1 part peat moss, and 2 parts compost. Select all of the mixtures below that are in a proportional relationship with this blend.

   ☐ 5 ft³ minerals, 5 ft³ peat moss, 10 ft³ compost

   ☐ 10 ft³ minerals, 15 ft³ peat moss, 15 ft³ compost

   ☐ 12 ft³ minerals, 12 ft³ peat moss, 24 ft³ compost

   ☐ 20 ft³ minerals, 20 ft³ peat moss, 40 ft³ compost

   ☐ 100 ft³ minerals, 100 ft³ peat moss, 200 ft³ compost

   ☐ 50 ft³ minerals, 50 ft³ peat moss, 50 ft³ compost

Apply

8. Melanie is making lemonade and finds a recipe that calls for 1 part lemon juice, 2 parts sugar, and 8 parts water. She juices 2 lemons to obtain 6 tablespoons of lemon juice. How much sugar and water will she need to make lemonade with the same ratio of ingredients as the recipe? (*Hint*: 1 cup equals 16 tablespoons)

9. The pizza dough recipe shown makes one batch of dough. Charlie wants to make a half batch. She has 1 cup of flour. How much more flour does she need?

Ingredient	Amount
Flour	3 c
Salt	$\frac{1}{2}$ tsp
Yeast	2 tsp
Water	1 c
Olive Oil	4 tsp

10. **MP Identify Structure** Half of an orange juice mixture is orange concentrate. Explain why the ratio of orange concentrate to water is 1 : 1.

11. **MP Find the Error** One cleaning solution uses 1 part vinegar with 2 parts water. Another cleaning solution uses 2 parts vinegar with 3 parts water. A student says that this represents a proportional relationship because, in each solution, there is one more part of water than vinegar. Find the error and correct it.

12. Patrick made a simple sugar solution using 3 parts sugar and 4 parts water. Thomas made a sugar solution using 6 parts sugar and 7 parts water. Whose solution was more sugary? Explain.

13. **MP Be Precise** How can you use a unit rate to determine if a relationship is proportional?

Tables of Proportional Relationships

I Can... determine whether two quantities shown in a table are in a proportional relationship by testing for equivalent ratios.

Explore Ratios in Tables

Online Activity You will explore how to determine if the ratios between two quantities are equivalent.

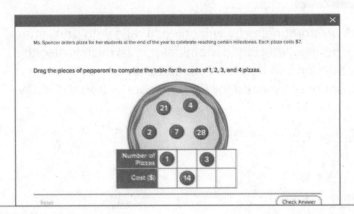

Learn Proportional Relationships and Tables

Two quantities are **proportional** if the ratios comparing them are equivalent.

In the table, all of the ratios comparing the cost to the number of pizzas are equivalent and have a unit rate of $\frac{\$7}{1\ \text{pizza}}$. So, the cost of the order is proportional to the number of pizzas ordered.

Number of Pizzas	1	2	3	4
Cost ($)	7	14	21	28
Cost per Pizza ($)	7	7	7	7

In relationships where these ratios are not equivalent, the two quantities are **nonproportional**.

In the table below, the ratios comparing the cost to the number of pizzas are different. So, the relationship represented by this table is nonproportional.

Number of Pizzas	1	2	3	4
Cost ($)	9	16	23	30
Cost per Pizza ($)	9	8	7.67	7.50

What Vocabulary Will You Learn?
constant of
  proportionality
nonproportional
proportional

Talk About It!
How can you use ratios to determine if a relationship is proportional?

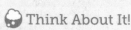
🌐 **Example 1** Proportional Relationships and Tables

Carrie earns $8.50 per hour babysitting.

Is the amount of money she earns proportional to the number of hours she spends babysitting?

Complete the table with the amount of money she earns for babysitting 1, 2, 3, and 4 hours.

Number of Hours	1	2	3	4
Amount Earned ($)				

You can check proportionality by writing each ratio with the same denominator. The most efficient denominator is 1, which is also the unit ratio. The relationship between the amount earned, in dollars, and the hours spent babysitting for each number of hours is shown.

$$\frac{8.5}{1} = \frac{8.5}{1} \qquad\qquad \frac{17}{2} = \frac{8.5}{1}$$

$$\frac{25.5}{3} = \frac{8.5}{1} \qquad\qquad \frac{34}{4} = \frac{8.5}{1}$$

So, the relationship is proportional because the ratios between the quantities are equivalent and have a unit rate of $8.50 per hour.

Check

An adult elephant drinks about 225 liters of water each day. Is the amount of water proportional to the number of days that have passed? Use the table provided to help answer the question.

Time (days)	1	2	3	4
Water (L)	225	450	675	900

Show your work here

🌐 **Go Online** You can complete an Extra Example online.

🌐 **Example 2** Proportional Relationships and Tables

A ticket agency charges a $6 service fee on any order. Each ticket for a concert costs $25.

Is the cost of an order proportional to the number of tickets ordered?

The table shows the relationship between the total cost of an order and the number of tickets ordered.

Number of Tickets	1	2	3	4
Total Cost of Order ($)	31	56	81	106

The cost of each additional ticket is an additional $25 because the service fee is only charged once per order.

For each number of tickets, write the relationship of the total cost to the number of tickets as a ratio with a denominator of 1. The first two are done for you.

$$\frac{31}{1} = \frac{31}{1}$$ $$\frac{56}{2} = \frac{28}{1}$$

$$\frac{81}{3} = \frac{\boxed{}}{1}$$ $$\frac{106}{4} = \frac{\boxed{}}{1}$$

Because the ratios between the quantities are not the same, the cost of an order is not proportional to the number of tickets ordered.

Check

The table shows how long it took Maria to run laps around the school track. Is the number of laps she ran proportional to the time it took her? Explain.

Laps	2	4	6
Time (s)	150	320	580

Show your work here

🌐 **Go Online** You can complete an Extra Example online.

Learn Identify the Constant of Proportionality

You have learned that two quantities are proportional if the ratios comparing them are equivalent or constant. The constant ratio is called the **constant of proportionality**. The constant of proportionality has the same value as the unit rate.

Creators of a stop motion animation can film 24 frames per second. The table shows the number of frames captured over 1, 2, 3, and 4 seconds.

Number of Seconds	1	2	3	4
Number of Frames	24	48	72	96

What is the constant of proportionality? _____

What is the unit rate? _____

Because the constant ratio is $\frac{24}{1}$, the constant of proportionality is 24, and the unit rate is 24 frames per second.

Talk About It!

What do you notice about the constant of proportionality and the unit rate?

Think About It!

How are the constant of proportionality and the unit rate related?

Talk About It!

If the relationship was not proportional, would there be a constant of proportionality? Explain.

🌐 Example 3 Identify the Constant of Proportionality

The winner of a jump rope competition jumped 124 times in 20 seconds and 186 times in 30 seconds.

What is the constant of proportionality?

Write the equivalent ratios so that each ratio has a denominator of 1.

$$\frac{124 \text{ jumps}}{\boxed{}} = \frac{\boxed{}}{1 \text{ second}}$$

$$\frac{\boxed{}}{30 \text{ seconds}} = \frac{\boxed{}}{1 \text{ second}}$$

Because the constant ratio is $\frac{6.2}{1}$, the constant of proportionality is 6.2, and the unit rate is 6.2 jumps per second.

Check

The cost of a birthday party at a skating rink is proportional to the number of guests. The skating rink charges $82.50 for 10 guests.

The choices show the number of guests and the total cost for four different parties that were hosted at different locations. Select each party that was hosted with the same constant of proportionality, or unit rate, as the skating rink's unit rate.

☐ 12 guests for $99.00

☐ 8 guests for $52.00

☐ 15 guests for $97.50

☐ 6 guests for $49.50

Show your work here

🐦 **Go Online** You can complete an Extra Example online.

🌐 **Example 4** Identify the Constant of Proportionality

After a volcanic eruption, lava flows down the slopes of the volcano. The distance the lava flows is proportional to the time.

Time (min)	Distance (m)
5	9
10	18
15	27
20	36

What is the constant of proportionality of the flow of lava?

For each time, find the ratio $\frac{\text{number of meters}}{\text{number of seconds}}$ and rewrite it with a denominator of 1.

$$\frac{9}{5} = \frac{\square}{1} \qquad \frac{18}{10} = \frac{\square}{1}$$

$$\frac{27}{15} = \frac{\square}{1} \qquad \frac{36}{20} = \frac{\square}{1}$$

Because the constant ratios are $\frac{1.8}{1}$, the constant of proportionality is 1.8, and the unit rate is 1.8 meters per second.

Copyright © McGraw-Hill Education

💭 **Think About It!**

Why is a table a good way to organize the information?

 Talk About It!

What does the constant of proportionality, 1.8, mean in the context of the problem?

Check

The tables show the amount that three friends earn during a bake sale. Write the correct unit rate that each friend earned in the spaces provided.

Allie

Earnings ($)	17.00	34.00	51.00
Time (h)	1	2	3

Ben

Earnings ($)	9.00	18.00	27.00
Time (h)	0.75	1.5	2.25

Sri

Earnings ($)	40.00	80.00	120.00
Time (h)	2.5	5	7.5

Allie	Ben	Sri

Show your work here

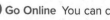 **Go Online** You can complete an Extra Example online.

Copyright © McGraw-Hill Education

Math History Minute

Erika Tatiana Camacho (1974-) is a Mexican-American mathematical biologist and Associate Professor at Arizona State University. In 2014, she won the Presidential Award for Excellence in Science, Mathematics, and Engineering Mentoring. Her high school teacher and mentor was Jaime Escalante, the subject of the 1988 movie *Stand and Deliver*.

🌐 **Apply** Sales Tax

Jalen went shopping for school clothes. The table shows the sales tax for various purchase amounts. Is the sales tax proportional to the purchase amount? What is the total cost, in dollars, for a purchase amount of $84?

Purchase Amount ($)	12	24	36	48
Sales Tax ($)	0.60	1.20	1.80	2.40

1 What is the task?

Make sure you understand exactly what question to answer or problem to solve. You may want to read the problem three times. Discuss these questions with a partner.

First Time Describe the context of the problem, in your own words.
Second Time What mathematics do you see in the problem?
Third Time What are you wondering about?

2 How can you approach the task? What strategies can you use?

3 What is your solution?

Use your strategy to solve the problem.

4 How can you show your solution is reasonable?

✍ **Write About It!** Write an argument that can be used to defend your solution.

💬 **Talk About It!**

If the relationship was not proportional, could you solve this problem? Explain.

Check

The table shows three membership options at a fitness center.

Membership	Cost
Basic	$20 per class
Fit Plus	$60 per month plus $10 per class
Fit Extreme	$75 per month plus $30 enrollment fee

Logan chooses the membership that represents a proportional relationship between the number of classes and the monthly cost. Which membership did Logan choose? How much will he spend if he takes 12 classes in a month?

Show your work here

Go Online You can complete an Extra Example online.

Foldables It's time to update your Foldable, located in the Module Review, based on what you learned in this lesson. If you haven't already assembled your Foldable, you can find the instructions on page FL1.

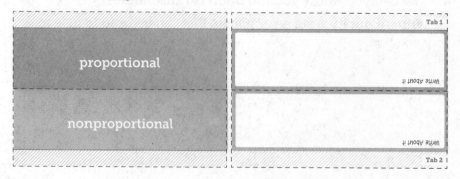

Practice

🔵 **Go Online** You can complete your homework online.

For each situation, complete the table given. Does the situation represent a proportional relationship? Explain.

1. The cost of a school lunch is $2.50. (Example 1)

Lunches Bought	1	2	3	4
Total Cost ($)				

2. Anna walks her dog at a constant rate of 12 blocks in 8 minutes. (Example 1)

Number of Blocks	12	24	36	48
Number of Minutes				

3. Fun Center rents popcorn machines for $20 per hour. In addition to the hourly charge, there is a rental fee of $35. (Example 2)

Hours	1	2	3	4
Cost ($)				

4. Jean has $280 in her savings account. Starting next week, she will deposit $30 in her account every week. (Example 2)

Weeks	1	2	3	4
Savings ($)				

5. Rocko paid $12.50 for 25 game tickets. Louisa paid $17.50 for 35 game tickets. What is the constant of proportionality? (Example 3)

6. A baker, in 70 minutes, iced 40 cupcakes and, in 49 minutes, iced 28 cupcakes. What is the constant of proportionality? (Example 3)

7. The table shows the amount of dietary fiber in bananas. Use the table to find the constant of proportionality. (Example 4)

Dietary Fiber (g)	9.3	18.6	27.9	37.2
Bananas	3	6	9	12

Test Practice

8. **Open Response** The table shows the distance traveled by a runner. Use the table to find the constant of proportionality.

Distance (mi)	4.55	13.65	22.75	31.85
Time (h)	0.5	1.5	2.5	3.5

Apply

9. The table shows the amount a restaurant is donating to a local school based on various dinner bills. Is the amount of the donation proportional to the dinner bill? If so, what would be the donation for a dinner bill of $50? If not, explain.

Donations	
Dinner Bill ($)	**Donation ($)**
25	4.50
30	5.40
35	6.30
40	7.20

10. The table shows the cost to mail various letters based on different weights. Is the cost of mailing a letter proportional to the weight? If so, what would be the cost of mailing a 6-ounce letter? If not, explain.

Mailing Costs	
Weight (oz)	**Cost ($)**
1	0.47
2	0.68
3	0.89
4	1.10

11. There are 8 fluid ounces in one cup. If you double the amount of fluid ounces, will the amount of cups also double? Write an argument that can be used to defend your solution.

12. Determine whether the cost of renting equipment is *sometimes*, *always*, or *never* proportional. Explain.

13. **MP** **Justify Conclusions** Noah ran laps around the school building. The table shows his times. He thinks the number of laps is proportional to his time. Explain how Noah may have come to that conclusion.

Laps	5	10	15
Time (min)	4	6	8

14. **Multiple Representations** Represent the proportional relationship $2 for 5 ears of corn and $4 for 10 ears of corn using another representation.

Graphs of Proportional Relationships

I Can... determine if a relationship is proportional by analyzing its graph and explain what the points (0, 0) and (1, *r*) mean on the graph of a proportional relationship.

Explore Proportional Relationships, Tables, and Graphs

Online Activity You will use Web Sketchpad to explore the graphs of proportional and nonproportional linear relationships.

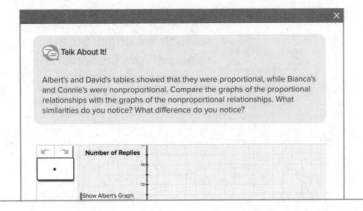

Learn Proportional Relationships and Graphs

A graph shows a proportional relationship if it is a straight line through the origin. A graph shows a nonproportional relationship if it is not a straight line, or is a straight line that does not pass through the origin.

Determine if the relationship shown in each graph is *proportional* or *nonproportional*.

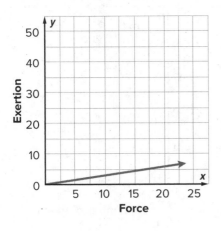

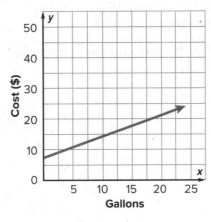

Copyright © McGraw-Hill Education

Talk About It!

Why does the line of the graph of a proportional relationship need to be straight and pass through the origin?

🌐 **Example 1** Proportional Relationships and Graphs

A rabbit challenges a tortoise to a race. The table shows the distance that the tortoise moved after 0, 1, 2, and 3 minutes.

Time (min)	Distance (ft)
0	0
1	6
2	12
3	18

Determine whether the number of feet the tortoise moves is proportional to the number of minutes by graphing the relationship on the coordinate plane.

Part A Graph the relationship.

Distance Traveled

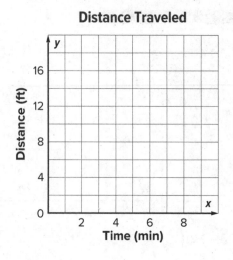

Part B Describe the relationship.

The line is straight and passes through the origin. So, the relationship is proportional.

💬 **Talk About It!**
If the graph of a proportional relationship must be a straight line through the origin, how does the table illustrate that same information?

Check

In a pack of snacks, one piece has 5 Calories, two pieces have 10 Calories, and three pieces have 15 Calories.

Part A Graph the relationship for 0, 1, 2, and 3 pieces on the coordinate plane.

Part B Describe the relationship.

Calories per Piece

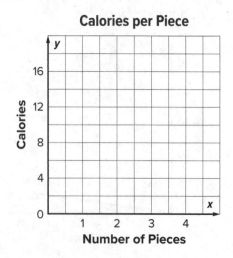

🌐 **Go Online** You can complete an Extra Example online.

🌐 Example 2 Proportional Relationships and Graphs

A rabbit challenges a tortoise to a race. The table shows the distance that the rabbit moved after 0, 1, 2, and 3 minutes.

Time (min)	Distance (ft)
0	0
1	8
2	8
3	15

Determine if the distance the rabbit moves is proportional to the time by graphing the relationship on the coordinate plane.

Part A Graph the relationship.

Part B Describe the relationship.

The relationship between time and distance is not proportional because the graph is not a straight line.

💬 **Talk About It!**

Is there another way to determine if the relationship is proportional or nonproportional?

Check

The table shows the account balance in a savings account at the end of each week.

Determine if the account balance is proportional to the time by graphing the relationship on the coordinate plane.

Time (wk)	Account Balance ($)
0	10
1	12
2	14
3	16

💬 **Talk About It!**

What part of the scenario makes the relationship nonproportional?

Part A
Graph the relationship.

Part B
Describe the relationship.

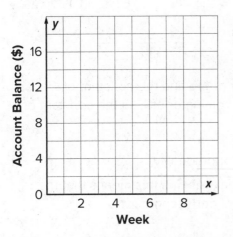

🅖 **Go Online**
You can complete an Extra Example online.

Learn Find the Constant of Proportionality from Graphs

When a proportional relationship is graphed, you can determine the constant of proportionality using any point on the line other than the origin. The constant of proportionality is the ratio of $\frac{y}{x}$ for any point on the line, except $(0, 0)$, when $x = 1$.

Write each ordered pair and write the ratio $\frac{y}{x}$, when $x = 1$, for each.

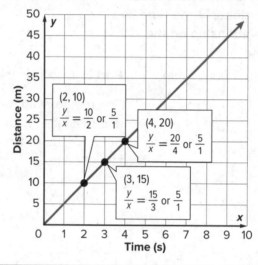

Example 3 Find the Constant of Proportionality from Graphs

In a 100-meter race, assuming the runner's rate is constant, the distance run is proportional to the time spent running. One runner's data are shown on the graph.

Find the constant of proportionality and describe what it means.

Part A Find the constant of proportionality.

The ratios $\frac{y}{x}$ or $\frac{distance}{time}$ for all of the given points are shown.

$$\frac{y}{x} = \frac{31}{5} = \frac{6.2}{1} \qquad \frac{y}{x} = \frac{62}{10} = \frac{6.2}{1} \qquad \frac{y}{x} = \frac{93}{15} = \frac{6.2}{1}$$

Write the ratios with a denominator of 1 to find the constant of proportionality. Each ratio has a denominator of 1, so the constant of proportionality is 6.2.

Part B Describe the constant of proportionality.

Because the constant of proportionality is 6.2, this means that the runner moves at a rate of 6.2 meters per second.

 Think About It!

How would you begin solving the problem?

 Talk About It!

Based on the constant of proportionality, 6.2, how far will the runner travel after 20 seconds? Explain your reasoning.

Check

Briana decides to save money each week for her family vacation. Use the graph to find the constant of proportionality. Then describe what it means.

Part A

The constant of proportionality is _____ .

Part B

The constant of proportionality means that Briana saves _____ each week.

Go Online You can complete an Extra Example online.

[Graph: Money Saved ($) vs Number of Weeks, showing points (0, 0), (2, 40), and (4, 80)]

Explore Analyze Points

Online Activity You will explore and analyze the points (0, 0) and (1, r) on a graph of a proportional relationship.

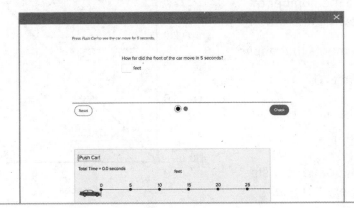

Learn Analyze Points on a Graph

When two quantities are proportional, you can use a graph to find the constant of proportionality and to interpret the point (0, 0).

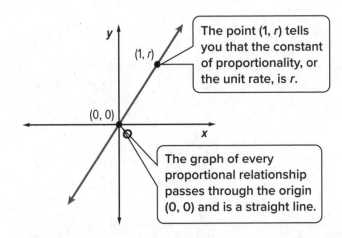

The point (1, r) tells you that the constant of proportionality, or the unit rate, is r.

The graph of every proportional relationship passes through the origin (0, 0) and is a straight line.

Talk About It!

What is the significance of (0, 0) on the graph of a proportional relationship?

Example 4 Analyze Points on a Graph

Copyright © McGraw-Hill Education

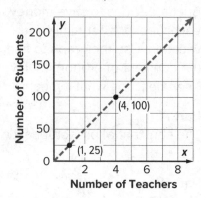

The number of students on a school trip is proportional to the number of teachers as shown in the graph. The line representing the relationship is a dashed line because the number of teachers can only be a whole number.

What do the points (0, 0) and (1, 25) represent?

The point (0, 0) means that, for zero teachers, there are zero students.

The point (1, 25) means that, for one teacher, there are _____ students. This also means that the constant of proportionality is 25 and the unit rate is 25 students for every 1 teacher.

Check

The cost of a smoothie is proportional to the number of ounces as shown in the graph.

Select all statements that apply.

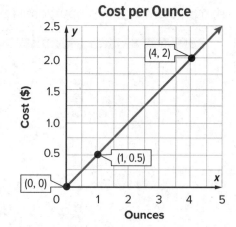

☐ The unit rate is 0.

☐ The unit rate is 0.5.

☐ The unit rate is 1.

☐ For each smoothie, it costs $1.50 for every 1 ounce.

☐ For each smoothie, it costs $2 for every 1 ounce.

☐ For each smoothie, it costs $1 for every 2 ounces.

☐ For each smoothie, it costs $2.00 for every 4 ounces.

🔖 **Go Online** You can complete an Extra Example online.

💭 Think About It!

In the ordered pairs, what does the *x*-coordinate represent? the *y*-coordinate?

💬 Talk About It!

Can you find the unit rate from any point on a graph that shows a proportional relationship? Explain your reasoning.

🌐 Apply Fundraising

Michele and Angelo are participating in a school fundraiser. The number of items each student sells after 1, 2, and 3 days is shown in the table. Which ordered pair represents a unit rate for the number of items sold per day?

Day	0	1	2	3
Michele's Number of Items	0	3	6	9
Angelo's Number of Items	0	1	4	5

1 What is the task?

Make sure you understand exactly what question to answer or problem to solve. You may want to read the problem three times. Discuss these questions with a partner.

First Time Describe the context of the problem, in your own words.
Second Time What mathematics do you see in the problem?
Third Time What are you wondering about?

2 How can you approach the task? What strategies can you use?

3 What is your solution?

Use your strategy to solve the problem.

🗨 Talk About It!

How could you have determined that Angelo's graph was not proportional before graphing the relationship on the coordinate plane?

4 How can you show your solution is reasonable?

🖊 **Write About It!** Write an argument that can be used to defend your solution.

Check

The heights of two plants are recorded after 1, 2, and 3 weeks. The data are shown in the table. Which ordered pair represents a unit rate for the number of inches grown per week?

Time (weeks)	Plant 1 Height (in.)	Time (weeks)	Plant 2 Height (in.)
0	0	0	0
1	5	1	3
2	6	2	6
3	7	3	9

🔘 **Go Online** You can complete an Extra Example online.

📖 **Foldables** It's time to update your Foldable, located in the Module Review, based on what you learned in this lesson. If you haven't already assembled your Foldable, you can find the instructions on page FL1.

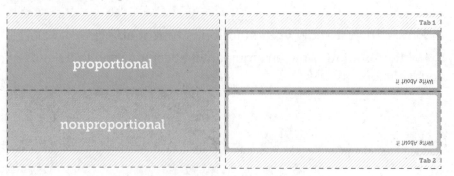

proportional

nonproportional

Tab 1

Write About It

Write About It

Tab 2

Practice

🔁 **Go Online** You can complete your homework online.

1. The cost of pumpkins is shown in the table. Determine whether the cost of a pumpkin is proportional to the number bought by graphing the relationship on the coordinate plane. Explain. **(Example 1)**

Number of Pumpkins	0	1	2	3	4
Cost ($)	0	4	8	12	16

Cost of Pumpkins

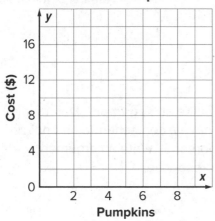

2. The table shows temperatures in degrees Celsius and their equivalent temperatures in degrees Fahrenheit. Determine whether the temperature in degrees Fahrenheit is proportional to the temperature in degrees Celsius by graphing the relationship on the coordinate plane. Explain. **(Example 2)**

Celsius (degrees)	0	5	10	15	20
Fahrenheit (degrees)	32	41	50	59	68

Temperature

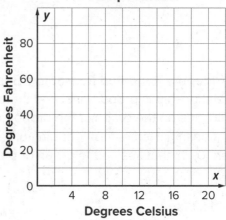

3. The total cost of online streaming is proportional to the number of months. What is the constant of proportionality? **(Example 3)**

Online Streaming of TV Shows/Movies

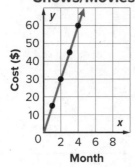

Test Practice

4. **Open Response** The cost per slice of pizza is proportional to the number of slices as shown in the graph. What do the points (0, 0) and (1, 2) represent? **(Example 4)**

Pizza Slices Cost

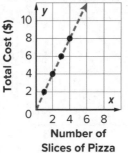

Apply

5. The table shows the number of Calories José and Natalie burned after 1, 2, and 3 minutes of running. Graph the relationship between the number of minutes running and the number of Calories burned for each person. By which ordered pair is a unit rate represented?

	Calories Burned	
Time (min)	José	Natalie
0	0	0
1	8	5
2	15	10
3	23	15

Calories Burned

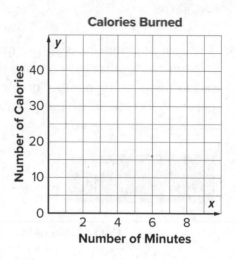

6. **MP Identify Repeated Reasoning** Suppose a relationship is proportional and the point (4, 10) lies on the graph of the proportional relationship. Name another point, other than (0, 0), that lies on the graph of the line.

7. **MP Make an Argument** Determine if a line can have a constant rate and not be proportional. Write an argument to defend your response.

8. **MP Find the Error** Karl said the point (1, 1) represents the constant of proportionality for the graph shown. Find his error and correct it.

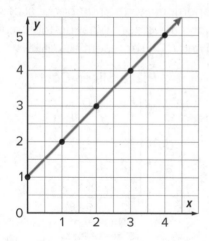

9. **Create** The graph of a proportional relationship is shown. Describe a real-world situation that could be represented by the graph. Be sure to include the meaning of the constant of proportionality.

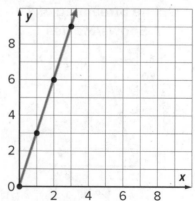

Equations of Proportional Relationships

I Can... write equations to represent proportional relationships and identify the constant of proportionality in the equation representing a proportional relationship.

Explore Proportional Relationships and Equations

Online Activity You will explore the equations of proportional relationships.

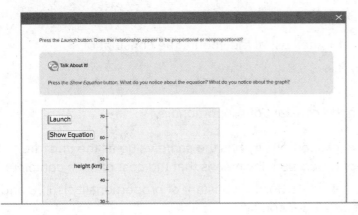

Learn Identify the Constant of Proportionality in Equations

Two quantities are proportional if the ratios comparing them are equivalent. This ratio is called the constant of proportionality. Proportional relationships can be represented by an equation in the form $y = kx$, where k is the constant of proportionality.

Words	Symbols
A linear relationship is proportional when the ratio of y to x is a constant k.	$y = kx$, where $k \neq 0$

Example	Graph
$y = 3x$	

Talk About It!

Can you write an equation in the form $y = kx$ for a line that does not pass through the origin? Explain.

🌐 **Example 1** Identify the Constant
of Proportionality in Equations

Olivia bought six containers of yogurt for $7.68. The equation
$y = 1.28x$ can be used to represent this situation, where y represents
the total cost of the yogurt and x represents the number of containers
bought.

**Identify the constant of proportionality. Then explain what it
represents.**

Part A Identify the constant of proportionality.

Compare the two equations.

$y = kx$, where k is the constant of proportionality

$y = 1.28x$

In the equation for the cost of the yogurt, the constant of
proportionality k is _____ .

Part B Explain the constant of proportionality.

The constant of proportionality has the same value as the unit rate.
The constant of proportionality means that the cost of each container
of yogurt is $ _____ . So, the constant of proportionality is 1.28 and
the unit rate is $1.28 per container.

💬 **Talk About It!**

How much would Olivia
pay for 10 yogurts at
this same rate?

Check

An airplane travels 780 miles in 4 hours. The equation $y = 195x$
models this situation.

Part A What is the constant of proportionality?

Part B What does the constant of proportionality represent in the
context of the problem?

Ⓐ An airplane travels 390 miles per hour.

Ⓑ An airplane travels 1 mile per 4 hours.

Ⓒ An airplane travels 1 mile per 585 hours.

Ⓓ An airplane travels 195 miles per hour.

🌐 **Go Online** You can complete an Extra Example online.

Learn Proportional Relationships and Equations

You can use the constant of proportionality, or unit rate, to write an equation in the form $y = kx$ that represents a proportional relationship.

You can find k by writing the ratio comparing y to x. To write the ratio, solve the equation $y = kx$ for k.

$y = kx$ Write the equation.

$\dfrac{y}{x} = \dfrac{kx}{x}$ Divide each side by x.

$\dfrac{y}{x} = k$ Simplify.

The constant of proportionality is the ratio $\dfrac{y}{x}$, where $x \neq 0$.

Example 2 Proportional Relationships and Equations

Jaycee bought 8 gallons of gas for $31.12.

Write an equation relating the total cost y to the number of gallons of gas x if it is a proportional relationship.

Step 1 Find the constant of proportionality between cost and gallons.

$$\frac{\text{total cost (\$)}}{\text{number of gallons}} = \frac{31.12}{8}$$

$$= \frac{3.89}{1}$$

The constant of proportionality is _____.

Step 2 Write the equation.

The total cost is $3.89 multiplied by the number of gallons. Let y = total cost and x = number of gallons.

So, the total cost for any number of gallons can be found using the equation $y = 3.89x$.

Check

A mechanic charges $270 for 5 hours of work on a car. Write an equation that compares the total cost to the number of hours he works on a car if it is a proportional relationship.

Show your work here

Go Online You can complete an Extra Example online.

Talk About It!
What is the equation for a proportional relationship where the constant of proportionality is 3.25?

Talk About It!
How can you use the equation $y = 3.89x$ to determine the cost for 15 gallons of gas?

Example 3 Proportional Relationships and Equations

The perimeter of a square is proportional to the length of one side. A square with 10-inch sides has a perimeter of 40 inches.

Write an equation relating the perimeter of the square to its side length. Then find the perimeter of a square with a 6-inch side.

Part A Write an equation.

The perimeter of a square with 10-inch sides is 40 inches. Write a ratio that compares the perimeter to the side length. Find the constant of proportionality. Find an equivalent ratio with a denominator of 1.

$$\frac{\text{perimeter}}{\text{side length}} = \frac{\square}{\square} = \frac{\square}{1}$$

So, the constant of proportionality is _____.

The perimeter is 4 times the length of a side, so the equation is written $y = 4x$.

Part B Use the equation to find the perimeter of a square with a side length of 6 inches.

$y = 4x$	Write the equation.
$y = 4(6)$	Replace x with 6.
$y = 24$	Multiply.

So, the perimeter of a square with sides of 6 inches is _____ inches.

Check

Jack is in charge of the 120,000-gallon community swimming pool. Each spring, he drains the pool in order to clean it. When finished, he refills the pool with fresh water. Jack can fill the pool with 500 gallons of water in 5 minutes.

Part A
Write an equation that represents the relationship between the total number of gallons y and the number of minutes spent filling the pool x.

Part B
How long will it take to become completely filled? _____

 Go Online You can complete an Extra Example online.

Think About It!
What do you need to find before you can write the equation?

Talk About It!
Suppose you know the perimeter of a square is 52 inches. Explain how you could use the equation $y = 4x$ to find the length of the side length of the square. Then find the length of the side.

🌐 Apply Running

Hugo can run 4 miles in 25 minutes. How many more miles can Hugo run in 90 minutes than in 25 minutes? Assume the relationship is proportional and he runs at a constant rate.

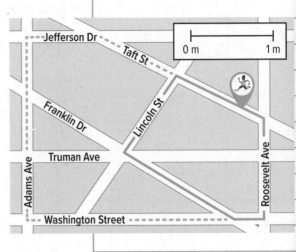

1 What is the task?

Make sure you understand exactly what question to answer or problem to solve. You may want to read the problem three times. Discuss these questions with a partner.

First Time Describe the context of the problem, in your own words.
Second Time What mathematics do you see in the problem?
Third Time What are you wondering about?

2 How can you approach the task? What strategies can you use?

3 What is your solution?

Use your strategy to solve the problem.

4 How can you show your solution is reasonable?

✏️ **Write About It!** Write an argument that can be used to defend your solution.

> 💬 **Talk About It!**
> Can you always assume that Hugo runs at a constant rate for any amount of time predicted? Explain your reasoning.

Check

Emily can ride her bike 15 kilometers in 45 minutes. How many more kilometers can she bike in two hours than in 45 minutes? Assume the relationship is proportional and she always bikes at the same rate.

Go Online You can complete an Extra Example online.

Foldables It's time to update your Foldable, located in the Module Review, based on what you learned in this lesson. If you haven't already assembled your Foldable, you can find the instructions on page FL1.

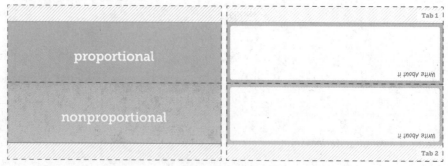

Practice

Go Online You can complete your homework online.

1. Liv earns $9.50 for every two bracelets she sells. The equation $y = 4.75x$, where x represents the number of bracelets and y represents the total cost in dollars earned, represents this situation. What is the constant of proportionality? What does the constant of proportionality represent in the context of the problem? (Example 1)

2. John ran 3 miles in 25.5 minutes. The equation $y = 8.5x$, where x represents the number of miles and y represents the total time in minutes, represents this situation. What is the constant of proportionality? What does the constant of proportionality represent in the context of the problem? (Example 1)

3. Lincoln bought 3 bottles of an energy drink for $4.50. Write an equation relating the total cost y to the number of energy drinks bought x. (Example 2)

4. The total cost of renting a cotton candy machine for 4 hours is $72. What equation can be used to model the total cost y for renting the cotton candy machine x hours? (Example 2)

5. Marley used 7 cups of water to make 4 loaves of French bread. What equation can be used to model the total cups of water needed y for making x loaves of French bread? How many cups of water do you need for 6 loaves of French bread? (Example 3)

6. Mrs. Henderson used $6\frac{3}{4}$ yards of fabric to make 3 elf costumes. What equation can be used to model the total number of yards of fabric y for x costumes? How many yards of fabric do you need for 7 elf costumes? (Example 3)

Test Practice

7. **Multiselect** The table shows the cost of 4 movie tickets at two theaters. Select the statements that are true about the situation.

Theater	Cost ($)
Movies Galore	30
Star Cinema	31

☐ The equation $y = 7.75x$ models the cost for tickets at Star Cinema.

☐ The equation $y = 30x$ models the cost for tickets at Movies Galore.

☐ The total cost of 9 tickets at Star Cinema would cost $69.75.

☐ The total cost for 1 ticket at Movies Galore is $30.

Apply

8. Roman can type 3 pages in 60 minutes. How many more pages can Roman type in 90 minutes than in 60 minutes? Assume the relationship is proportional and he types at a constant rate.

9. On average, Asia makes 14 out of 20 free throws. Assuming the relationship is proportional, how many more free throws is she likely to make if she shoots 150 free throws?

10. Evan earned $26 for 4 hours of babysitting. What equation can be used to model his total earnings y for babysitting x hours? Then graph the equation on the coordinate plane. What is the unit rate? How is that represented on the graph?

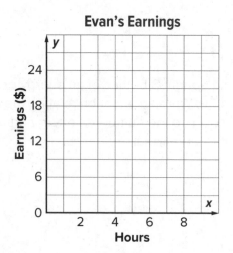

Evan's Earnings

11. 🔖 **Persevere with Problems** The Diaz family spent $38.25 on 3 large pizzas. What is the cost of one large pizza? Assume the situation is proportional. Explain how you solved.

12. 🔖 **Use a Counterexample** Determine whether the statement is *true* or *false*. If false, give a counterexample.
    The constant of proportionality in an equation can never be 0.

13. 🔖 **Justify Conclusions** A recipe for homemade modeling clay includes $\frac{1}{3}$ cup of salt for every cup of water. If there are 6 cups of salt, how many gallons of water are needed? Identify the constant of proportionality. Explain your reasoning.

Solve Problems Involving Proportional Relationships

I Can... solve problems involving proportional relationships by making a table, using a graph, or writing an equation.

Learn Proportions

A **proportion** is an equation stating that two ratios are equivalent. Suppose a recipe indicates a ratio of 3 cups of milk for every 4 cups of flour. The ratio 3 cups of milk to 4 cups of flour is equivalent to 6 cups of milk to 8 cups of flour. This relationship can be written as a proportion.

cups of milk ⟶ $\dfrac{3}{4} = \dfrac{6}{8}$ ⟵ cups of milk
cups of flour ⟶ ⟵ cups of flour

The equals sign means that the ratios are equivalent and the original ratio of 3 cups of milk for every 4 cups of flour is maintained. The unit rate is $\dfrac{3}{4}$ cup of milk for every cup of flour.

You can use any representation to solve a problem involving a proportional relationship. The multiple representations table shows different methods for finding the number of cups of milk needed if 3 cups of flour are used.

Words	Ratio Table
The unit rate is $\dfrac{3}{4}$ cup of milk for every cup of flour. If you use 3 cups of flour, you need $3\left(\dfrac{3}{4}\right)$ or $2\dfrac{1}{4}$ cups of milk.	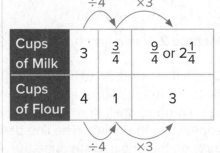 

Graph	Example
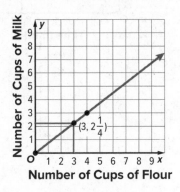	The unit rate is $\dfrac{3}{4}$ cup of milk to 1 cup of flour, so the constant of proportionality is $\dfrac{3}{4}$. Let y represent the number of cups of milk, and x represent the number of cups of flour. $$y = \dfrac{3}{4}x$$ $$y = \dfrac{3}{4}(3), \text{ or } 2\dfrac{1}{4}$$

Number of Cups of Milk (graph y-axis)
Number of Cups of Flour (graph x-axis)
(3, 2$\tfrac{1}{4}$)

Copyright © McGraw-Hill Education

What Vocabulary Will You Learn?
proportion

🗨 Talk About It!
How does each representation show the unit rate?

Copyright © McGraw-Hill Education

Think About It!

Is the expected wait time for 240 people less than, greater than, or equal to 40 minutes? How do you know?

Talk About It!

How does the ratio table illustrate the unit rate, or constant of proportionality?

Talk About It!

In this situation, how accurate is the use of a graph when finding the x- or y-coordinate?

🌐 Example 1 Solve Problems Involving Proportional Relationships

The wait time to ride a roller coaster is 20 minutes when 160 people are in line.

How long is the expected wait time when 240 people are in line? Assume the relationship is proportional.

Method 1 Use a table.

There is no whole number by which you can multiply 160 by to obtain 240.

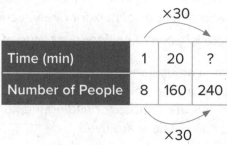

Scale back to find the unit rate. Because 20 ÷ 20 = 1, divide 160 by 20 to obtain 8. When 8 people are in line, the expected wait time is 1 minute.

Then scale forward to find the expected wait time when 240 people are in line. Because 8(30) = 240, multiply 1 by 30 to obtain 30.

When 240 people are in line, the expected wait time is 30 minutes.

Method 2 Use a graph.

Graph the points (0, 0) and (20, 160). Because the relationship is assumed to be proportional, draw a dotted line connecting the points. The line passes through the origin. Determine the corresponding x-coordinate for when the y-coordinate is 240.

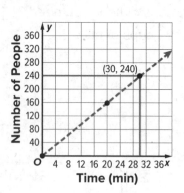

The corresponding x-coordinate when the y-coordinate is 240 appears to be 30. So, the expected wait time when 240 people are in line is 30 minutes.

(continued on next page)

Method 3 Use an equation.

The equation that represents a proportional relationship is of the form $y = kx$, where k is the constant of proportionality, or unit rate.

Find the unit rate, the number of people in line when the expected wait time is 1 minute. Let y represent the number of people in line and x represent the number of minutes.

$y = kx$	Write the equation.
$160 = k(20)$	Replace y with 160 and x with 20.
$\dfrac{160}{20} = \dfrac{k(20)}{20}$	Divide each side by 20 to find the value of k.
$8 = k$	Simplify. The unit rate, or constant of proportionality, is 8.

When 8 people are in line, the expected wait time is 1 minute. Use the equation $y = 8x$ to find the expected wait time when 240 people are in line.

$y = 8x$	Write the equation. The constant of proportionality is 8.
$240 = 8x$	Replace y with 240.
$\dfrac{240}{8} = \dfrac{8x}{8}$	Divide each side by 8 to find the value of x.
$30 = x$	Simplify.

So, the expected wait time when 240 people are in line is 30 minutes.

Check

Matthew paid $49.45 for five used video games of equal cost. The relationship between the number of video games and the total cost is proportional. What is the total cost for 11 used video games? Use any strategy.

Show your work here

Go Online You can complete an Extra Example online.

 Talk About It!

Suppose a classmate drew the double number line shown to represent and solve this problem. Is this a valid method? Explain. Does this method show the unit rate? Is the unit rate necessary to solve the problem? Explain.

Time (min)	0	10	20	30
Number of People	0	80	160	240

After two hours, the outside air temperature had risen 6°F. The temperature is forecasted to continue to increase at this same rate for the next several hours.

At this rate, how many hours will it take the temperature to rise an additional 13°F?

Choose a strategy for solving this problem. For this problem, using an equation is an advantageous strategy because calculations with fractional values will be involved. You know this because there is no whole number by which you can multiply 6 to obtain 13.

The equation that represents a proportional relationship is of the form $y = kx$, where k is the constant of proportionality, or unit rate.

Find the unit rate, the rise in temperature in degrees Fahrenheit per hour. Let y represent the rise in temperature in degrees Fahrenheit and x represent the number of hours.

$y = kx$	Write the equation.
$6 = k(2)$	Replace y with 6 and x with 2.
$\dfrac{6}{2} = \dfrac{k(2)}{2}$	Divide each side by 2 to find the value of k.
$3 = k$	Simplify. The unit rate, or constant of proportionality, is 3.

Each hour, the temperature is forecasted to rise 3°F.

Use the equation $y = 3x$ to find the number of hours it is expected to take the temperature to rise an additional 13°F.

$y = 3x$	Write the equation. The constant of proportionality is 3.
$13 = 3x$	Replace y with 13.
$\dfrac{13}{3} = \dfrac{3x}{3}$	Divide each side by 3 to find the value of x.
$4\dfrac{1}{3} = x$	Simplify.

So, after $4\dfrac{1}{3}$ hours, or 4 hours and 20 minutes, the temperature is forecasted to rise an additional 13°F.

Check

Brooke bought 8 bottled teas for $13.52. Assume the relationship between the number of bottled teas and total cost, in dollars, is proportional. How much can Brooke expect to pay for 12 bottled teas?

🌐 **Go Online** You can complete an Extra Example online.

🌐 Apply Blood Drives

At a recent statewide blood drive, the ratio of Type O to non-Type O donors was 37 : 43. Suppose there are 300 donors at a local blood drive. About how many are Type O?

1 What is the task?

Make sure you understand exactly what question to answer or problem to solve. You may want to read the problem three times. Discuss these questions with a partner.

First Time Describe the context of the problem, in your own words.
Second Time What mathematics do you see in the problem?
Third Time What are you wondering about?

2 How can you approach the task? What strategies can you use?

Record your observations here

3 What is your solution?

Use your strategy to solve the problem.

Show your work here

4 How can you show your solution is reasonable?

⚡ **Write About It!** Write an argument that can be used to defend your solution.

🔺 **Go Online** watch the animation online.

💬 **Talk About It!**
How can you use the ratio 37 : 43 to solve the problem?

Check

The ratio of girls to boys in a school is 2 : 3. The relationship between the number of girls and the number of boys is proportional. How many boys are there if there are 345 students in the school?

Show your work here

 Go Online You can complete an Extra Example online.

Pause and Reflect

Compare and contrast the different methods presented in this lesson for solving problems involving proportional relationships.

Record your observations here

Practice

⬥ **Go Online** You can complete your homework online.

For each problem, use any method. Assume each relationship is proportional. (Examples 1 and 2)

1. For every three girls taking classes at a martial arts school, there are 4 boys who are taking classes. If there are 236 boys taking classes, predict the number of girls taking classes at the school.

2. A grading machine can grade 96 multiple choice tests in 2 minutes. If a teacher has 300 multiple choice tests to grade, predict the number of minutes it will take the machine to grade the tests.

3. A 6-ounce package of fruit snacks contains 45 pieces. How many pieces would you expect in a 10-ounce package?

4. Of the 50 students in the cafeteria, 7 have red hair. If there are 750 students in the school, predict the number of students who have red hair.

Test Practice

5. The wait times for two different rides are shown in the table. If there are 120 people in line for the swings, how long can you expect to wait to ride the ride?

Ride	Wait Times
Carousel	6 minutes for 48 people in line
Swings	12 minutes for 75 people in line

6. **Open Response** Ingrid types 3 pages in the same amount of time that Tanya types 4.5 pages. If Ingrid and Tanya start typing at the same time and continue at their respective rates, how many pages will Tanya have typed when Ingrid has typed 11 pages?

Apply

7. The ratio of kids to adults at a school festival is 11 : 7. Suppose there are a total of 810 kids and adults at the festival. How many adults are at the festival?

8. The ratio of laptops to tablets in the stock room of a store is 13 : 17. If there are a total of 90 laptops and tablets in the stock room, how many laptops are in the stock room?

9. (MP) **Persevere with Problems** Lisa is painting the exterior surfaces at her home. A gallon of paint will cover 350 square feet. How many gallons of paint will Lisa need to paint one side of her fence? Explain how you solved.

Item to Paint	Length (ft)	Width (ft)
Fence	26	7
Barn Door	11	6

10. (MP) **Find the Error** The rate of growth for a plant is 0.2 centimeter per 0.5 day. A student found the number of days for the plant to grow 3.6 centimeters to be 1.44 days. Find the error and correct it.

11. **Create** Write a real-world problem involving a proportional relationship. Then solve the problem.

12. (MP) **Be Precise** When is it more beneficial to solve a problem involving a proportional relationship using an equation than using a graph?

Foldables Use your Foldable to help review the module.

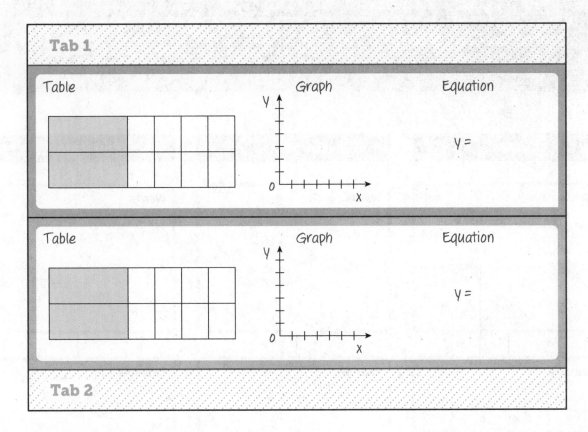

Tab 1

Table   Graph   Equation

y =

Table   Graph   Equation

y =

Tab 2

Rate Yourself!

Complete the chart at the beginning of the module by placing a checkmark in each row that corresponds with how much you know about each topic after completing this module.

Write about one thing you learned.

_____

_____

_____

_____

_____

_____

Write about a question you still have.

_____

_____

_____

_____

_____

Reflect on the Module

Use what you learned about proportional relationships to complete the graphic organizer.

Essential Question

What does it mean for two quantities to be in a proportional relationship?

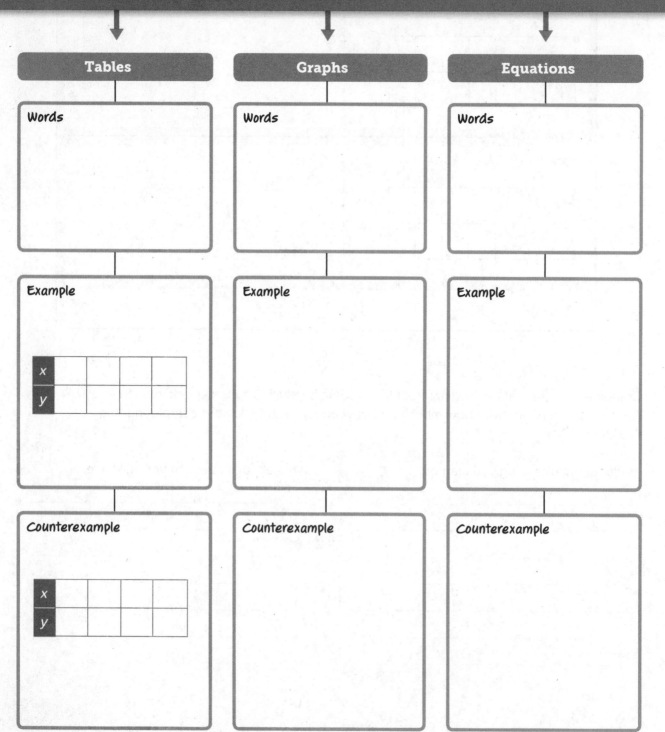

Tables	Graphs	Equations
Words	**Words**	**Words**
Example	**Example**	**Example**

Example (Tables)

x			
y			

| **Counterexample** | **Counterexample** | **Counterexample** |

Counterexample (Tables)

x			
y			

Test Practice

1. Multiple Choice Noreen can walk $\frac{1}{3}$ mile in 12 minutes. What is her average speed in miles per hour? (Lesson 1)

(A) 36 miles per hour

(B) 12 miles per hour

(C) $\frac{1}{4}$ mile per hour

(D) $1\frac{2}{3}$ miles per hour

2. Open Response The table shows the amount spent on tomatoes at three different stands at a farmer's market. Which stand sold their tomatoes at the least expensive price per pound, and what is that price? Round to the nearest cent, if necessary. (Lesson 1)

Stand	Weight (lb)	Cost ($)
A	$\frac{3}{4}$	2.00
B	$1\frac{1}{4}$	3.45
C	$5\frac{1}{2}$	14.85

3. Multiselect One recipe for homemade playdough calls for 4 parts flour, 1 part salt, and 2 parts water. Select all of the mixtures below that are in a proportional relationship with this recipe. (Lesson 2)

☐ 8 cups flour, 2 cups salt, 4 cups water

☐ 2 cups flour, $\frac{1}{2}$ cup salt, $\frac{1}{2}$ cup water

☐ 6 cups flour, $1\frac{1}{2}$ cups salt, 3 cups water

☐ 10 cups flour, 1 cup salt, 2 cups water

4. Open Response The ratio of Braydon's number of laps he ran to the time he ran is 6 : 2. The ratio of Monique's number of laps she ran to the time she ran is 10 : 4. Explain why these ratios are not in a proportional relationship. (Lesson 2)

5. Open Response One month, Miko bought two books at a used book store for a total of $2. Over the next few months, she bought four books for a total of $5, six books for a total of $7, and eight books for a total of $10. Is the cost proportional to the number of books purchased? Explain. (Lesson 3)

6. Table Item The table shows the time and distance Hector walked in a 5-kilometer-long walk for charity. The distance is proportional to the time walked. Complete the table to show the times of his last three walks. (Lesson 3)

Distance (km)	Time (min)
1.5	13.5
2	
4.5	
5	

7. Multiselect The relationship between the number of slices of pizza purchased and the number of students served is shown in the graph. Select all of the statements that are true. (Lesson 4)

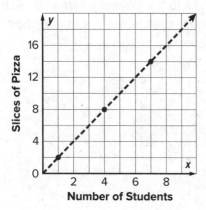

☐ The relationship is proportional.

☐ The point (9, 18) satisfies this relationship.

☐ The constant of proportionality is $\frac{1}{2}$.

☐ The constant of proportionality is 2.

8. Grid The cost of dance lessons is $12 for 1 lesson, $22 for 2 lessons, and $32 for 3 lessons. (Lesson 4)

A. Graph the ordered pairs on the coordinate plane.

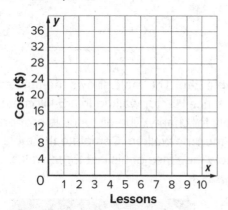

B. Determine whether the cost is proportional to the number of lessons. Explain your reasoning.

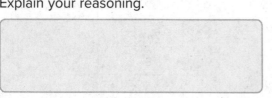

9. Equation Editor Mrs. Jameson paid $202.50 for a group of 9 students to visit an amusement park. (Lesson 5)

A. Write an equation relating the total cost y and the number of students x attending the park.

B. What would be the total cost if four more students wanted to join the group?

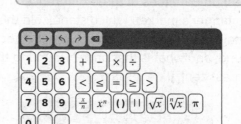

10. Equation Editor A homeowner whose house was assessed at $120,000 pays $1,800 in taxes. At the same rate, what is the tax on a house assessed at $135,000? (Lesson 6)

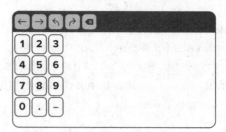

Solve Percent Problems

℮ Essential Question

How can percent describe the change of a quantity?

What Will You Learn?

Place a checkmark (✓) in each row that corresponds with how much you already know about each topic **before** starting this module.

KEY	Before			After		
▢ — I don't know. ◈ — I've heard of it. ★ — I know it!	▢	◈	★	▢	◈	★
finding percent of change						
solving problems involving taxes						
solving problems involving tips and markups						
solving problems involving discounts						
solving problems involving interest						
solving problems involving commissions and fees						
finding percent error						

📖 Foldables Cut out the Foldable and tape it to the Module Review at the end of the module. You can use the Foldable throughout the module as you learn about solving problems involving percent.

What Vocabulary Will You Learn?

Check the box next to each vocabulary term that you may already know.

☐ amount of error ☐ markdown ☐ principal

☐ commission ☐ markup ☐ sales tax

☐ discount ☐ percent error ☐ selling price

☐ fee ☐ percent of change ☐ simple interest

☐ gratuity ☐ percent of decrease ☐ tip

☐ interest ☐ percent of increase ☐ wholesale cost

Are You Ready?

Study the Quick Review to see if you are ready to start this module.
Then complete the Quick Check.

Quick Review

Example 1
Multiply with decimals.

Multiply 240 × 0.03 × 5.

240 × 0.03 × 5

$\quad$ = 7.2 × 5 $\qquad$ Multiply 240 by 0.03.

$\quad$ = 36 $\qquad$ Multiply 7.2 by 5.

Example 2
Write decimals as fractions and percents.

Write 0.35 as a fraction.

$0.35 = \frac{35}{100}$ $\qquad$ 0.35 means *thirty-five hundredths*

$\quad = \frac{7}{20}$ $\qquad$ Find an equivalent ratio.

Write 0.35 as a percent.

$0.35 = \frac{35}{100}$ $\qquad$ 0.35 means *thirty-five hundredths*

$\quad = 35\%$ $\qquad$ Definition of percent

Quick Check

1. Suppose Nicole saves $2.50 every day. How much money will she have in 4 weeks?

2. Approximately 0.92 of a watermelon is water. What percent and what fraction, in simplest form, represent this decimal?

How Did You Do?

Which exercises did you answer correctly in the Quick Check?
Shade those exercise numbers at the right.

Percent of Change

I Can... use proportional relationships to solve percent of change problems.

Explore Percent of Change

Online Activity You will use bar diagrams to determine how a percent can be used to describe a change when a quantity increases or decreases.

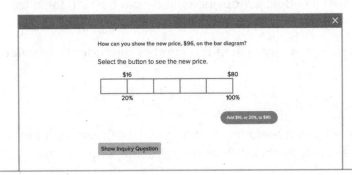

Learn Percent of Increase

A **percent of change** is a ratio, written as a percent, which compares the change in quantity to the original amount. If the original amount increased, then it is called a **percent of increase**.

The price of a monthly internet plan increased from $36 to $45. You can use a bar diagram to determine the percent of increase.

Draw a bar to represent the original price, $36. Because the original price is the whole, label the length of the bar 100%.

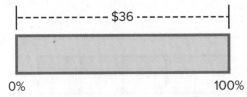

The price increased from $36 to $45, which is an increase of $9. Because $36 divided by $9 is 4, divide the bar representing $36 into four equal-size sections of $9 each. Each section represents 25% of the whole, $36.

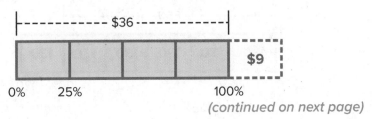

(continued on next page)

Copyright © McGraw-Hill Education

What Vocabulary Will You Learn?

percent of change

percent of decrease

percent of increase

Talk About It!

When finding percent of change, why is it important that the two quantities have the same unit of measure?

Each $9 section is 25% of the whole, $36. So, the price increased by 25%.

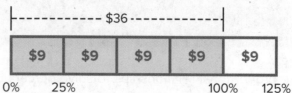

Note that the new price, $45, is 125% of the original price, but that the price increased by 25%.

The enrollment at a middle school increased from 650 students to 780 students in five years.

What is the percent of increase in the number of students enrolled at the school?

Method 1 Use a bar diagram.

Draw a bar to represent the original enrollment, 650. Because the original number of students is the whole, label the length of the bar 100%.

Enrollment increased from 650 to 780 students, which is an increase of 130 students. Divide the bar into equal parts so that the amount of increase, 130, is represented by one of the parts. Because 650 divided by 130 is 5, divide the bar representing 650 into five equal-size sections of 130 students each. Each section represents 20% of the whole, 650.

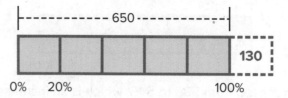

Each section that represents 130 students is 20% of the whole, 650. So, the enrollment increased by 20%.

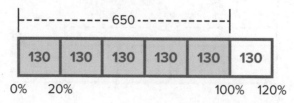

(continued on next page)

By how many students did the enrollment increase?

Copyright © McGraw-Hill Education

64 **Module 2** • Solve Percent Problems

Method 2 Use equivalent ratios.

Step 1 Identify the part and the whole.

original amount = 650   This is the whole.

new amount = 780   This is the whole plus the part.

amount of increase = 130   This is the part.

Step 2 Find the percent of increase.

$$\frac{part}{whole} = \frac{130}{650}$$ Write the part-to-whole ratio. The part is 130. The whole is 650.

$$= 0.20$$ Divide.

$$= \frac{20}{100}$$ Write an equivalent ratio, as a rate per 100.

$$= 20\%$$ Definition of percent

So, using either method, the percent of increase in student enrollment is 20%.

Check

A certain brand of shampoo contains 12 ounces in a bottle. The company decides to redesign the shape of the shampoo bottle. The new type of bottle contains 15 ounces of shampoo. What is the percent of increase in the amount of shampoo the new bottle contains? Use any strategy.

Go Online You can complete an Extra Example online.

Pause and Reflect

Describe some examples of where you might see percent of change in your everyday life.

Example 2 Percent of Increase

The average cost of gas in 2000 was about $1.50 per gallon. Suppose you purchase gas for $3.30 per gallon.

What is the percent of increase in the cost of a gallon of gas?

Step 1 Identify the part and the whole.

original amount = $1.50   This is the whole.

new amount = $3.30   This is the whole plus the part.

amount of increase = $1.80   This is the part.

Step 2 Find the percent of increase.

$$\frac{part}{whole} = \frac{1.80}{1.50}$$ Write the part-to-whole ratio. The part is 1.80. The whole is 1.50.

$$= 1.20$$ Divide.

$$= \frac{120}{100}$$ Write an equivalent ratio, as a rate per 100.

$$= 120\%$$ Definition of percent

So, the percent of increase in the price per gallon of gas is 120%.

Check

Frankie calculated that she would need 9.5 yards of wallpaper to decorate her room. The wallpaper she chose had a large pattern, so instead she needed 13 yards of that paper. What is the percent of increase, rounded to the nearest percent, in the amount of wallpaper she needs?

Show your work here

 Go Online You can complete an Extra Example online.

Learn Percent of Decrease

A percent of change is a ratio, written as a percent, which compares the change in quantity to the original amount. If the original amount is decreased, then the ratio is called a **percent of decrease.**

Jordan trained for a 5K race and his time decreased from 35 minutes to 28 minutes. You can use a bar diagram to determine the percent of decrease.

Draw a bar to represent the original time, 35 minutes. Because the original time is the whole, label the length of the bar 100%.

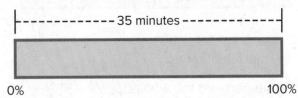

Jordan's time decreased from 35 minutes to 28 minutes, which is a decrease of 7 minutes. Because 35 divided by 7 is 5, divide the bar into five equal-size sections of 7-8 minutes each. Each section represents 20% of the whole, 35 minutes.

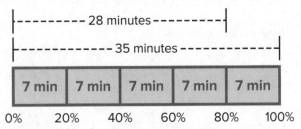

Each 7-minute section is 20% of the whole, 35. So, Jordan's time decreased by 20%.

Note that Jordan's new time is 80% of his original time, but that his time decreased by 20%.

Talk About It!

The percent of decrease is 20%. Explain why it is not 80%.

🌐 Example 3 Percent of Decrease

At the beginning of a chemistry experiment, the volume of liquid in a container was 25.2 milliliters. During the experiment, the volume dropped to 18.9 milliliters.

What is the percent of decrease in the volume of liquid?

Method 1 Use a bar diagram.

Draw a bar to represent the original volume, 25.2 milliliters. Because the original volume is the whole, label the length of the bar 100%.

Think About It!

Before you can find the percent of decrease, what value do you need to calculate?

(continued on next page)

The volume decreased from 25.2 milliliters to 18.9 milliliters, which is a decrease of 6.3 milliliters. Because 25.2 divided by 6.3 is 4, divide the bar into four equal-size sections of 6.3 milliliters each. Each section represents 25% of the whole, 25.2 milliliters. So, the percent of decrease in the volume of the liquid is 25%.

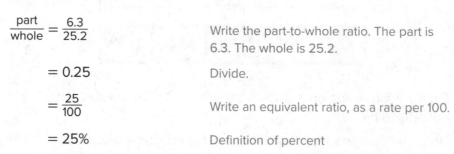

Method 2 Use equivalent ratios.

Step 1 Identify the part and the whole.

original amount = 25.2   This is the whole.

new amount = 18.9   This is the whole minus the part.

amount of decrease = 6.3   This is the part.

Step 2 Find the percent of increase.

$$\frac{part}{whole} = \frac{6.3}{25.2}$$ Write the part-to-whole ratio. The part is 6.3. The whole is 25.2.

$$= 0.25$$ Divide.

$$= \frac{25}{100}$$ Write an equivalent ratio, as a rate per 100.

$$= 25\%$$ Definition of percent

So, the percent of decrease in the volume of the liquid is 25%.

Check

When resting, a spring measures 73.3 millimeters. When the spring is compressed, it measures 51.29 millimeters. Find the percent of decrease in the length of the spring. Round to the nearest percent if necessary. Use any strategy.

 Go Online You can complete an Extra Example online.

> 💬 Talk About It!
>
> How can you use estimation to know that your answer is reasonable?

🌐 **Apply** Movies

The first known motion picture was filmed in 1888 and lasted for only 2.11 seconds. Today, we watch movies that last an average of about two hours. What is the percent of change in the times from 1888 to today? Round your answer to the nearest whole percent if necessary.

1 What is the task?

Make sure you understand exactly what question to answer or problem to solve. You may want to read the problem three times. Discuss these questions with a partner.

First Time Describe the context of the problem, in your own words.
Second Time What mathematics do you see in the problem?
Third Time What are you wondering about?

2 How can you approach the task? What strategies can you use?

Record your observations here

3 What is your solution?

Use your strategy to solve the problem.

Show your work here

4 How can you show your solution is reasonable?

✎ **Write About It!** Write an argument that can be used to defend your solution.

💬 Talk About It!
Why is the percent of change such a large number?

Check

The graph shows Alexa's bank account balance over the past four months. Between which consecutive months is the percent of increase the least?

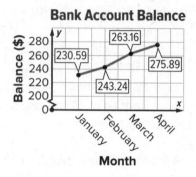

Bank Account Balance

Show your work here

Go Online You can complete an Extra Example online.

Foldables It's time to update your Foldable, located in the Module Review, based on what you learned in this lesson. If you haven't already assembled your Foldable, you can find the instructions on page FL1.

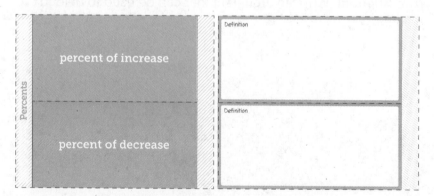

Practice

Go Online You can complete your homework online.

Find each percent of change. Identify it as a percent of increase or decrease. (Examples 1–3)

1. 8 feet to 10 feet

2. 62 trees to 31 trees

3. 136 days to 85 days

4. Last month, the online price of a powered ride-on car was $250. This month, the online price is $330. What is the percent of increase for the price of the car? (Example 1)

5. At end of the first half of a football game, Nathan had carried the ball for 50.5 yards. By the end of the game, he carried the ball for a total of 75 yards. Find the percent of increase in the number of yards he carried. Round to the nearest whole tenth if necessary. (Example 1)

6. A music video website received 5,000 comments on a new song they released. The next day, the artist performed the song on television and an additional 1,500 comments were made on the website. What was the percent of increase? (Example 1)

7. When Ricardo was 9 years old, he was 56 inches tall. Ricardo is now 12 years old and he is 62 inches tall. Find the percent of increase in Ricardo's height to the nearest tenth. (Example 1)

8. At a garage sale, Petra priced her scooter for $15.50. She ended up selling it for $10.75. Find the percent of decrease in the price of the scooter. Round to the nearest tenth if necessary. (Example 2)

9. At the beginning of a baking session, there were 2.26 kilograms of flour in the bag. By the end of the baking session, there was 0.98 kilogram of flour in the bag. What is the percent of decrease, rounded to the nearest tenth, for the amount of flour? (Example 2)

Test Practice

10. Open Response The table shows the number of candid pictures of students for the yearbook for two consecutive years. What was the percent of decrease in the number of candid student pictures from 2015 to 2016, rounded to the nearest tenth?

Year	Number of Photos
2015	236
2016	214

11. The side length of the square shown is tripled. Which percent of increase is greater: the percent of increase for the perimeter of the square or the percent of increase for the area? How much greater?

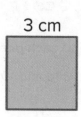

3 cm

12. The Keatings have several bird feeders in their yard. They started with a 10.5-pound of birdseed. After 2 months, 12 ounces remained. What is the percent of change in the amount of birdseed after two months? Round your answer to the nearest tenth of a percent if necessary.

13. **Create** Write and solve a real-world problem involving a percent of increase with decimals.

14. (MP) **Justify Conclusions** Each of Mrs. White's two children received $50 for their savings account. The original amounts in their savings accounts were $500 and $300, respectively. Without calculating, which savings account had a greater percent of increase? Explain.

15. (MP) **Reason Abstractly** Determine if the following statement is *true* or *false*. Explain.

    When the percent of change is a decrease, the original amount will be greater than the new amount.

16. (MP) **Reason Abstractly** Can a percent of change be greater than 100%? Explain.

Tax

I Can... use proportional relationships to find the amount of tax charged for an item.

What Vocabulary Will You Learn?
sales tax

Explore Sales Tax

Online Activity You will use Web Sketchpad to explore how sales tax changes the total cost to purchase an item.

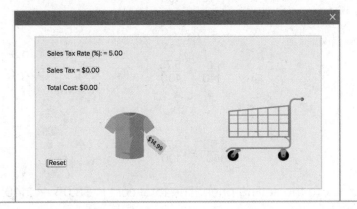

Sales Tax Rate (%): = 5.00

Sales Tax = $0.00

Total Cost: $0.00

$14.99

Reset

Learn Sales Tax

A tax is an amount of money added to the price of certain goods and services. Tax is usually calculated as a percentage of the cost of the item. Some common forms of tax are sales tax, income tax, property tax, and hotel tax. Tax revenue is used to pay for government-provided services.

Sales tax is a state or local tax that is added to the price of an item or service. The total cost to purchase an item is the selling price plus the sales tax.

Sales tax varies depending on the city or state in which you live. Consider the cost of a $10 T-shirt with a sales tax rate of 7.5%. You can use equivalent ratios, and the percentage written as a rate per 100, to determine the amount of the tax. Let t represent the amount of tax.

sales tax on T-shirt ⟶ $\left.\dfrac{t}{10} = \dfrac{7.5}{100} \right\}$ Percent
cost of T-shirt ⟶

÷10

$\dfrac{0.75}{10} = \dfrac{7.5}{100}$

Because 100 ÷ 10 is 10, divide 7.5 by 10 to find t. The amount of tax is $0.75 or 75 cents.

÷10

Talk About It!

How can you find the total cost of the T-shirt, including the tax?

Before you can find the total Carie paid, what quantity do you need to find?

🗨 Talk About It!

Compare and contrast the two methods used for finding the total cost of the equipment.

🌐 **Example 1** Sales Tax

Carie wants to buy sports equipment that costs $140. The sales tax in her city is 5.75%.

What is the total cost of the equipment?

Method 1 Use ratio reasoning.

Write a proportion relating the two equivalent ratios. Let t represent the amount of sales tax. Then solve using ratio reasoning.

sales tax on sports equipment ⟶ $\dfrac{t}{140} = \dfrac{5.75}{100}$ } Percent
cost of sports equipment ⟶

$$\overset{\times 1.40}{\overgroup{\dfrac{t}{140} = \dfrac{5.75}{100}}}$$

Because $100 \times 1.40 = 140$, multiply 5.75 by 1.40 to find the value of t.

$$\underset{\times 1.40}{\undergroup{}}$$

$$\dfrac{8.05}{140} = \dfrac{5.75}{100}$$ $5.75 \times 1.40 = 8.05$, so, $t = 8.05$.

Add the sales tax to the selling price. The total cost is $8.05 + $140, or $148.05.

Method 2 Use properties of operations. Let t represent the amount of sales tax.

$\dfrac{t}{140} = \dfrac{5.75}{100}$ Write the proportion.

$\dfrac{t}{140} = 0.0575$ Divide 5.75 by 100. A one-step equation results.

$140 \cdot \left(\dfrac{t}{140}\right) = (0.0575) \cdot 140$ Multiplication Property of Equality; Notice the tax is equal to the product of the percent, written as a decimal, and the cost.

$t = 8.05$ Simplify.

Add the sales tax to the selling price. The total cost is $8.05 + $140, or $148.05.

So, using either method, the total cost of the sports equipment is $148.05.

Check

Ashley is buying a laptop that sells for $749. The sales tax rate in her city is $8\frac{1}{4}$%. What is the total cost for the laptop? Round your answer to the nearest cent. Use any strategy.

Go Online You can complete an Extra Example online.

🌐 Example 2 Hotel Tax

The cost of a hotel room rented for 2 nights is $280. There is also a 12% hotel room tax.

What is the total cost of the hotel room?

Method 1 Use ratio reasoning.

Write a proportion. Let t represent the amount of tax. Then solve using ratio reasoning.

tax on hotel room ⟶ $\dfrac{t}{280} = \dfrac{12}{100}$ ⎫ Percent
cost of hotel room ⟶   ⎭

$$\overset{\times 2.80}{\dfrac{t}{280} = \dfrac{12}{100}}\underset{\times 2.80}{}$$

Because $100 \times 2.80 = 280$, multiply 12 by 2.80 to find the value of t.

$$\dfrac{33.60}{280} = \dfrac{12}{100}$$ $12 \times 2.80 = 33.60$, so, $t = 33.60$.

Add the hotel tax to the cost of the hotel room. The total cost is $280 + $33.60, or $313.60.

(continued on next page)

Copyright © McGraw-Hill Education

> 💬 **Think About It!**
>
> Is the tax less than, greater than, or equal to $28? How do you know?

In Method 2, how could you use the steps in solving the equation to find the tax rate, or percentage, of any value?

Method 2 Use properties of operations. Let t represent the amount of tax.

$$\frac{t}{280} = \frac{12}{100}$$ Write the proportion.

$$\frac{t}{280} = 0.12$$ Divide 12 by 100. A one-step equation results.

$$280 \cdot \left(\frac{t}{280}\right) = (0.12) \cdot 280$$ Multiplication Property of Equality; Notice the tax is equal to the product of the percent, written as a decimal, and the cost.

$$t = 33.60$$ Simplify.

Add the hotel tax to the cost of the hotel room. The total cost is $280 + $33.60, or $313.60.

So, using either method, the total cost of the hotel room is $313.60.

Check

The cost of a hotel room for 5 nights is $610. There is a 9.5% hotel tax. What is the total cost of the hotel room? Use any strategy.

 Go Online You can complete an Extra Example online.

Pause and Reflect

Use the Internet, or another source, to research the sales tax rate for your city or state. Describe an item you may want to purchase and research the cost of the item. Trade with a partner and use the sales tax rate to find the total cost of the item, including sales tax.

🌐 **Example 3** Sales Tax

Henry purchases $56.00 worth of clothing at the store. The sales tax in his city is 7.5%.

What is the total cost of the clothing?

Method 1 Find the sales tax.

Use a proportion to find the amount of tax. Let t represent the amount of tax. Then add the tax to the cost of the clothing.

$$\frac{t}{56} = \frac{7.5}{100} \qquad \text{Write the proportion.}$$

$$\frac{t}{56} = 0.075 \qquad \text{Divide 7.5 by 100.}$$

$$56 \cdot \left(\frac{t}{56}\right) = (0.075) \cdot 56 \qquad \text{Multiplication Property of Equality; Notice the tax is equal to the product of the percent, written as a decimal, and the cost.}$$

$$t = 4.20 \qquad \text{Simplify.}$$

Add the sales tax to the selling price.

$$\$4.20 + \$56.00 = \$60.20 \qquad \text{The total cost is \$60.20.}$$

Method 2 Add the sales tax percent to 100%.

Because 100% represents the selling price, add the sales tax percent to 100%. The total percent is 100% + 7.5% or 107.5%.

$$\frac{c}{56} = \frac{107.5}{100} \qquad \text{Write the proportion. Let } c \text{ represent the total cost.}$$

$$\frac{c}{56} = 1.075 \qquad \text{Divide 107.5 by 100.}$$

$$56 \cdot \left(\frac{c}{56}\right) = (1.075) \cdot 56 \qquad \text{Multiplication Property of Equality; Notice the total cost is equal to the product of the total percent, written as a decimal, and the cost.}$$

$$c = 60.20 \qquad \text{Simplify.}$$

So, the total cost of the clothing is $60.20.

(continued on next page)

🫧 **Think About It!**
How would you begin solving the problem?

Copyright © McGraw-Hill Education

Talk About It!

If you are at a store and need to quickly calculate the total with sales tax, what method would you use? Explain.

Method 3 Write an equation.

The total cost is the cost of the clothing plus the sales tax.
Let c represent the total cost.

$c = 56 + 0.075(56)$	Write an equation. Write 7.5% as 0.075.
$= 56(1 + 0.075)$	Distributive Property
$= 56(1.075)$	Add.
$= 60.20$	Multiply.

Increasing the price by 7.5% is the same as multiplying the price by 1.075. So, using any method, the total cost of the clothing is $60.20.

Check

Jimena purchased $24 worth of crafting supplies. The tax rate in her city is 6.25%. What is the total cost of the crafting supplies? Use any strategy.

Go Online You can complete an Extra Example online.

Pause and Reflect

Compare what you have learned in this lesson to what you previously learned about proportional relationships.

🌐 **Apply** Shopping

Davit goes shopping at his local grocery store and buys the items shown in the table. In the city he lives, there is a 7.3% sales tax on all items, except food. How much money does Davit spend at the grocery store? Round to the nearest cent if necessary.

Item	Cost ($)
Chicken	6.25
Carrots	1.99
Potatoes	2.55
Paper Plates	5.50
Napkins	6.15

1 What is the task?

Make sure you understand exactly what question to answer or problem to solve. You may want to read the problem three times. Discuss these questions with a partner.

First Time Describe the context of the problem, in your own words.
Second Time What mathematics do you see in the problem?
Third Time What are you wondering about?

2 How can you approach the task? What strategies can you use?

Record your observations here

3 What is your solution?

Use your strategy to solve the problem.

Show your work here

Copyright © McGraw-Hill Education

4 How can you show your solution is reasonable?

🖋 **Write About It!** Write an argument that can be used to defend your solution.

> 💬 Talk About It!
>
> How can you use estimation to determine if your answer is reasonable?

Check

The table shows the items that Victoria purchased at the market. In her city, there is an 8% sales tax on all items, except for food. How much money does Victoria spend at the market? Round to the nearest cent if necessary.

Item	Cost ($)
Cereal	3.55
Milk	3.10
Laundry Detergent	8.95
Yogurt	5.35
Dish Soap	2.95

Go Online You can complete an Extra Example online.

Pause and Reflect

How can you use mental math to estimate your total including sales tax?

Practice

🧭 **Go Online** You can complete your homework online.

Find the total cost to the nearest cent. (Examples 1–3)

1. $18 breakfast; 7% tax

2. $24 shirt; 6% tax

3. $49.95 pair of shoes; 5% tax

4. Emily wants to buy new boots that cost $68. The sales tax rate in her city is $5\frac{1}{2}$%. What is the total cost for the boots? (Example 1)

5. Jack wants to buy a coat that costs $74.95. The sales tax rate in his city is $6\frac{1}{2}$%. What is the total cost for the coat? (Example 1)

6. Mr. Phuong stayed in a hotel room for 2 nights that cost $210. The hotel room tax rate in the city is 12%. What is the total cost for the hotel room? (Example 2)

7. The cost of a hotel room during Lacy's trip is $325. The hotel room tax in the city she is in is 10.5%. What is the total cost of the hotel room? (Example 2)

Test Practice

8. Robert spends $30.45, before tax, at the bookstore. If the sales tax rate in his city is 7.25%, what is the total cost of his purchase? (Example 3)

9. Multiple Choice Anya purchased $124.35 worth of home improvement items at the hardware store. If the sales tax rate in her city is 6.75%, what is the total cost of her purchase? (Example 3)

Ⓐ $131.10

Ⓑ $8.39

Ⓒ $115.96

Ⓓ $132.74

Apply

10. Jen purchased the items shown in the table. In the city she lives in, the sales tax rate is 7.15%. In another city, the sales tax rate is 6.35%. How much more is she spending if she purchases the items in the city where she lives? Round to the nearest cent.

Item	Cost ($)
Shirt	11.99
Shoes	35.50
Belt	6.75
Socks	3.00

11. Shamir is trying to decide between two cities to travel to for a weekend trip. The prices for a hotel room for two nights and the hotel tax rate are listed in the table. What is the difference in cost between the two cities for a weekend trip? Round to the nearest cent.

City	Cost ($)	Tax Rate
A	250	12.5%
B	215	14.75%

12. MP **Find the Error** A student is finding the total cost c, including sales tax, of a book that costs $9.95. The sales tax rate is 9%. Find the student's mistake and correct it.

$c = 1.9 \times 9.95$
$c \approx 18.91$
$c \approx \$18.91$

13. MP **Identify Structure** Write two different expressions to find the total cost of an item that costs $\$a$ if the sales tax is 6%. Explain why the expressions give the same result.

14. What are some similarities between sales tax and hotel tax?

15. MP **Reasoning** A dollhouse is on sale for $160. The tax rate is $5\frac{1}{4}$%. Without calculating, will the tax be greater than or less than $8? Write an argument that can be used to defend your solution.

Tips and Markups

I Can... use proportional relationships to find the amount to pay for a tip and the amount of markup on items.

Learn Tips

A **tip**, or **gratuity**, is an additional amount of money paid in return for a service. This amount is sometimes a percent of the service cost. The total amount paid is the cost of the service plus the tip.

Example 1 Tips

The bill for a group of eight people at a restaurant was $125 before the tip was added. The group wants to add an 18% tip.

What will be the total bill including the tip?

Method 1 Use ratio reasoning.

Write a proportion. Then solve using ratio reasoning. Let t represent the amount of the tip.

amount of tip ⟶ $\dfrac{t}{125} = \dfrac{18}{100}$ } Percent
amount of bill ⟶

$$\overset{\times 1.25}{\overset{\frown}{\dfrac{t}{125}} = \dfrac{18}{100}}$$ Because $100 \times 1.25 = 125$, multiply 18 by
$$\underset{\times 1.25}{\underset{\smile}{}}$$ 1.25 to find the value of t.

$$\dfrac{22.50}{125} = \dfrac{18}{100}$$ $18 \times 1.25 = 22.50$, so, $t = 22.50$.

Add the tip to the bill. The total cost is $125 + $22.50, or $147.50.

Method 2 Use properties of operations. Let t represent the amount of the tip.

$$\dfrac{t}{125} = \dfrac{18}{100}$$ Write the proportion.

$$\dfrac{t}{125} = 0.18$$ Divide 18 by 100. A one-step equation results.

$$125 \cdot \left(\dfrac{t}{125}\right) = (0.18) \cdot 125$$ Multiplication Property of Equality

$$t = 22.50$$ Simplify.

Add the tip to the bill. The total cost is $125 + $22.50, or $147.50.
So, using either method, the total cost of the bill is $147.50.

What Vocabulary Will You Learn?

gratuity

markup

selling price

tip

wholesale cost

Think About It!

What is a good estimate for the solution? Explain how you calculated the solution.

Talk About It!

How could you solve the problem another way?

Check

Amy wants to tip her hairstylist 20% for a haircut that costs $48. What is her total bill with tip? Use any strategy.

(Show your work here)

⊘ **Go Online** You can complete an Extra Example online.

Learn Markup

In order to make a profit, stores typically sell items for more than what they pay for them. The amount the store pays for an item is called the **wholesale cost**. The amount of increase is called the **markup**. The **selling price** is the amount the customer pays for an item. The selling price is equal to the wholesale cost plus the markup.

$$\text{selling price} = \text{wholesale cost} + \text{markup}$$

🌐 Example 2 Markup

The wholesale cost for each shirt at a clothing store is $17. The store manager plans to mark up the shirts by 125%.

What will be the selling price for each shirt?

Method 1 Use ratio reasoning.

Write a proportion. Then solve using ratio reasoning. Let x represent the selling price.

amount of markup ⟶ $\dfrac{x}{17} = \dfrac{125}{100}$ ⎫ Percent
wholesale cost ⟶

$\times 0.17$

$\dfrac{x}{17} = \dfrac{125}{100}$ Because $100 \times 0.17 = 17$, multiply 125 by 0.17 by to find the value of x.

$\times 0.17$

$\dfrac{21.25}{17} = \dfrac{125}{100}$ $125 \times 0.17 = 21.25$, so, $x = 21.25$.

Add the markup to the wholesale cost. The selling price is $17 + $21.25, or $38.25.

(continued on next page)

Method 2 Use properties of operations.

$$\frac{x}{17} = \frac{125}{100}$$ Write the proportion. Let *x* represent the selling price.

$$\frac{x}{17} = 1.25$$ Divide 125 by 100. A one-step equation results.

$$17 \cdot \left(\frac{x}{17}\right) = (1.25) \cdot 17$$ Multiplication Property of Equality

$$x = 21.25$$ Simplify.

Add the markup to the wholesale cost. The selling price is $17 + $21.25, or $38.25.

So, using either method, the selling price is $38.25.

Check

The wholesale cost for a basketball backboard is $32. If the markup is $85\frac{1}{2}$%, what is the selling price? Use any strategy.

 Go Online You can complete an Extra Example online.

Example 3 Markup

Ben's family is shopping for a new car. The selling price of a car is $24,199.50. Ben researches to find that the wholesale cost of the car is $22,100.00.

What is the percent of markup?

Finding the percent of markup is the same as finding the percent of increase.

Step 1 Identify the part and the whole.

original amount = $22,100.00 This is the whole.

new amount = $24,199.50 This is the whole plus the part.

amount of increase = $2,099.50 This is the part.

(continued on next page)

Talk About It!

Compare the wholesale cost with the selling price. How do you know the selling price is reasonable?

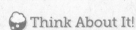 **Think About It!**

What is a good estimate for the solution? Explain how you calculated that estimate.

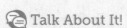

Step 2 Find the percent of increase.

$$\frac{\text{part}}{\text{whole}} = \frac{2{,}099.50}{22{,}100.00}$$ Write the part-to-whole ratio. The part is 2,099.50. The whole is 22,100.00.

$$= 0.095$$ Divide.

$$= \frac{9.5}{100}$$ Write an equivalent ratio, as a rate per 100.

$$= 9.5\%$$ Definition of percent

So, the percent of markup for the wholesale price of the car is 9.5%.

Check

Mikka is making jewelry for a craft show. The wholesale cost of a bracelet is $12.50. If she sells them for $20, what is the percent of markup?

⟲ **Go Online** You can complete an Extra Example online.

Pause and Reflect

Compare and contrast tips and markups. Where have you seen or used tips and markups in your everday life?

Talk About It!

How is finding the percent of markup different than finding the selling price of an item?

🌐 **Apply** Dining Out

Kerry and three friends went out for dinner. They split a large pizza, and each person had a salad and a soda. They want to leave a 15% tip on the cost of the food, and the sales tax is $8\frac{1}{4}$%. How much will each person pay if they split the bill evenly?

Pizza	$18.60
Salad	$2.50
Soda	$2.25

1 What is the task?

Make sure you understand exactly what question to answer or problem to solve. You may want to read the problem three times. Discuss these questions with a partner.

First Time Describe the context of the problem, in your own words.
Second Time What mathematics do you see in the problem?
Third Time What are you wondering about?

2 How can you approach the task? What strategies can you use?

Record your observations here

3 What is your solution?

Use your strategy to solve the problem.

Show your work here

💬 **Talk About It!**

What steps should you take before splitting the bill?

4 How can you show your solution is reasonable?

🖊 **Write About It!** Write an argument that can be used to defend your solution.

Check

Brian has $24 worth of pizza delivered to his house. He pays the bill plus a 15% tip and 7% sales tax. He also pays a $3 delivery fee that is charged after the tax and tip. How much change does he receive, if he pays with two $20 bills?

Show your work here

Go Online You can complete an Extra Example online.

Pause and Reflect

Explain how tips and markups are percents of increase.

Record your observations here

Practice

Find the total cost to the nearest cent. Use any strategy. (Examples 1 and 2)

1. $20 haircut; 10% tip

2. $24 lunch; 15% tip

3. $185 TV; 5% markup

4. Vera went to the local salon to get a haircut. The cost was $24. Vera tipped the hair stylist 18%. What was the total cost of haircut including the tip? Round to the nearest cent. (Example 1)

5. The Gomez family ordered $39.50 worth of pizza and subs. They gave the delivery person a 20% tip. What was the total cost of the food and tip? Round to the nearest cent. (Example 1)

6. The wholesale cost of a bicycle is $98.75. The markup for the bicycle is 33.3%. Find the selling price of the bicycle. Round to the nearest cent. (Example 2)

7. The wholesale cost for a purse in a department store is $12.50. The store plans to mark up the purse by 140%. What will be the selling price of the purse? Round to the nearest cent. (Example 2)

8. Keri is making doll clothes for a holiday craft show. The wholesale cost of the materials for one outfit is $9.38. If she sells an outfit for $15, what is the percent of markup? Round to the nearest percent. (Example 3)

9. A pet store sells a large dog kennel for $98.50. The wholesale cost of the kennel is $63.55. What is the percent of markup? Round to the nearest percent. (Example 3)

Test Practice

10. Open Response An elementary school wants to purchase a new swing set. The table shows the selling price of the swing sets they are interested in buying. The markup for both swing sets is $20\frac{1}{4}$%. The school decides to buy the Adventurers swing set. What is the selling price of the swing set they are buying?

Swing Set	Wholesale Price ($)
Adventurers	3,056
Thunder Ridge	4,125

Apply

11. Louise bought a frame and canvas online for $125. Then she bought the supplies shown in the table and paid a 7% sales tax on these supplies only. She took the amount that she spent on the canvas, frame, and supplies, and marked up the price by 115%. How much did she charge for her painting? Round to the nearest cent.

Item	Cost ($)
Brushes	12.50
Paint	19.50

12. Brendan has $65 worth of balloons and flowers delivered to his mother. He pays the bill plus an 8.5% sales tax and an 18% tip on the total cost including tax. He also pays a $10 delivery fee that is charged after the tax and tip. How much change does he receive if he pays with two $50 bills? Round to the nearest cent.

13. Toru takes his dog to be groomed. The fee to groom the dog is $70 plus a 10% tip. Is $80 enough to pay for the service and tip? Explain your reasoning.

14. **MP** **Use a Counterexample** Determine if the following statement is *true* or *false*. If false, provide a counterexample.

It is impossible to increase the cost of an item by less than 1%.

15. **MP** **Identify Structure** Write two different expressions to find the total cost of a service that costs x if the tip is 18%. Explain why the expressions give the same result.

16. **Create** Write and solve a real-world problem in which you find sales tax and a tip.

Discounts

I Can... use proportional relationships to find the amount of discount or markdown.

Learn Discounts

Discount or **markdown** is the amount by which the regular price of an item is reduced. The sale price is the original price minus the discount. The percent of discount is a percent of decrease.

$$\text{sale price} = \text{original price} - \text{discount}$$

🌐 Example 1 Discounts

Summer Sports is having a sale. A volleyball has an original price of $59. It is on sale for 65% off the original price.

What is the sale price of the volleyball?

Method 1 Find the amount of the discount. Let d represent the amount of the discount.

Write a proportion. Then solve using ratio reasoning.

$$\begin{array}{r}\text{amount of discount} \longrightarrow \\ \text{original price} \longrightarrow\end{array} \left.\dfrac{d}{59} = \dfrac{65}{100}\right\} \text{ Percent}$$

$$\overset{\times 0.59}{\overset{\frown}{\dfrac{d}{59} = \dfrac{65}{100}}}_{\times 0.59}$$

Because $100 \times 0.59 = 59$, multiply 65 by 0.59 to find the value of d.

$$\dfrac{38.35}{59} = \dfrac{65}{100}$$

$65 \times 0.59 = 38.35$, so, $d = 38.35$.

Subtract the discount from the original price. The sale price is $59 − $38.35, or $20.65.

(continued on next page)

What Vocabulary Will You Learn?
discount

markdown

💬 Talk About It!
In Method 1, why is the answer not $38.35?

Method 2 Subtract the discount percent from 100%.

Because 100% represents the original price, subtract the percent of the discount from 100%. The sale price is 100% − 65% or 35% of the original price. Then write a proportion and solve using ratio reasoning. Let x represent the sale price.

sale price ⟶
original price ⟶ $\dfrac{x}{59} = \dfrac{35}{100}$ } Percent

$$\overset{\times 0.59}{\overset{\frown}{\underset{\underset{\times 0.59}{\smile}}{\dfrac{x}{59} = \dfrac{35}{100}}}}$$

Because $100 \times 0.59 = 59$, multiply 35 by 0.59 to find the value of x.

$$\dfrac{20.65}{59} = \dfrac{35}{100}$$ $35 \times 0.59 = 20.65$, so, $x = 20.65$.

So, using either method, the sale price is $20.65.

Check

A restaurant decreased their prices for a day to their prices from 1964. A pizza that usually sells for $15.40 was marked down 85%. What was the price of the pizza in 1964? Use any strategy.

Show your work here

 Go Online You can complete an Extra Example online.

🌐 **Example 2** Combined Discounts

During a clearance sale at an electronics store, certain tablets were marked down 20%. One day, an additional 30% was taken off already-reduced prices. A tablet originally sold for $375.

What was the final price after both discounts were applied?

Step 1 Find the price after the first discount.

Because 100% represents the original price, subtract the percent of the original discount from 100%. The sale price is 100% − 20% or 80% of the original price. Then write a proportion and solve using ratio reasoning. Let x represent the sale price after the first discount.

(continued on next page)

sale price $\longrightarrow$
original price $\longrightarrow$ $\left.\dfrac{x}{375} = \dfrac{80}{100}\right\}$ Percent

×3.75

$\dfrac{x}{375} = \dfrac{80}{100}$ Because $100 \times 3.75 = 375$, multiply 80 by 3.75 to find the value of x.

×3.75

$\dfrac{300}{375} = \dfrac{80}{100}$ $80 \times 3.75 = 300$, so, $x = 300$. The price after the first discount is $300.

Step 2 Find the final price after the additional discount.

Because 100% now represents the price after the first discount, subtract the percent of the additional discount from 100%. The sale price is 100% − 30% or 70% of the price after the first discount. Then write a proportion and solve using ratio reasoning. Let x represent the sale price after the additional discount.

sale price $\longrightarrow$
original price $\longrightarrow$ $\left.\dfrac{x}{300} = \dfrac{70}{100}\right\}$ Percent

×3

$\dfrac{x}{300} = \dfrac{70}{100}$ Because $100 \times 3 = 300$, multiply 70 by 3 to find the value of x.

×3

$\dfrac{210}{300} = \dfrac{70}{100}$ $70 \times 3 = 210$, so, $x = 210$.

So, the final price of the tablet after both discounts were applied was $210.

Check

A clothing store marked all of their summer clothes down 50%. A sign in the store indicates that an additional 20% is to be taken off clearance prices. What is the final price of a top that originally sold for $58?

Show your work here

> **Talk About It!**
> The final price had a discount of 20% followed by a discount of 30%. Is this the same as finding 20% + 30% or 50% of the original price? Use the values in the Example to justify your reasoning.

Go Online You can complete an Extra Example online.

Think About It!

Are you trying to find the part, percent, or whole?

Talk About It!

When writing the proportion, why is the percent 75 out of 100 instead of 25 out of 100?

 Example 3 Find the Original Price

Sandy has a 25% discount coupon for athletic equipment. She buys hockey equipment for a final price of $172.50.

What is the original price?

If the amount of the discount is 25%, the sale price is 100% − 25%, or 75%, of the original price. So, 75% of the original price is $172.50.

Write a proportion and solve using ratio reasoning. Let x represent the original price.

$$\left.\begin{array}{l}\text{sale price} \longrightarrow \\ \text{original price} \longrightarrow\end{array}\frac{172.50}{x} = \frac{75}{100}\right\} \text{ Percent}$$

$$\overset{\times 2.3}{\frac{172.50}{x} = \frac{75}{100}}_{\times 2.3}$$

Because 75 × 2.3 = 172.50, multiply 100 by 2.3 to find the value of x.

$$\frac{172.50}{230} = \frac{75}{100}$$

100 × 2.3 = 230, so, x = 230.

So, the original price of the hockey equipment is $230.

Check

A pair of running shoes is on sale for $125.99. If the sale price is discounted 9% from the original price, what is the original price? Round to the nearest cent.

Show your work here

 Go Online You can complete an Extra Example online.

🌐 Apply Shopping

The Wares want to buy a new computer. Store A has a regular price of $1,300 and is offering a discount of 20%. Store B has a regular price of $1,089 with no discount. They want to purchase the computer that is less expensive. If there is a $7\frac{1}{4}$% sales tax in their city, at which store should they buy their computer, and how much money will they save if they buy at that store instead of the other store?

🖱 **Go Online** watch the animation.

1 What is the task?

Make sure you understand exactly what question to answer or problem to solve. You may want to read the problem three times. Discuss these questions with a partner.

First Time Describe the context of the problem, in your own words.
Second Time What mathematics do you see in the problem?
Third Time What are you wondering about?

2 How can you approach the task? What strategies can you use?

3 What is your solution?

Use your strategy to solve the problem.

4 How can you show your solution is reasonable?

✍ **Write About It!** Write an argument that can be used to defend your solution.

💬 **Talk About It!**

How can you quickly determine which computer will be cheaper before the sales tax is applied?

Copyright © McGraw-Hill Education

Check

A bike shop has two road bikes on sale. The Road Warrior has a regular price of $379 and is discounted 25%. The Road Racer has a regular price of $320 and is discounted 12%. If the sales tax rate is 6.5%, which bike has the less expensive sale price? How much will you save by buying that bike?

Show your work here

🔘 **Go Online** You can complete an Extra Example online.

Pause and Reflect

Write a paragraph explaining why a 30% discount is not the same as a 20% discount plus an additional 10% discount. Which is the better discount?

Record your observations here

Practice

Go Online You can complete your homework online.

Find the sale price to the nearest cent. Use any strategy. (Example 1)

1. $140 coat; 10% discount

2. $80 boots; 25% discount

3. $325 tent; 15% discount

4. A toy store is having a sale. A video game system has an original price of $99. It is on sale for 40% off the original price. Find the sale price of the game system. Round to the nearest cent. (Example 1)

5. A yearly coffee club subscription costs $65. Avery received an offer for 62% off the subscription cost. What is the sale price of the subscription? Round to the nearest cent. (Example 1)

6. During a clearance sale at a sporting goods store, skateboards were marked down 30%. On Saturday, an additional 25% was taken off already reduced prices of skateboards. If a skateboard originally cost $119.50, what was the final price after all discounts had been taken? Round to the nearest cent. (Example 2)

7. At an electronics store, a smart phone is on sale for 35% off the original price of $679. If you use the store credit card, you can receive an additional 15% off the sale price. What is the final price of the smart phone if you use the store credit card? Round to the nearest cent. (Example 2)

8. Gary had a 40% discount for new tires. The sale price of a tire was $96.25. What was the original price of the tire? Round to the nearest cent. (Example 3)

9. A swimsuit is on sale for $45.50. If the sale price is discounted 5% from the original price, what was the original price? Round to the nearest cent. (Example 3)

Test Practice

10. Open Response A shoe store is having a clearance sale on their summer shoes. All summer shoes are marked 55% off. A sign states you can take an additional 10% off the clearance sale prices. Kelly is deciding between two pairs of sandals shown in the table. If she buys the blue sandals, what is the final price Kelly will pay? Round to the nearest cent.

Shoes	Original Price
Blue Sandals	$75
Tan Sandals	$68

Apply

11. A recreational outlet has two trampolines on sale. The table shows the original prices. The Skye Bouncer is discounted 15% and the Ultimate is discounted 13%. If the sales tax rate is 7.5%, which trampoline has the lower sale price? How much will you save by buying that trampoline? Round to the nearest cent.

Trampoline Model	Original Price ($)
Skye Bouncer	1,480
Ultimate	1,450

12. Pets Plus and Pet Planet are having a sale on the same aquarium. At Pets Plus the aquarium is on sale for 30% off the original price and at Pet Planet it is discounted by 25%. If the sales tax rate is 8%, which store has the lower sale price? How much will you save by buying the aquarium there? Round to the nearest cent.

Store	Original Price of Aquarium ($)
Pets Plus	118
Pet Planet	110

13. (MP) **Persevere with Problems** Suppose an online store has an item on sale for 10% off the original price. By what percent does the store have to increase the price of the item in order to sell it for the original amount? Explain.

14. (MP) **Identify Structure** A shoe store buys packs of socks wholesale for $5 each and marks them up by 40%. The store decides to discount the packs of socks by 40%. Is the discounted price $5? Explain your reasoning.

15. A model car is for sale online. The owner discounts the price by 5% each day until it sells. On the third day the car sells. If the original price of the model car was $40, how much did the car sell for? What percent discount does this represent from the original price?

16. Describe a real-world problem where an item is discounted by 75%. Then find the original price.

Interest

I Can... use the simple interest formula to find the amount of interest earned for a given principal, at a given interest rate, for a given period of time.

Explore Interest

Online Activity You will use eTools to explore the simple interest formula.

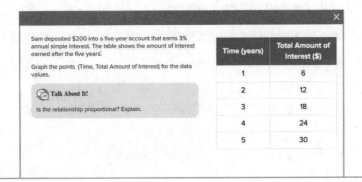

Sam deposited $200 into a five-year account that earns 3% annual simple interest. The table shows the amount of interest earned after the five years.

Graph the points (Time, Total Amount of Interest) for the data values.

Talk About It!

Is the relationship proportional? Explain.

Time (years)	Total Amount of Interest ($)
1	6
2	12
3	18
4	24
5	30

Learn Simple Interest

If you borrow money or deposit money, the **principal** is the amount of money borrowed or deposited. **Interest** is the amount paid or earned for the use of the principal. **Simple interest** is calculated using specified periods of time.

Percents are used to calculate interest. You can earn interest by letting the bank use the money you deposit in a bank account. You will pay interest if you borrow money from the bank or use a credit card.

Go Online Watch the animation to find the interest earned for the information in the table.

principal	$430
interest rate	2% or 0.02
time	5 years

The animation shows that the steps for finding the interest are:

$I = prt$ Interest formula

$I = 430 \cdot 0.02 \cdot 5$ principal is $430, rate is 0.02, and time is 5 years

$I = 43$ Simplify.

So, the interest is $43. This means that, in 5 years, you will have $430 + $43, or $473, in your bank account if you do not deposit or withdraw in those 5 years.

(continued on next page)

Words	Symbols
Simple Interest *I* is the product of the principal *p*, interest rate *r*, and the length of time *t*, expressed in years.	$I = prt$

🌐 Example 1 Find Simple Interest

Magdalena put $580 into a savings account. The account pays 2.5% simple interest each year.

If she neither adds nor withdraws money from the account, how much interest will she earn after 2 years?

The period of time is annual or yearly, so $t =$ _____.
The rate is 2.5% or, as a decimal, _____.

Use the Interest formula to find the amount of interest Magdalena will earn.

$I = prt$ Interest formula

$I = \boxed{} \cdot \boxed{} \cdot \boxed{}$ The principal is 580, the rate is 0.025, and the time is 2.

$I = 29$ Multiply.

So, Magdalena earned $ _____ in interest in 2 years.

Check

Curtis deposits $550 into a savings account. The account pays 1.75% simple interest on an annual basis. If he does not add or withdraw money from the account, how much interest will he earn after 3 years?

Show your work here

🅦 **Go Online** You can complete an Extra Example online.

🌐 Example 2 Find Simple Interest

Suppose Magdalena opens a savings account for only 6 months.
She puts $580 in that account and earns 3.5% each year.

How much interest will she earn?

Because the period of time is annual or yearly, 6 months equals $\frac{6}{12}$ or

_____ year.

What is the rate 3.5% written as a decimal? _____

Use the Interest formula to find the amount of interest Magdalena
will earn.

$I = prt$ Interest formula

$I = \boxed{} \cdot \boxed{} \cdot \boxed{}$ The principal is 580, the rate is
0.035, and the time is 0.5.

$I = 10.15$ Multiply.

So, Magdalena earned $ _____ in interest in _____ months.

Check

Carrie invests $430 into a savings account. The account pays
2.5% simple interest on an annual basis. If she does not add or
withdraw money from the account, how much interest will she
earn after 15 months?

🅡 **Go Online** You can complete an Extra Example online.

 Copyright © McGraw-Hill Education

💭 Think About It!

What is the simple
interest formula? Are
you trying to find the
interest, principal, rate,
or *time*?

💬 Talk About It!

How is the process of
finding simple interest
different when the
time is given in
months?

Example 3 Find Simple Interest

🌐 **Example 3** Find Simple Interest

Rondell's parents are starting a real estate business. They borrow $99,400 from the bank to finance their start-up operations. The simple interest rate is $4\frac{3}{4}\%$ per year, and they plan to take 15 years to repay the loan.

How much simple interest will they pay?

The period of time is annual or yearly, so $t =$ _____.

What is the rate $4\frac{3}{4}\%$ written as a decimal? _____

Use the Interest formula to find the amount of interest they will pay.

$I = prt$ Interest formula

$I =$ ☐ · ☐ · ☐   The principal is 99,400, the rate is 0.0475, and the time is 15.

$I = 70,822.50$ Multiply.

So, Rondell's parents will pay $ _____ in interest on the loan.

Check

Abbie borrows $3,216 at a rate of $6\frac{4}{5}\%$ per year. How much simple interest will she owe if it takes 2 years to repay the loan?

Show
your work
here

🌐 **Go Online** You can complete an Extra Example online.

Pause and Reflect

Why is it important to pay off loans as soon as possible?

Record your
observations
here

🌐 Apply Car Shopping

Alex is buying a car that costs $18,000. The tax rate is 7%, and he plans to make a down payment of $2,000. The sales tax is added to the price of the car before the down payment is made. He is considering a five-year loan with a simple interest rate of 6.2% each year. What will be his monthly payment?

1 What is the task?

Make sure you understand exactly what question to answer or problem to solve. You may want to read the problem three times. Discuss these questions with a partner.

First Time Describe the context of the problem, in your own words.
Second Time What mathematics do you see in the problem?
Third Time What are you wondering about?

2 How can you approach the task? What strategies can you use?

3 What is your solution?

Use your strategy to solve the problem.

4 How can you show your solution is reasonable?

🔺 **Write About It!** Write an argument that can be used to defend your solution.

💬 Talk About It!

How can the amount of a down payment affect the monthly payment?

Check

Donte is buying a car that costs $8,000. He is deciding between a 3-year loan and a 4-year loan. Use the rates in the table to determine how much money he will save if he chooses the 3-year loan instead of the 4-year loan.

Time (y)	Simple Interest Rate (%)
3	2.25
4	2.5
5	3

Go Online You can complete an Extra Example online.

Pause and Reflect

Have you ever wondered when you might use the concepts you learn in math class? What are some everyday scenarios in which you might use what you learned today?

Practice

🧭 **Go Online** You can complete your homework online.

Find the simple interest earned, to the nearest cent, for each principal, interest rate, and time. (Example 1)

1. $530, 6%, 1 year

2. $1,200, 3.5%, 2 years

3. $750, 7%, 3 years

4. Elena's father put $460 into a savings account for her. The account pays 2.5% simple interest each year. If he neither adds nor withdraws money from the account, how much interest will the account earn after 4 years? Round to the nearest cent. (Example 1)

5. Ethan put $1,250 into a savings account. The account pays 4.5% simple interest on an annual basis. If he does not add or withdraw money from the account, how much interest will he earn after 2 years? Round to the nearest cent. (Example 1)

6. Marc deposits $840 into a savings account. The account pays 2% simple interest on an annual basis. If he does not add or withdraw money from the account, how much interest will he earn after 6 months? Round to the nearest cent. (Example 2)

7. Nina's grandmother deposits $3,000 into a savings account for her. The account pays 5.5% simple interest on an annual basis. If she does not add or withdraw money from the account, how much interest will she earn after 21 months? Round to the nearest cent. (Example 2)

8. Jack borrows $2,700 at a rate of 8.2% per year. How much simple interest will he owe if it takes 3 months to repay the loan? Round to the nearest cent. (Example 3)

9. Liliya's parents borrow $1,400 from the bank for a new washer and dryer. The interest rate is 7.5% per year. How much simple interest will they pay if they take 18 months to repay the loan? Round to the nearest cent. (Example 3)

Test Practice

10. Open Response The table shows the interest rates for auto repair loans based on how long it takes to pay off the loan. Jin borrows $3,600 and plans to pay the loan off in 18 months. How much simple interest will he owe if it takes 18 months to repay the loan? Round to the nearest cent.

Time	Rate (%)
6 months	3.5
12 months	4.0
18 months	4.25

Apply

11. Jarvis is buying a boat that costs $6,000. He has $500 for a down payment. He is deciding between a 2-year loan and a 3-year loan. Use the rates in the table to determine how much money he will save if he chooses the 2-year loan instead of the 3-year loan. Round to the nearest cent.

Time (years)	Simple Interest Rate (%)
1	1.5
2	2.25
3	3.0

12. Evelyn is buying a motorcycle that costs $14,000. The tax rate is 6.75%, and she plans to make a down payment of $1,500. The sales tax is added before the down payment is applied. She is considering a three-year loan. What will be her monthly payments? Round to the nearest cent.

Time (years)	Simple Interest Rate (%)
1	3.25
3	6.75
5	8.5

13. (MP) **Persevere with Problems** Suppose Serena invests $2,500 for 3 years and 6 months and earns $328.13. What was the rate of interest? Explain how you solved.

14. A student stated that an interest rate could not be less than 1%. Do you agree? Why or why not?

15. (MP) **Justify Conclusions** Suppose you earn 2% on $1,000 for 2 years. Explain how the simple interest is affected if the rate is doubled. What happens if the time is doubled?

16. Name a principal and interest rate where the amount of simple interest earned in 10 years would be $200. Justify your answer.

Commission and Fees

I Can... use proportional relationships to find the amount of commission earned on sales and the amount of fees for certain services.

Learn Commission and Fees

The following are four ways in which people are paid.

Hourly	These employees are paid a set amount each hour, such as $15 per hour.
Salary	These employees are paid a set amount, no matter how many hours they work, such as $45,000 per year.
Commission Only	These employees are paid only a percent of the amount that they sell.
Salary plus Commission	These employees are paid a small base salary, plus a percent of the amount that they sell.

Some employees, such as realtors, car salesmen, and stockbrokers, are paid a percent of the amount that they sell. This payment is called a **commission**.

If you pay a commission to a person or business, you are paying a **fee**. A fee is a payment for a service. It can be a fixed amount, a percent of the charge, or both.

A realtor may earn 5% commission on the sale of a home. Suppose the realtor sells a house for $140,000. You can use equivalent ratios, and the percentage written as a rate per 100, to determine the amount of commission. Let c represent the amount of commission.

amount of commission → $\frac{c}{140,000} = \frac{5}{100}$ } Percent
price of house →

$\times 1,400$

$\frac{7,000}{140,000} = \frac{5}{100}$ Because 100 × 1,400 is 140,000, multiply 5 by 1,400 to find c.

$\times 1,400$

The amount of commission is $7,000. This means the real estate agent will be paid $7,000 for selling a $140,000 home.

Copyright © McGraw-Hill Education

What Vocabulary Will You Learn?
commission
fee

🌎 Example 1 Find Commission

Angie works in a jewelry store and earns a 6.25% commission on every piece of jewelry she sells.

How much commission does she earn for selling a ring that costs $1,300?

Method 1 Use ratio reasoning.

Write a proportion and solve using ratio reasoning. Let c represent the amount of commission.

amount of commission ⟶
cost of ring ⟶ $\dfrac{c}{1{,}300} = \dfrac{6.25}{100}$ } Percent

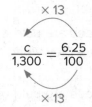

$\dfrac{c}{1{,}300} = \dfrac{6.25}{100}$ Because $100 \times 13 = 1{,}300$, multiply 6.25 by 13 to find the value of c.

$\dfrac{81.25}{1{,}300} = \dfrac{6.25}{100}$ $6.25 \times 13 = 81.25$, so, $c = 81.25$.

Method 2 Use properties of operations.

$\dfrac{c}{1{,}300} = \dfrac{6.25}{100}$ Write the proportion. Let c represent the amount.

$\dfrac{c}{1{,}300} = 0.0625$ Divide 6.25 by 100. A one-step equation results.

$1{,}300 \cdot \left(\dfrac{c}{1{,}300} \right) = (0.0625) \cdot 1{,}300$ Multiplication Property of Equality

$c = 81.25$ Simplify.

So, using either method, the amount of commission is $81.25.

Check

Nathan sells jewelry and earns a 9.75% commission on every piece he sells. How much commission would he earn on a bracelet that sold for $220? Use any strategy.

Show your work here

🅖 **Go Online** You can complete an Extra Example online.

Think About It!

What is a good estimate for the solution? Explain how you calculated the estimate.

Talk About It!

How does your answer compare to your estimate?

🌐 Example 2 Find the Amount of Sales

Caleb needs to earn $1,200 each month to cover his living expenses. He earns an 8% commission on everything he sells.

What is the minimum amount he needs to sell in order to earn $1,200?

Write a proportion. Then solve using ratio reasoning. Let a represent the amount he needs to sell.

amount of commission ⟶
amount of sales ⟶ $\dfrac{1,200}{a} = \dfrac{8}{100}$ } Percent

$\times 150$

$\dfrac{1,200}{a} = \dfrac{8}{100}$ Because $8 \times 150 = 1,200$, multiply 100 by 150 to find the value of a.

$\times 150$

$\dfrac{1,200}{15,000} = \dfrac{8}{100}$ $100 \times 150 = 15,000$, so, $a = 15,000$.

So, Caleb needs to sell $15,000 to earn $1,200 in commission.

Check

Quentin wants to earn at least $945 this month in commission. What is the minimum amount he needs to sell if he earns a 5.25% commission?

🌐 **Go Online** You can complete an Extra Example online.

Pause and Reflect

Did you make any errors when completing the Check exercise? What can you do to make sure you don't repeat the errors in the future?

 Example 3 Fees

Sybrina bought some paper supplies from an online retailer. The retailer charges a $6.95 shipping fee or 8% of the total purchase, whichever is greater.

If Sybrina's total purchase is $85.50, how much shipping will she pay?

Step 1 Find the percent of the total purchase.

Because the shipping fee may be 8% of the total purchase, find 8% of $85.50.

Write a proportion and solve using the properties of operations. Let x represent the potential shipping fee.

$$\frac{x}{85.50} = \frac{8}{100} \qquad \text{Write the proportion.}$$

$$\frac{x}{85.50} = 0.08 \qquad \text{Divide 8 by 100.}$$

$$85.50 \cdot \left(\frac{x}{85.50}\right) = (0.08) \cdot 85.50 \qquad \text{Multiplication Property of Equality}$$

$$x = 6.84 \qquad \text{Simplify.}$$

Step 2 Compare the two fees.

The 8% shipping fee of $6.84 is less than the $6.95 shipping fee. Sybrina must pay the greater amount.

So, Sybrina will pay $6.95 in shipping fees.

Check

Tickets for a concert are sold by online ticket resellers. One company charges a fee of 4% of the ticket price. Another company charges a flat fee of $2.50. The ticket you want to buy costs $60.50. If you buy the ticket online, which is the lesser fee?

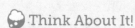 **Think About It!**

What is a good estimate for an 8% shipping fee? Explain how you calculated the estimate.

 **Talk About It!**

How can you solve the problem another way?

Go Online You can complete an Extra Example online.

🌐 Apply Personal Finance

David is starting a new job in sales. He can choose to earn only a 10% commission on sales each month, or earn a monthly base salary of $1,500 with a 3% commission on sales over $7,500. Which pay method would earn him more money if he has an average sales of $16,000 each month?

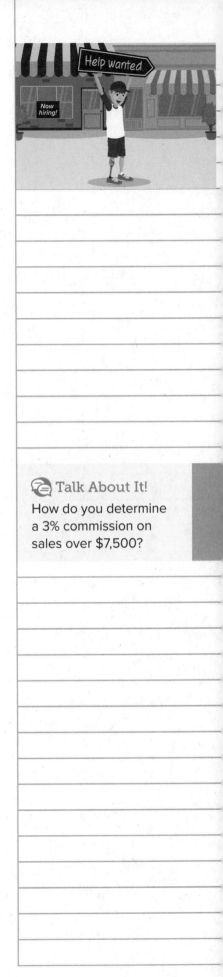

1 What is the task?

Make sure you understand exactly what question to answer or problem to solve. You may want to read the problem three times. Discuss these questions with a partner.

First Time Describe the context of the problem, in your own words.
Second Time What mathematics do you see in the problem?
Third Time What are you wondering about?

2 How can you approach the task? What strategies can you use?

Record your observations here

3 What is your solution?

Use your strategy to solve the problem.

Show your work here

💬 Talk About It!

How do you determine a 3% commission on sales over $7,500?

4 How can you show your solution is reasonable?

✏️ **Write About It!** Write an argument that can be used to defend your solution.

Check

Kelly has a job in sales that pays only a commission of 24% of her total monthly sales. She is offered a new job with a monthly salary of $3,750 with a commission of 9.5% on her total monthly sales over $12,500. If she estimates her total monthly sales will average $20,000, which statement is correct?

(A) The new job will pay more because she would earn $4,462.50, and her current job only pays $4,220.

(B) The new job will pay more because she would earn $4,742.75, and her current job only pays $4,220.

(C) Her current job will pay more because she will earn $4,800, and the new job would only pay $4,462.50.

(D) Her current job will pay more because she will earn $4,800, and the new job would only pay $4,742.75.

Show your work here

🔳 **Go Online** You can complete an Extra Example online.

Pause and Reflect

How are *parts* and *wholes* represented in fees and commissions? Give examples to support your answer.

Record your observations here

Practice

📡 **Go Online** You can complete your homework online.

Solve each problem. Use any strategy, such as a bar diagram, ratio table, or division.

1. Mrs. Hollern works in a jewelry store and earns an 8.5% commission on every piece of jewelry she sells. How much commission would she earn for selling a necklace that costs $4,600? **(Example 1)**

2. Booker's father sells computer software and earns a 4.25% commission on every software package he sells. How much commission would he earn on a software package that sold for $15,725? Round to the nearest cent. **(Example 1)**

3. Chase wants to earn at least $900 this month in commission. What is the minimum amount he needs to sell in order to earn $900 if he earns a 3.3% commission on everything he sells? Round to the nearest dollar. **(Example 2)**

4. Sophia needs to earn at least $1,800 each month to cover her living expenses. What is the minimum amount she needs to sell in order to earn at least $1,800 if she earns a 10.75% commission on everything she sells? Round to the nearest dollar. **(Example 2)**

5. Mrs. Jackson needs to earn $3,000 a month. How much does she need to sell if she earns 15% commission on everything she sells? **(Example 2)**

6. Raymond purchases concert tickets online for $43.50. There is a 3% processing fee. What is the total cost of the tickets? Round to the nearest cent. **(Example 3)**

Test Practice

7. The table shows the shipping fee options for two online marketplaces. If you sell a smart phone online for $115, which is the lesser fee? **(Example 3)**

Online Market	Shipping Fees
Deals	Flat fee of $12
Sell UR Stuff	11% of selling price

8. **Multiple Choice** Mai bought some party supplies from an online store. The shop charges a $7.95 shipping fee or 6.25% of the total purchase, whichever is greater. Suppose Mai's total purchase is $128. How much shipping will she pay?

Ⓐ $7.95

Ⓑ $8.00

Ⓒ $15.95

Ⓓ $16.00

Apply

9. Addison is considering two sales jobs. The job offers are shown in the table. She estimates her total monthly sales will average $18,000. Which job should Addison take if she wants to earn more money each month? How much more will she earn?

	Offer
Job 1	A commission of 24% of her total monthly sales.
Job 2	Salary of $4,000 with a commission of 12.25% on her total monthly sales over $14,500.

10. The table shows Frank and his brother's monthly salaries. One week, Frank and Daniel each have sales of $9,500. Who earned more money that week? How much more?

	Offer
Daniel	A commission of 15% on his weekly sales.
Frank	A weekly salary of $450 plus an 18% commission on his weekly sales over $4,000.

11. **MP** **Find the Error** A student is finding a 6.5% commission on $525.08 worth of sales. Find the student's mistake and correct it. Let x represent the amount of commission.

$$\frac{x}{525.08} = \frac{65}{100}$$

$$\frac{x}{525.08} = 0.65$$

$$x = 341.30$$

12. **Create** Write and solve a real-world problem in which you find the commission.

13. Determine if the following statement is *true* or *false*. Write an argument that can be used to defend your solution.

A commission can be less than 1%.

14. **MP** **Persevere with Problems** Natalie is a real estate agent and earns a 3% commission on all homes she sells. She estimates that she will earn $8,250 on the next house she sells. She actually earns $9,210. What was Natalie's estimate for the selling price of home? What was the actual selling price of the home?

Percent Error

I Can... use proportional relationships to solve percent error problems.

What Vocabulary Will You Learn?
amount of error
percent error

Explore Percent Error

 Online Activity You will use Web Sketchpad to explore percent error.

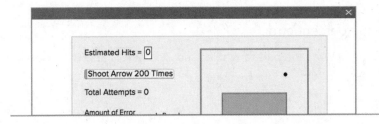

Learn Percent Error

Percent error is a ratio, written as a percent, that compares the inaccuracy of an estimate, or amount of error, to the actual amount.

The **amount of error** is the positive difference between the estimate and the actual amount. To find the amount of error, subtract the lesser amount from the greater amount.

Suppose 800 people are estimated to attend the high school football game. The actual attendance was 1,000 people. You can use a bar diagram to represent and find the percent of error.

Draw a bar to represent the actual attendance, 1,000 people. Because this is the whole, label the length of the bar 100%.

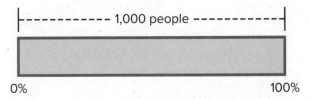

The amount of error is 1,000 − 800, or 200 people. Because 1,000 ÷ 200 = 5, divide the bar into 5 equal-size sections of 200.

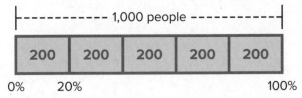

Each section represents 20% of the whole, 200 people. So, the percent error is 20%.

Talk About It!
Why do you think it is important to subtract the lesser amount from the greater amount when finding the amount of error?

Copyright © McGraw-Hill Education

Think About It!

Why might a bar diagram not be advantageous to help solve this problem?

Talk About It!

Suppose it actually took the contractor 19.5 hours. What part of solving the problem would stay the same? What part would change?

Example 1 Percent Error

A contractor estimates that it will take him 16 hours to complete a home improvement project. It actually takes him 12.5 hours.

What is the percent error of the contractor's estimate?

Step 1 Identify the part and the whole.

actual amount = 12.5   This is the whole.

estimated amount = 16   This is the whole plus the part.

amount of error = 16 − 12.5, or 3.5   This is the part.

Step 2 Find the percent error.

$$\frac{part}{whole} = \frac{3.5}{12.5}$$ Write the part-to-whole ratio. The part is 3.5. The whole is 12.5.

$$= 0.28$$ Divide.

$$= \frac{28}{100}$$ Write an equivalent ratio, as a rate per 100.

$$= 28\%$$ Definition of percent

So, the percent error is 28%.

Check

The Ramirez family was going on a trip. Their GPS system estimated that it would take them 4.25 hours to reach their destination. It actually took 5 hours because of stops. Find the percent error of the estimated time.

Show your work here

Go Online You can complete an Extra Example online.

🌐 **Apply** Sports

A school newspaper estimates that their basketball team will win 23 out of 25 games for the season. After 10 games, they have won 8. If the team continues winning at this rate, what will be the percent error of the newspaper's estimate once the season is over?

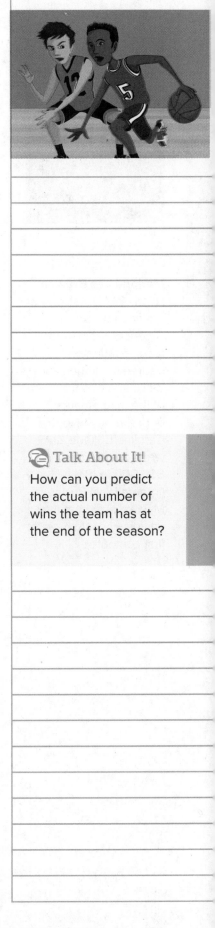

1 What is the task?

Make sure you understand exactly what question to answer or problem to solve. You may want to read the problem three times. Discuss these questions with a partner.

First Time Describe the context of the problem, in your own words.
Second Time What mathematics do you see in the problem?
Third Time What are you wondering about?

2 How can you approach the task? What strategies can you use?

Record your observations here

3 What is your solution?

Use your strategy to solve the problem.

Show your work here

🗨 **Talk About It!**
How can you predict the actual number of wins the team has at the end of the season?

4 How can you show that your solution is reasonable?

✎ **Write About It!** Write an argument that can be used to defend your solution.

Check

It is predicted that a softball team will win 35 out of their 50 games for their summer season. After 20 games, they have won 16. If the team continues to win at this rate, what will be the percent error of the prediction once the season is over?

Show your work here

Go Online You can complete an Extra Example online.

Pause and Reflect

Explain how ratio and proportional reasoning can be used to solve problems involving percent.

Record your observations here

Practice

Go Online You can complete your homework online.

Solve each problem.

1. Doug estimates that his soccer team will win 7 games this year. The team actually wins 10 games. What is the percent error of Doug's estimate? Round the answer to the nearest tenth percent, if necessary. (Example 1)

2. A mayor estimates that 4,000 people will attend the first day of the county fair. A total of 8,400 people actually attend the first day of the fair. What is the percent error of the mayor's estimate? Round the answer to the nearest tenth percent, if necessary. (Example 1)

3. Maya estimates that the wait time for her favorite roller coaster is 35 minutes. The actual wait time is 55.5 minutes. What is the percent error of Maya's estimate? Round the answer to the nearest tenth of a percent, if necessary. (Example 1)

4. Oliver estimates the weight of his cat to be 16 pounds. The actual weight of his cat is 14.25 pounds. What is the percent error of Oliver's estimate rounded to the nearest tenth of a percent? (Example 1)

5. A jar of marbles should contain 100 marbles. The jar actually has 99 marbles. What is the percent error to the nearest hundredth of a percent? (Example 1)

6. A cyclist estimates that he will bike 80 miles this week. He actually bikes 75.5 miles. What is the percent error of the cyclist's estimate rounded to the nearest hundredth of a percent? (Example 1)

Test Practice

7. The table shows the predicted and actual amount of snow for a local city. What is the percent error for the amount of snowfall? Round the answer to the nearest tenth of a percent if necessary. (Example 1)

Snowfall (inches)	
Predicted	6.75
Actual	10.25

8. **Multiple Choice** Jin's mother estimates there is a half gallon of fruit punch remaining in the container. There is actually 72 ounces of fruit punch in the container. What is the percent error of Jin's mother's estimate, rounded to the nearest tenth?

Ⓐ 11.1%

Ⓑ 36%

Ⓒ 88.9%

Ⓓ 144%

Apply

9. A school newspaper estimates that their academic team will win 25 out of 30 matches for the season. After 15 matches, they have won 12. If the team continues winning at this rate, what will be the percent error of the newspaper's estimate once the season is over? Round to the nearest percent.

10. A toy company that makes bubbles fills its 8-ounce bottles using a machine. To check that the machine fills the bottles with the proper amount, the company randomly checks bottles off the assembly line. A bottle passes inspection if the percent error of the amount is 2% or less. What is the range of values that a bottle could contain to pass inspection? Round to the nearest hundredth.

11. **Ⓜ️ Find the Error** A student is finding the percent error for an estimated length of 22 inches with an actual length of 25 inches. Find the student's mistake and correct it.

25 in. − 22 in. = 3 in.

$$\frac{3}{25} = 0.12$$
$$= 0.12\%$$

12. A student population of 1,200 was estimated to increase by 15% in the next five years. The population actually increased by 20%. Find the estimated and actual student populations and describe the percent error.

13. **Ⓜ️ Make an Argument** Make an argument for why you cannot find the percent error when the actual value is 0. Explain.

14. **Ⓜ️ Use a Counterexample** Determine if the following statement is *true* or *false*. If false, provide a counterexample.

Percent error can never be greater than 100%.

📖 **Foldables** Use your Foldable to help review the module.

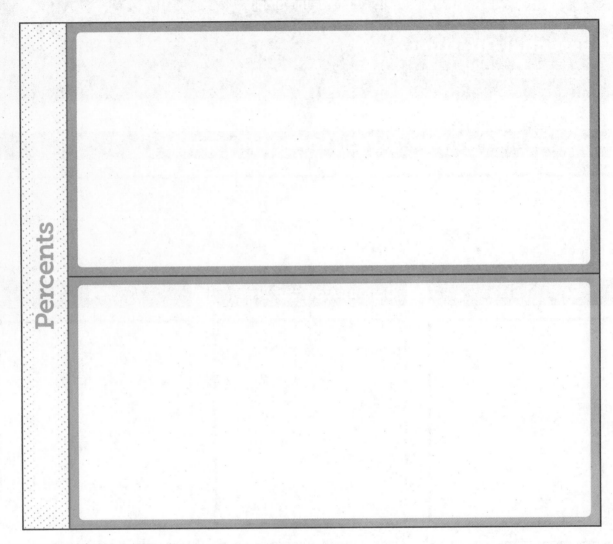

Percents

Rate Yourself! ⬛ ◈ ⭐

Complete the chart at the beginning of the module by placing a checkmark in each row that corresponds with how much you know about each topic after completing this module.

Write about one thing you learned.	Write about a question you still have.
_____	_____
_____	_____
_____	_____
_____	_____

Reflect on the Module

Use what you learned about percents to complete the graphic organizer.

e Essential Question

How can percent describe the change of a quantity?

How do you find the percent of change between two amounts?

Percent Error	Commission and Fees	Tax

Tips and Markups	Discounts	Interest

Test Practice

1. Open Response An ice cream shop had 316 customers the first weekend it opened. The second weekend it had 468. What is the percent of increase in the number of customers? Round to the nearest percent. (Lesson 1)

[   ]

2. Open Response Cynthia is training to run the 100-yard dash. Each week she records her best 100-yard dash time so that she can track her improvement. The table shows her best times after each of the first 3 weeks of training. (Lesson 1)

Week	Time (s)
1	14.0
2	13.5
3	13.1

A. Find the percent of change from Week 1 to Week 2, and from Week 2 to Week 3. Round to the nearest tenth percent if necessary.

[   ]

B. How much greater is the percent of change between the first two weeks than between the second and third weeks? Round to the nearest tenth percent.

[   ]

3. Multiple Choice The wholesale cost of a pair of sandals at a shoe store is $22.60. The markup for the sandals is 45%. What is the selling price of the sandals? (Lesson 3)

Ⓐ $10.17
Ⓑ $32.77
Ⓒ $41.15
Ⓓ $54.90

4. Equation Editor Julia wants to purchase a pair of headphones that are on sale for $42.79. The sales tax rate in her county is 6%. Calculate the amount of sales tax Julia will pay on the purchase. Round to the nearest cent if necessary. (Lesson 2)

[   ]

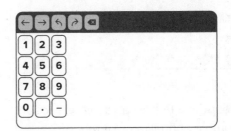

5. Table Item For parties of 6 or more people at a certain restaurant, a tip of 18% is automatically included on the bill.

Select *yes* or *no* to indicate whether or not each of the following tip amounts is correct for an 18% gratuity on each subtotal. (Lesson 3)

	yes	no
subtotal: $129.50 tip: $23.27		
subtotal: $98.40 tip: $17.71		
subtotal: $142.17 tip: $22.75		

6. Multiple Choice During a sale at a furniture store, sofas were marked down 30%. You have a coupon for an additional 15% off. What is the final price of a sofa that originally cost $550? (Lesson 4)

Ⓐ $302.50
Ⓑ $327.25
Ⓒ $396.00
Ⓓ $275.00

7. Equation Editor A set of curtains normally sells for $58.99. Roberta used a coupon good for 20% off the regular price. To the nearest cent, how much did Roberta pay for the curtains before sales tax? (Lesson 4)

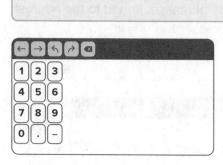

8. Open Response Rebecca's savings account pays a simple annual interest rate of 2.5%. Suppose she deposits $4,280 in the account and makes no additional deposits or withdrawals for 4 years. What will the total value of the account be after 4 years? (Lesson 5)

9. Multiselect There are 118 different elements in the periodic table. While giving a presentation, Clark says that there are 128 different elements. Select each true statement. (Lesson 7)

☐ There are actually 118 elements.

☐ The amount of error in Clark's estimate is 10.

☐ The amount of error in Clark's estimate is 118.

☐ To the nearest tenth percent, the percent error in Clark's estimate is 8.5%.

☐ To the nearest tenth percent, the percent error in Clark's estimate is 10%.

10. Equation Editor Manny is a realtor and earns 3% commission on the sale of a house. If the house sells for $155,000, how much commission would he earn? (Lesson 6)

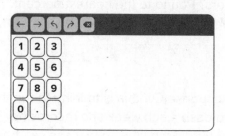

11. Open Response Mr. Gomez earns a 5.4% commission on his total sales for the month. If he wants to earn at least $3,500 in commission this month, what is the minimum amount he needs to sell? Round to the nearest dollar. Explain how you found your answer. (Lesson 6)

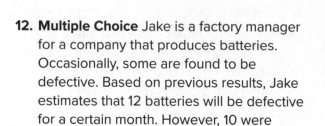

12. Multiple Choice Jake is a factory manager for a company that produces batteries. Occasionally, some are found to be defective. Based on previous results, Jake estimates that 12 batteries will be defective for a certain month. However, 10 were actually defective. What is the percent error of Jake's estimate? (Lesson 7)

Ⓐ 2%

Ⓑ 16%

Ⓒ 1.6%

Ⓓ 20%

Module 3

Operations with Integers and Rational Numbers

e Essential Question

How are operations with rational numbers related to operations with integers?

What Will You Learn?

Place a checkmark (✓) in each row that corresponds with how much you already know about each topic **before** starting this module.

KEY

⬛ — I don't know.　◆ — I've heard of it.　★ — I know it!

	Before			After		
	⬛	◆	★	⬛	◆	★
adding and subtracting integers						
finding the distance between two integers on a number line						
multiplying and dividing integers						
simplifying expressions using the order of operations						
evaluating algebraic expressions involving integers						
writing fractions as decimals						
writing repeating decimals as fractions and mixed numbers						
adding and subtracting rational numbers						
multiplying and dividing rational numbers						
simplifying expressions involving rational numbers using the order of operations						
evaluating algebraic expressions involving rational numbers						

📖 Foldables Cut out the Foldable and tape it to the Module Review at the end of the module. You can use the Foldable throughout the module as you learn about operations with rational numbers.

What Vocabulary Will You Learn?

Check the box next to each vocabulary term that you may already know.

☐ absolute value

☐ Additive Inverse Property

☐ additive inverses

☐ bar notation

☐ Distributive Property

☐ Multiplicative Identity Property

☐ multiplicative inverses

☐ Multiplicative Property of Zero

☐ opposites

☐ order of operations

☐ rational number

☐ repeating decimal

☐ terminating decimal

Are You Ready?

Study the Quick Review to see if you are ready to start this module.
Then complete the Quick Check.

Quick Review

Example 1

Write an equivalent fraction.

Write a fraction that is equivalent to $\frac{25}{100}$.

$$\frac{25}{100} = \frac{1}{4}$$

Divide the numerator and denominator by 25.

An equivalent fraction to $\frac{25}{100}$ is $\frac{1}{4}$.

Example 2

Graph mixed numbers on a number line.

Graph $3\frac{2}{3}$ on a number line.

Find the two whole numbers between which $3\frac{2}{3}$ lies.

$$3 < 3\frac{2}{3} < 4$$

Because the denominator is 3, divide each space into 3 sections.

Draw a dot at $3\frac{2}{3}$.

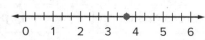

Quick Check

1. Write a fraction that is equivalent to $\frac{24}{36}$.

2. Write a fraction that is equivalent to $\frac{45}{50}$.

3. Graph $1\frac{1}{4}$ on a number line.

How Did You Do?

Which exercises did you answer correctly in the Quick Check?
Shade those exercise numbers at the right.

Add Integers

I Can... use different methods, including algebra tiles, number lines, or absolute value, to add integers.

Explore Use Algebra Tiles to Add Integers

🔾 **Online Activity** You will use algebra tiles to model addition of integers, and make a conjecture about the sign of the sum of two integers.

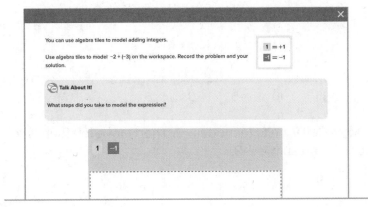

What Vocabulary
Will You Learn?
additive inverse
Additive Inverse
  Property
opposites

Learn Add Integers with the Same Sign

To add two integers with the same sign, you can use a horizontal or vertical number line.

The equation $-3 + (-4) = -7$ is modeled on the horizontal number line. Start at zero. Move left three units to model the negative integer, -3. Then, move left four units to model adding the negative integer -4. The sum is -7.

The equation $-5 + (-4) = -9$ is modeled on the vertical number line. Start at zero. Move down five units to model the negative integer -5. Then move down four units to model the negative integer -4. The sum is -9.

(continued on next page)

Copyright © McGraw-Hill Education

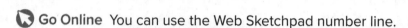 Talk About It!

How does a number line help show that the sum of two negative numbers will always be negative?

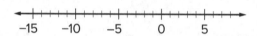 Talk About It!

Compare Method 1 and Method 2. Give an example of a situation where Method 1 might be more advantageous.

The number lines illustrate the rules for adding two integers with the same sign. To add two integers with the same sign, add their absolute values. The sum is:

- _____ if both integers are positive

- _____ if both integers are negative

Example 1 Add Integers with the Same Sign

Find −7 + (−2).

Method 1 Use a number line.

 Go Online You can use the Web Sketchpad number line.

$$\begin{array}{ccccccc} & | & | & | & | & | \\ -15 & -10 & -5 & 0 & 5 \end{array}$$

Start at zero. Move left 7 units to model −7. Then move left 2 units to model adding −2. The sum is −9.

So, −7 + (−2) = −9.

Method 2 Use the absolute value.

Because the integers have the same sign, find the sum of the absolute values.

$$|-7| = \boxed{} \text{ and } |-2| = \boxed{}$$

The sum of their absolute values is 7 + 2 or 9.

Because both integers are negative, the sum will be negative.

So, the solution to −7 + (−2) is −9.

Check

Find the sum of −5 + (−11).

Show your work here

 Go Online You can complete an Extra Example online.

🌐 Example 2 Add Integers with the Same Sign

Allie borrowed $139 from her parents to purchase an ebook reader. During the first month, she purchased $47 in apps, games, and movies which was added to the amount she already owed her parents.

What integer represents the amount of money Allie had at the end of the first month?

The addition expression $-139 + (-47)$ represents the amount of money Allie had at the end of the month.

Because the integers have the same sign, find the sum of the absolute values.

$$|-139| = \boxed{} \text{ and } |-47| = \boxed{}$$

The sum of their absolute values is $139 + 47$ or _____ .

Because both integers are negative, the sum will be _____ .

So, the amount of money Allie has is $-\$186$.

Check

A contestant on a game show has $-1,500$ points. He loses another 750 points. What is his new score?

Show your work here

🌐 **Go Online** You can complete an Extra Example online.

💭 Think About It!

Does a positive or negative integer represent borrowing and spending money?

💬 Talk About It!

What does the value $-\$186$ mean in the problem?

Learn Find Additive Inverses

The integers 4 and −4 are opposites. **Opposites** have the same absolute value but different signs. Two integers that are opposites are called **additive inverses** and their sum is zero.

The **Additive Inverse Property** can be used to find additive inverses.

Words	Example	Algebra
The sum of any number and its additive inverse is zero.	$4 + (-4) = 0$	$a + (-a) = 0$

Complete the table showing integers and their additive inverses.

Integer	Additive Inverse	Sum
−1	1	.
2		0
	−3	0
−4		0
	5	0

Talk About It!

Which number is its own additive inverse? Explain.

Pause and Reflect

Are you ready to move on to the next Example? If yes, what have you learned that you think will help you? If no, what questions do you still have? How can you get those questions answered?

Record your observations here

🌐 Example 3 Find Additive Inverses

A hiking trail begins at an elevation of 150 feet above sea level. It leads down to the shore of the ocean, which has an elevation of 0 feet above sea level.

What integer represents the change in the elevation of the trail from beginning to end?

The trail begins at a positive height above the shore. To reach sea level at 0 feet, the change in elevation would have to be equal to the additive inverse of 150 feet. The additive inverse of 150 feet is −150 feet.

So, the integer that represents the change in elevation of the trail from beginning to end is _____.

Check

A puffin is flying at 29 feet above sea level. What is the elevation, in feet, that it will have to fly to reach sea level?

Show your work here

🌀 **Go Online** You can complete an Extra Example online.

Learn Add Integers with Different Signs

To add two integers with different signs, you can use a horizontal or vertical number line.

The horizontal number line models the equation $5 + (−3) = 2$. Start at zero. Move right five units to model the positive integer 5. Then move left three units to model adding the negative integer −3. The sum is 2.

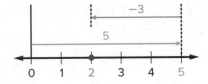

Predict how you can use a vertical number line to add the integers. Turn the page to check your prediction.

(continued on next page)

> 🔵 Think About It!
>
> Does the hiking trail ascend or descend?

> 💬 Talk About It!
>
> What do you notice about the signs of a pair of additive inverses?

Copyright © McGraw-Hill Education

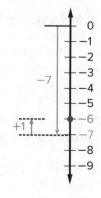

The vertical number line models the equation $-7 + 1 = -6$. Start at zero. Move down seven units to model the negative integer -7. Then move up one unit to model the positive integer 1. The sum is -6.

The number lines illustrate the rules for adding two integers with different signs. To add integers with different signs, subtract their absolute values. The sum is:

- positive if the positive integer's absolute value is greater

- negative if the negative integer's absolute value is greater

Example 4 Add Integers with Different Signs

Find $11 + (-4)$.

Method 1 Use a number line.

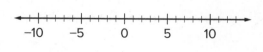 **Go Online** You can use the Web Sketchpad number line.

Start at zero. Move right 11 units to model 11. Then move left 4 units to model adding -4. The sum is 7.

So, $11 + (-4) = $ ⬚.

Method 2 Use the absolute value.

Because the integers have different signs, find the difference of the absolute values.

$$|11| = \boxed{} \text{ and } |-4| = \boxed{}$$

The difference in their absolute values is $11 - 4$ or 7.

Because $|11| > |-4|$, the sum will have the same sign as _____.

So, $11 + (-4) = 7$.

Check

Find the sum of $10 + (-22)$.

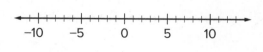 **Go Online** You can complete an Extra Example online.

Talk About It!

Give an example of adding integers with different signs. Does your example reinforce the statements about the sign of a sum?

Talk About It!

How could you change the second addend so that the sum is negative?

Example 5 Add Integers with Different Signs

A whale swam at a depth of 275 feet below the surface of the ocean. After 10 minutes, it rose 194 feet.

What integer represents the location of the whale, relative to the ocean's surface, after 10 minutes? Interpret the integer within the context of the problem.

Because the integers have different signs, find the difference of the absolute values.

$|-275| = $ [ ] and $|194| = $ [ ]

The difference of their absolute values is $275 - 194$ or _____.

Because $|-275| > |194|$, the sum will have the same sign as

_____.

So, the integer that describes the whale's location, in feet, is -81. This means the whale is 81 feet below the ocean's surface.

Check

At 5:00 P.M., a thermometer shows an outside temperature of 2°F. Then, over the next three hours, the temperature drops 11°F. Which integer represents the thermometer reading, in degrees, at 8:00 P.M.?

 (Show your work here)

 Go Online You can complete an Extra Example online.

Pause and Reflect

When you first saw this Example, what was your reaction? Did you think you could solve the problem? Did what you already know help you solve the problem?

(Record your observations here)

Copyright © McGraw-Hill Education

Talk About It!

Why would an answer of a positive integer not make sense in the context of the problem?

Talk About It!

Can you think of a change that could be made to the problem where a positive integer answer would make sense?

Example 6 Add Three or More Integers

Find −26 + 74 + (−14).

Method 1 Add the numbers in order.

−26 + 74 + (−14)   Write the expression.

= [ ] + (−14)   Add −26 + 74.

= [ ]   Add 48 + (−14).

Method 2 Group like signs together.

−26 + 74 + (−14)   Write the expression.

= −26 + ([ ]) + [ ]   Use the Commutative Property.

= [ ] + 74   Add −26 + (−14).

= [ ]   Add.

So, the sum of −26 + 74 + (−14) is 34.

Check

Find −14 + 8 + (−6).

Show your work here

Go Online You can complete an Extra Example online.

 Think About It!

Can you group the negative addends together to add those first? What property allows you to do that?

 Talk About It!

Compare and contrast Method 1 and Method 2.

🌐 Example 7 Add Three or More Integers

A roller coaster starts at point *A*, 43 feet above the ground. It ascends 55 feet, descends 80 feet, then ascends 110 feet to point *B*.

What is the height of the roller coaster at point *B* in relation to the ground?

Ascending can be represented by a positive integer and descending can be represented by a negative integer. So, the addition expression $43 + 55 + (-80) + 110$ models the situation.

Simplify the expression.

$43 + 55 + (-80) + 110$ Write the expression.

$= 43 + 55 + \boxed{} + \left(\boxed{}\right)$ Commutative Property of Addition

$= \boxed{} + 110 + (-80)$ Add $43 + 55$.

$= \boxed{} + (-80)$ Add $98 + 110$.

$= \boxed{}$ Add $208 + (-80)$.

So, point *B* is 128 feet above the ground.

Copyright © McGraw-Hill Education

🗨 Think About It!
What addition expression could represent this problem?

🗨 Talk About It!
Describe another way you can solve the problem.

Check

An unmanned submarine is doing tests along the ocean bottom. The following table shows the change in the depth of the submarine each minute for 4 minutes.

Minute	Change in Depth (ft)
1	−150
2	162
3	−175
4	−180

What is the total change in depth of the submarine after 4 minutes? Write your answer as an integer.

Show your work here

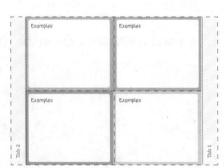

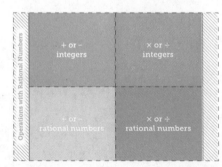

Go Online You can complete an Extra Example online.

Foldables It's time to update your Foldable, located in the Module Review, based on what you learned in this lesson. If you haven't already assembled your Foldable, you can find the instructions on page FL1.

Practice

Go Online You can complete your homework online.

Add. (Examples 1, 4, and 6)

1. −3 + (−8)

−11

2. −11 + (−13)

−24

3. 9 + (−35)

−26

4. −28 + 14

14

5. −22 + (−10) + 15

−17

6. 18 + (−12) + 5

11

7. Roger owes his father $15. He borrows another $25 from him. What integer represents the balance that he owes his father? (Example 2)

8. A football team lost 14 yards on their first play then lost another 7 yards on the next play. What integer represents the total change in yards for the two plays? (Example 2)

9. Kwan's beginning account balance was $20. His ending balance is $0. What integer represents the change in his account balance from beginning to end? (Example 3)

10. Lucy's dog lost 6 pounds. How much weight does her dog need to gain in order to have a net change of 0 pounds? (Example 3)

11. The table shows Jewel's scores for the first 9 holes and the second 9 holes of her game of golf. What integer represents her score for the entire game? (Example 5)

Holes	Score
1–9	3 over par
10–18	4 under par

12. At 4:00 A.M., the outside temperature was −28°F. By 4:00 P.M. that same day, it rose 38 degrees. What integer represents the temperature at 4:00 P.M.? (Example 5)

Test Practice

13. In 20 seconds, a roller coaster goes up a 100-meter hill, then down 72 meters, and then back up a 48-meter rise. How much higher or lower from the start of the ride is the coaster after the 20 seconds? (Example 7)

14. Open Response Joe opened a bank account with $80. He then withdrew $35 and deposited $115. What is his account balance after these transactions?

Apply

15. The table shows the transactions for one week for Tasha and Jamal. Who has the greater account balance at the end of the week? How much greater?

Transactions	Tasha	Jamal
Initial Deposit	$250	$200
Withdrawals	$20	$60
Deposits	$65	$135
Debit Card Purchases	$46	$27

16. A hot air balloon rises 340 feet into the air. Then it descends 130 feet, goes up 80 feet, and then down another 45 feet. How many feet will the balloon need to travel to return to the ground? Represent this amount as an integer. Explain.

17. What value of x would result in the numerical value of zero for each expression?

a. $-10 + 11 + x$

b. $7 + x + (-10)$

c. $x + 1 + (-1)$

18. (MP) **Find a Counterexample** Patrick stated that the sum of a positive integer and a negative integer is always negative. Find a counterexample that illustrates why this statement is not true.

19. Explain how you know that the sum of 2, 3, and -2 is positive without computing.

20. Create Write and solve a real-world problem where you add three integers and the sum is negative.

Subtract Integers

I Can... use different methods, including algebra tiles, number lines, or the additive inverse, to subtract integers.

Explore Use Algebra Tiles to Subtract Integers

Online Activity You will use algebra tiles to model subtraction of integers, and draw conclusions about the sign of the difference of the two integers.

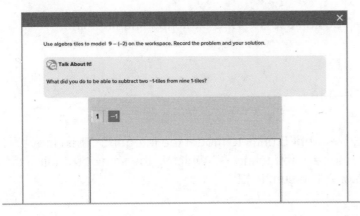

Learn Subtract Integers

To subtract integers, you can use a horizontal or vertical number line.

The horizontal number line models the equation $4 - 9 = -5$. Start at zero. Move right four units to model the integer 4. Then move left nine units to model subtracting 9. The difference is -5.

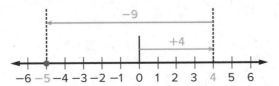

The vertical number line models the equation $4 - 9 = -5$. Start at zero. Move up four units to model the integer 4. Then move down nine units to model subtracting 9. The difference is -5.

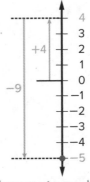

(continued on next page)

Talk About It!
The Commutative Property is true for addition. For example, $7 + 2 = 2 + 7$. Is the Commutative Property true for subtraction? Does $7 - 2 = 2 - 7$? Explain your reasoning using a number line.

The number lines illustrate the rules for subtracting two integers.

Words	Symbols	Example
To subtract an integer, add the additive inverse of the integer.	$p - q = p + (-q)$	$4 - 9 = 4 + (-9) = -5$

Example 1 Subtract Integers

Find $5 - (-7)$.

Method 1 Use a number line.

 Go Online You can use the Web Sketchpad number line.

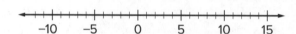

Start at zero. Move right 5 units to model the integer 5. Then move right 7 units to model subtracting −7, which is the same as adding the additive inverse, 7. The sum is 12.

So, $5 - (-7) = 12$.

Method 2 Use the additive inverse.

$5 - (-7) = 5 + 7$ To subtract −7, add the additive inverse of −7.

$= \boxed{}$ Add 5 + 7.

So, $5 - (-7) = 12$.

Check

Find $11 - (-15)$.

Show your work here

 Go Online You can complete an Extra Example online.

Think About It!

Will you use a number line or will you add the additive inverse to solve this problem?

Talk About It!

Compare and contrast Method 1 and Method 2.

Example 2 Subtract Integers

Find −24 − (−17).

$-24 - (-17) = -24 + 17$ To subtract −17, add its additive inverse.

$= -7$ Add.

So, $-24 - (-17) =$ _____ .

Check

Find $-39 - (-24)$.

Example 3 Subtract Expressions

Evaluate $x - y$ if $x = -23$ and $y = 19$.

$x - y = -23 - 19$ Replace x with −23 and y with 19.

$= -23 + (-19)$ To subtract 19, add its additive inverse.

$= -42$ Add $-23 + (-19)$.

So, when $x = -23$, and $y = 19$, $x - y =$ _____ .

Check

Evaluate $p - q$ if $p = -21$ and $q = 37$.

Go Online You can complete an Extra Example online.

Explore Find Distance on a Number Line

Online Activity You will calculate distance traveled by using a number line to find the difference of the two integers.

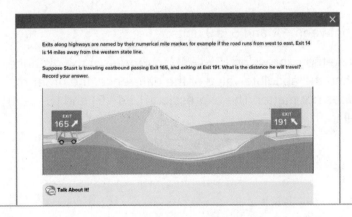

Think About It!

Predict the sign of the difference between the two integers.

Talk About It!

Describe a situation where the difference between two numbers is greater than either number. Then explain why that happens.

Copyright © McGraw-Hill Education

Learn Find the Distance Between Integers

Find the distance between −4 and 5.

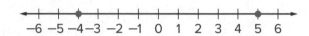 **Go Online** Watch the animation to learn how to find the distance between two integers.

Method 1 Use a number line.

Step 1 Plot the integers on a number line.

The animation shows two points at −4 and 5.

Step 2 Count the number of units between the two integers.

There are 9 units between −4 and 5.

Method 2 Use an expression.

The distance between two integers is equal to the absolute value of their difference.

distance = |difference of integers|

Step 1 Write an expression for the distance.

$|-4 - 5|$

Step 2 Simplify the expression.

$|-4 - 5| = \left|\right|$

$= \boxed{}$

The distance between −4 and 5 is 9 units.

You can also use the expression |5−(−4)| to represent the distance. Because you find the absolute value of the difference, the order of the integers does not matter. The expressions |−4−5| and |5−(−4)| are both equal to 9.

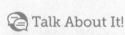 **Talk About It!**

Why do we take the absolute value of the difference?

Example 4 Find the Distance Between Integers

Find the distance between −9 and 8.

Method 1 Use a number line.

 Go Online You can use the Web Sketchpad number line.

Start at −9. Move right until you reach 8.

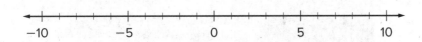

There are _____ units between −9 and 8.

Method 2 Use the absolute value.

To find the distance between integers, you can find the absolute value of their difference.

$|-9 - 8| = |-9 + (-8)|$ Add the additive inverse of 8.

$= \left|\ \ \ \right|\ $ or $\ \left|\ \ \ \right|$ Simplify.

So, the distance between −9 and 8 is 17 units.

Check

Find the distance between −5 and 9 on the number line.

Show your work here

 Go Online You can complete an Extra Example online.

Pause and Reflect

When finding the distance between integers with different signs, which method would you choose to use? Explain.

Record your observations here

Think About It!

What subtraction expression could be used to find the distance?

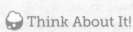 **Example 5** Find the Distance Between Integers

The highest point in California is Mount Whitney with an elevation of 14,494 feet. The lowest point is Death Valley with an elevation of −282 feet.

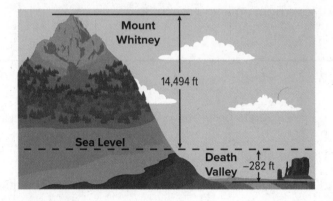

What is the distance between the height of Mount Whitney and the depth of Death Valley?

$|14{,}494 - (-282)| = |14{,}494 + 282|$ To subtract −282, add its additive inverse.

$= |14{,}776|$ Add.

$=$ ☐   Find the absolute value.

So, the distance between the two points is 14,776 feet.

Check

The top of an iceberg is 55 feet above sea level, while the bottom is 385 feet below sea level. What is the distance between the top and bottom of the iceberg?

 Show your work here

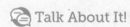 **Think About It!**

Will the distance be greater or less than 14,494 feet?

Talk About It!

Is it reasonable to have a negative answer? Why or why not?

 Go Online You can complete an Extra Example online.

🌐 Apply The Solar System

The table shows the minimum and maximum temperatures on various celestial objects in the solar system.

Celestial Object	Minimum Temperature (°F)	Maximum Temperature (°F)
Moon	−387	253
Mars	−225	70
Mercury	−279	801
Venus	864	864

Scientists want to send a probe to study the celestial object with the greatest variation in temperature. To which celestial object should they send the probe?

1 What is the task?

Make sure you understand exactly what question to answer or problem to solve. You may want to read the problem three times. Discuss these questions with a partner.

First Time Describe the context of the problem, in your own words.
Second Time What mathematics do you see in the problem?
Third Time What are you wondering about?

2 How can you approach the task? What strategies can you use?

3 What is your solution?

Use your strategy to solve the problem.

4 How can you show your solution is reasonable?

✏️ **Write About It!** Write an argument that can be used to defend your solution.

🖱 **Go Online** Watch the animation.

💬 Talk About It!

On which celestial object from the table would it be most reasonable to live? Explain.

Check

The table shows the highest and lowest points of elevation, in relation to sea level, in four countries. Which country in this list has the greatest variation in elevation? the least?

Country	Highest Point (ft)	Lowest Point (ft)
Jordan	6,083	−1,404
United Kingdom	4,406	−13
Sweden	6,903	−8
Ireland	3,406	−10

Go Online You can complete an Extra Example online.

Foldables It's time to update your Foldable, located in the Module Review, based on what you learned in this lesson. If you haven't already assembled your Foldable, you can find the instructions on page FL1.

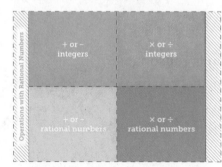

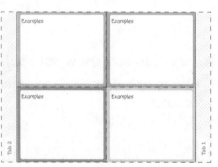

Practice

Go Online You can complete your homework online.

Subtract. (Examples 1 and 2)

1. $9 - (-2)$

2. $-20 - 10$

3. $13 - (-63)$

4. $28 - 14$

5. $-10 - 0$

6. $-33 - 33$

7. $-18 - (-12)$

8. $-28 - (-13)$

9. $-18 - (-40)$

10. Evaluate $a - b$ if $a = 10$ and $b = -7$.
(Example 3)

11. Evaluate $x - y$ if $x = -11$ and $y = 26$.
(Example 3)

12. Find the distance between -6 and 7 on a number line. (Example 4)

13. Find the distance between -14 and 5 on a number line. (Example 4)

14. The highest and lowest recorded temperatures for the state of Texas are 120° Fahrenheit and -23° Fahrenheit. Find the range of these extreme temperatures. (Example 5)

Test Practice

15. Open Response The table shows the starting and ending elevations of a hiking trail. How much greater is the elevation of the ending point than the starting point for the trail?

Point on Trail	Elevation
Starting Point	180 ft below sea level
Ending Point	260 ft above sea level

Apply

16. The table shows the maximum and minimum account balances for three college students for one month. Giovanni claimed that he had the least variation (from maximum to minimum) in his account balance that month. Is he correct? Write a mathematical argument to justify your solution.

Student	Maximum Balance ($)	Minimum Balance ($)
Jordan	145	−25
Giovanni	168	15
Elisa	152	−10

17. The table shows the record high and record low temperatures for certain U.S. states. Which state in the list had the greatest variation in temperature? the least?

State	Record High Temperature (°F)	Record Low Temperature (°F)
Alaska	100	−80
Idaho	118	−60
Nevada	125	−50
Utah	117	−69

18. (MP) **Use a Counterexample** Determine if each statement is *true* or *false*. If false, provide a counterexample.

   a. Distance is always positive.

   b. Change is always positive.

19. (MP) **Find the Error** A student is finding $4 - (-2)$. Find the student's mistake and correct it.

$$4 - (-2) = 4 - 2$$
$$= 2$$

20. **Create** Write a subtraction expression with a positive and negative integer whose difference is negative. Then find the difference.

21. If you subtract two negative integers, will the difference *always, sometimes,* or *never* be negative? Explain using examples to justify your solution.

Multiply Integers

I Can... use number lines and mathematical properties to multiply integers.

Explore Use Algebra Tiles to Multiply Integers

Online Activity You will use algebra tiles to model integer multiplication, and make a conjecture about the sign of the product of the two integers.

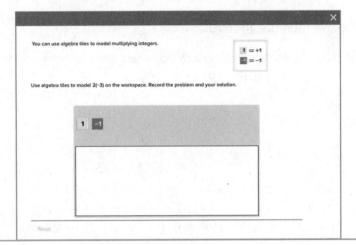

Learn Multiply Integers with Different Signs

Because multiplication is repeated addition, you can use a number line to show that 3(−6) means that −6 is used as an addend 3 times.

The number line models 3(−6).

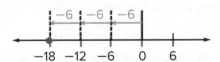

The number line illustrates the rule for multiplying integers with different signs.

Words	Examples
The product of two integers with different signs is negative.	3(−6) = −18 −3(6) = −18

Think About It!

Predict the sign of the two integers.

 Talk About It!

How could you represent −7(5) as an addition expression?

 Think About It!

What multiplication expression could you use to solve Example 2?

 Talk About It!

In the expression, why is 90 negative? Why would a positive integer not make sense as the product in this situation?

Example 1 Multiply Integers with Different Signs

Find −7(5).

Method 1 Use a number line.

Using the Commutative Property of Multiplication, you can rewrite −7(5) as 5(−7). Use a number line to show five groups of −7.

−7(5) = ☐

Method 2 Use the multiplication rule.

−7(5) = ☐   Multiply. Integers with different signs result in a negative product.

So, −7(5) = −35.

Check

Find 9(−13).

Show your work here

🔴 **Go Online** You can complete an Extra Example online.

🌐 Example 2 Multiply Integers with Different Signs

A submarine is diving from the surface of the water and descends at a rate of 90 feet per minute.

What integer represents the submarine's location, in feet, after 11 minutes?

The submarine descends at a rate of 90 feet per minute for 11 minutes. The expression 11(−90) represents the situation.

The signs of the integers are different. The product is negative.

11(−90) = ☐

So, after 11 minutes, the location of the submarine will be 990 feet below the surface.

Check

A helicopter descends at a rate of 275 feet per minute. What integer represents the change in the helicopter's altitude, in feet, after 7 minutes?

Show your work here

 Go Online You can complete an Extra Example online.

Learn Multiply Integers with the Same Sign

 Go Online Watch the animation to learn how to multiply integers with the same sign.

The animation shows that you can use a pattern to show that the product of two negative numbers is positive. Notice that when you decrease the first factor by 1, the product increases by 3. You can continue the pattern to negative numbers.

$(2)(-3) = -6$

$(1)(-3) = -3$

$(0)(-3) = 0$

$(-1)(-3) = 3$

$(-2)(-3) = 6$

The pattern shows that the product of two negative numbers is positive.

When multiplying two integers with the same sign, such as 5 and 6, or −5 and −6, the sign of the product is always positive.

Words	Examples
The product of two integers with the same sign is positive.	$6(5) = 30$ $-6(-5) = 30$

Math History Minute

Negative numbers were not widely recognized by mathematicians until the 1800s, with a few exceptions. In 6th century India, negative numbers were introduced to represent debts and Indian mathematician **Brahmagupta (598–668)** stated rules for adding, subtracting, multiplying, and dividing negative numbers.

Example 3 Multiply Integers with the Same Sign

Find −8(−9).

$-8(-9) = \boxed{}$ The signs of the integers are the same. The product is positive.

So, the product of −8(−9) is 72.

Check

Find −5(−13).

Example 4 Multiply Integers with the Same Sign

Evaluate xy if $x = -14$ and $y = -7$.

$xy = -14(-7)$ Replace x with −14 and y with −7.

$= \boxed{}$ The signs of the integers are the same. The product is positive.

So, the value of the expression is 98.

Check

Evaluate pq if $p = -16$ and $q = -7$.

 Go Online You can complete an Extra Example online.

Example 5 Multiply Three or More Integers

Find −4(−7)(−2).

$-4(-7)(-2) = [-4(-7)](-2)$ Associative Property

$= \boxed{}(-2)$ Multiply (−4)(−7).

$= \boxed{}$ Multiply 28(−2).

So, the product of −4(−7)(−2) is −56.

Check

Find −3(−11)(−3).

Example 6 Multiply Three or More Integers

Evaluate ab^2c when $a = -5$, $b = 4$, and $c = -9$.

$ab^2c = -5(4)^2(-9)$ Substitute −5 for a, 4 for b, and −9 for c.

$= -5\left(\boxed{}\right)(-9)$ Multiply 4 · 4.

$= [-5(16)](-9)$ Associative Property

$= \boxed{}(-9)$ Multiply −5(16).

$= \boxed{}$ Multiply −80(−9).

So, the value of the expression is 720.

Check

Evaluate $pqrs$ if $p = -7$, $q = 15$, $r = 1$, and $s = -2$.

Go Online You can complete an Extra Example online.

Think About It!

What do you notice about the expression?

Talk About It!

How can you tell, without multiplying, if the product will be positive or negative?

Learn Use Properties to Multiply Integers

 Talk About It!

How can the properties of multiplication help you multiply integers?

In mathematics, properties can be used to justify statements you make while verifying or proving another statement. Some of the properties of mathematics are listed below.

Additive Inverse Property	$a + (-a) = 0$
Distributive Property	$a(b + c) = ab + ac$
Multiplicative Identity Property	$1 \cdot a = a$
Multiplicative Property of Zero	$a \cdot 0 = 0$

 Go Online Watch the animation online to learn how to use properties of multiplication to multiply integers.

The animation shows that $2(-1) = -2$ using properties of multiplication, beginning with the true statement $2(0) = 0$.

$2(0) = 0$ Multiplicative Property of Zero

$2[1 + (-1)] = 0$ Additive Inverse Property

$2(1) + 2(-1) = 0$ Distributive Property

$2 + 2(-1) = 0$ Multiplicative Identity Property

In order for $2 + 2(-1)$ to equal 0, $2(-1)$ must equal -2, based on the Additive Inverse Property. This shows, using properties, that the product of two integers with different signs is negative.

You can also use properties to show that the product of two negative integers is positive.

Show that $(-2)(-1) = 2$ by writing the correct property for each step.

$0 = -2(0)$ _____

$0 = -2[1 + -1]$ _____

$0 = -2(1) + (-2)(-1)$ _____

$0 = -2 + (-2)(-1)$ _____

$2 = (-2)(-1)$ _____

Apply Agriculture

Two farms begin the year with $1,500 of extra savings each in case they lose money at the year's end. The table shows the amount of money each farm earned, or revenue, and the amount of money each farm spent, or expenses, for one month. If these results are consistent with the revenue and expenses for each month of the year, which farm will have enough savings to continue to operate for the whole year and how much will they have left over?

	Farm 1		Farm 2	
	Revenue ($)	Expenses ($)	Revenue ($)	Expenses ($)
Farm Supplies		134		211
Water		44		248
Maintenance		152		147
Potatoes Sold	308		476	

1 What is the task?

Make sure you understand exactly what question to answer or problem to solve. You may want to read the problem three times. Discuss these questions with a partner.

First Time Describe the context of the problem, in your own words.
Second Time What mathematics do you see in the problem?
Third Time What are you wondering about?

2 How can you approach the task? What strategies can you use?

Record your observations here

3 What is your solution?

Use your strategy to solve the problem.

Show your work here

4 How can you show your solution is reasonable?

⬛ **Write About It!** Write an argument that can be used to defend your solution.

🔁 Talk About It!

To the nearest integer, how much would each farm have to begin with in order for them both to have savings left over? Explain.

Check

James started a lawn care business in his community. He thinks that he should be able to operate his business for 20 weeks before it gets too cold. The table shows the expenses and revenue after one week.

Item	Expenses ($)	Revenue ($)
Gasoline	32	
Lawn Mower Maintenance	12	
Grass Seed and Weed Killer	17	
Income ($25 per lawn)		125

How much money will he earn in 20 weeks?

Show your work here

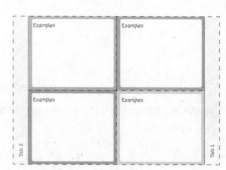 **Go Online** You can complete an Extra Example online.

Foldables It's time to update your Foldable, located in the Module Review, based on what you learned in this lesson. If you haven't already assembled your Foldable, you can find the instructions on page FL1.

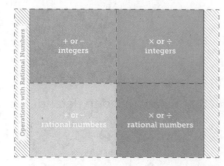

Practice

🡒 **Go Online** You can complete your homework online.

Multiply. (Examples 1, 3, and 5)

1. $4(-7)$

2. $-14(5)$

3. $9(-12)$

4. $-6(-8)$

5. $-10(-10)$

6. $-11(-13)$

7. $7(-5)(4)$

8. $(-8)(-7)(3)$

9. $-2(-12)(-8)$

10. Evaluate ab if $a = -16$ and $b = -5$. (Example 4)

11. Evaluate xy if $x = -10$ and $y = -7$. (Example 4)

12. Evaluate xyz^2 if $x = -2$, $y = 7$, and $z = -4$. (Example 6)

13. Evaluate a^2bc if $a = 3$, $b = -14$, and $c = -6$. (Example 6)

14. Mrs. Rockwell lost money on an investment at a rate of $4 per day. What is the change in her investment, due to the lost money, after 4 weeks? (Example 2)

Test Practice

15. Open Response The table shows the number of questions answered incorrectly by each player on a game show. If each missed question is worth −7 points, what is the change in Olive's score due to the incorrect questions?

Player	Incorrect Questions
Laura	8
Olive	9

Apply

16. Payton starts a lemonade stand for the summer. She thinks that she should be able to operate her business for 14 weeks. The table shows the expenses and revenue after 1 week. Based on this, how much money will Payton make during the 14 weeks?

Item	Expenses	Revenue
Cups	$5	
Lemonade	$6	
Ice	$7	
Income		$45

17. Rakim's goal is to have at least $500 in his checking account at the end of the year. The table shows his activity for the month of January. Will Rakim make his goal if the month of January's activity is representative of how much he will save and spend each month? Explain.

Transaction	Amount
Debit Card Purchases	$500
ATM Withdraws	$750
Deposits	$1,300

18. **MP Reason Inductively** The product of two integers is −24. The difference between the two integers is 14. The sum of the two integers is 10. What are the two integers?

19. **MP Identify Structure** Name the property illustrated by the following.

   a. $-x \cdot 1 = -x$

   b. $x \cdot (-y) = (-y) \cdot x$

20. If you multiply three negative integers, will the product *always, sometimes,* or *never* be negative. Explain.

21. **MP Identify Structure** Name all the values of x if $6|x| = 48$.

Divide Integers

I Can... use a related multiplication sentence to divide integers.

Explore Use Algebra Tiles to Divide Integers

Online Activity You will use algebra tiles to model integer division, and check solutions using multiplication.

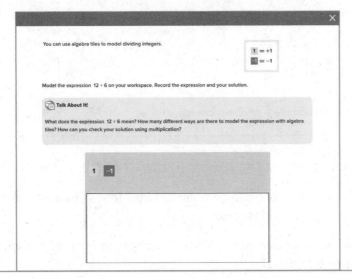

Learn Divide Integers with Different Signs

Division is the inverse operation of multiplication, the same way that subtraction is the inverse of addition. You can find a quotient by using a related multiplication sentence.

$$36 \div (-3) = -12 \qquad \rightarrow \qquad -3 \cdot (-12) = 36$$

What number multiplied by -3 results in 36?

So, $36 \div (-3)$ is -12.

The table shows the rules for dividing integers with different signs.

Words	Examples
The quotient of two integers with different signs is negative.	$-36 \div 3 = -12$ $36 \div (-3) = -12$

Talk About It!

What do you think is the quotient of $-36 \div 3$? Explain your reasoning.

Your Notes ↘

 Think About It!

Predict the sign of the quotient.

 **Talk About It!**

A friend stated that the quotient was positive. How can you use multiplication to show that the quotient is negative?

Talk About It!

What does the solution −95 represent in the context of the problem?

Example 1 Divide Integers with Different Signs

Find 90 ÷ (−10).

90 ÷ (−10) = −9   The signs of the integers are different.
                              The quotient is negative.

So, 90 ÷ (−10) = _____.

Check

Find 72 ÷ (−9).

Show your work here

🌐 **Go Online** You can complete an Extra Example online.

🌐 **Example 2** Divide Integers with Different Signs

The best underwater divers can dive almost 380 feet in four minutes, using only a rope as they descend.

At this rate, what integer represents the change, in feet, of a diver's position after one minute?

Step 1 Identify the distance descended.

What integer represents the change, in feet, in the diver's position after four minutes? ☐

Step 2 Write the division expression.

☐ ÷ ☐   Divide the total change in the diver's position by the number of minutes, 4.

Step 3 Divide −380 by 4 to find the change after one minute.

−380 ÷ 4 = ☐   The signs of the integers are different. The quotient is negative.

So, the integer −95 represents the change in the diver's position, in feet, after one minute.

Copyright © McGraw-Hill Education

Check

A neighbor is improving his dog's fitness level by working on agility training. Before beginning the training regiment, the dog weighed 65 pounds. After nine weeks of training, the dog weighed 47 pounds. What integer represents the change in weight that the dog averaged, in pounds per week, over the nine weeks?

Show your work here

Learn Divide Integers with the Same Sign

Division is the inverse operation of multiplication. You can find a quotient by finding a related multiplication sentence.

$$\underbrace{-20 \div (-4)}_{\text{same signs}} = \overset{\text{positive}}{\overset{\text{quotient}}{5}} \qquad \rightarrow \qquad -4\left(\boxed{}\right) = -20$$

The table shows the rules for dividing integers with the same sign.

Words	Examples
The quotient of two integers with the same sign is positive.	$20 \div 4 = 5$ $-20 \div (-4) = 5$

Talk About It!

How are division and multiplication of integers similar? How are they different?

Example 3 Divide Integers with the Same Sign

Find $-30 \div (-6)$.

$-30 \div (-6) = \boxed{}$ The signs of the integers are the same. The quotient is positive.

So, $-30 \div (-6) = 5$.

Check

Find $-84 \div (-12)$.

Show your work here

Go Online You can complete an Extra Example online.

Copyright © McGraw-Hill Education

Example 4 Divide Integers with the Same Sign

Evaluate $\frac{y}{x}$ if $y = -155$ and $x = -5$.

$\frac{y}{x} = \frac{-155}{-5}$ Replace y with -155 and x with -5.

$= \boxed{} \div \boxed{}$ Write as a division sentence.

$= \boxed{}$ The signs of the integers are the same. The quotient is positive.

So, the value of the expression is 31.

Check

Evaluate $\frac{y}{x}$ if $y = -162$ and $x = -6$.

Go Online You can complete an Extra Example online.

Pause and Reflect

Compare and contrast how to predict the sign of the integer when adding or subtracting, and when multiplying or dividing.

🌐 Apply Personal Finance

Natalie had $165 in her bank account at the beginning of the summer. Over the next 10 weeks, she worked at a summer camp and added $160 to her savings each week, while spending only $40 per week. Once she gets back to school, she plans to spend $105 per week. For how many weeks can she make withdrawals until her balance is $0?

1 What is the task?

Make sure you understand exactly what question to answer or problem to solve. You may want to read the problem three times. Discuss these questions with a partner.

First Time Describe the context of the problem, in your own words.
Second Time What mathematics do you see in the problem?
Third Time What are you wondering about?

2 How can you approach the task? What strategies can you use?

3 What is your solution?

Use your strategy to solve the problem.

4 How can you show your solution is reasonable?

✏️ **Write About It!** Write an argument that can be used to defend your solution.

 Talk About It!

If Natalie wants to be able to withdrawal $105 for 15 weeks, how much can she spend each week during the summer?

Check

You start a pool cleaning business in the neighborhood. You start your business with $1,000 of savings. The table shows the expenses and revenue after one week. At this rate, how many weeks will your savings last?

Item	Expenses	Revenue
Cleaning Chemicals	$31	
Brushes and Towels	$17	
Transportation	$10	
Income ($15 per pool)		$30
Flyers for Advertising	$12	

Show your work here

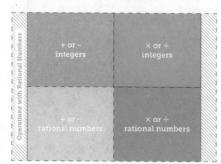

Go Online You can complete an Extra Example online.

Foldables It's time to update your Foldable, located in the Module Review, based on what you learned in this lesson. If you haven't already assembled your Foldable, you can find the instructions on page FL1.

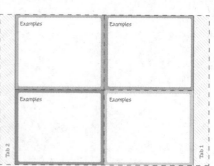

Practice

Go Online You can complete your homework online.

Divide. (Examples 1 and 3)

1. $22 \div (-2)$

2. $-110 \div 11$

3. $75 \div (-3)$

4. $-64 \div (-8)$

5. $-39 \div (-13)$

6. $-50 \div (-10)$

Evaluate each expression if $m = -32$, $n = 2$, and $p = -8$. (Example 4)

7. $\dfrac{m}{n}$

8. $\dfrac{m}{p}$

9. $\dfrac{p}{n}$

Evaluate each expression if $f = -15$, $g = 5$, and $h = -45$. (Example 4)

10. $\dfrac{f}{g}$

11. $\dfrac{h}{f}$

12. $\dfrac{h}{g}$

Test Practice

13. A submarine descends to a depth of 660 feet below the surface in 11 minutes. At this rate, what integer represents the change, in feet, of the submarine's position after one minute? (Example 2)

14. Equation Editor Aaron made 3 withdrawals last month. Each time, he withdrew the same amount. If Aaron withdrew a total of $375, what integer represents the change in his account after the first withdrawal?

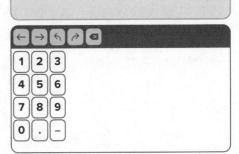

Apply

15. Over the summer, Paulo opens a dog-washing business and begins with $32. The table shows how much he earns and spends each week. He works for 8 weeks washing dogs. When he starts back at school, he budgets $60 to spend each week. How many weeks pass before he needs to wash more dogs?

Weekly Revenue and Expenses		
	Revenue	Expense
Sales	$124	
Supplies		$8

16. Nick had $100 in his savings account. Over the next 6 months, he worked at a seasonal store, where each month he earned $400 and spent $250. He put the remaining amount in his savings account each month. Now that the job is over, he plans to spend $200 per month. For how many months can he make withdrawals from his savings account until his balance is $0?

17. Ⓜ **Make an Argument** The Associative Property holds true for multiplication because $(-3 \times 4) \times (-2) = -3 \times [4 \times (-2)]$. Does the Associative Property hold true for division of integers? Explain.

18. Ⓜ **Justify Conclusions** Is the following statement *true* or *false*? Justify your response.

If n is a negative integer, $\frac{n}{n} = -1$.

19. Create Write and solve a real-world problem in which you divide a positive and negative integer.

20. Write a division sentence that divides a negative integer by a positive integer. Then write a multiplication sentence that proves your division sentence is correct.

Apply Integer Operations

I Can... use the order of integer operations to evaluate expressions.

Example 1 Order of Integer Operations

Find −4(3) + (−7).

$$-4(3) + (-7) = -12 + (-7) \qquad \text{Multiply } -4(3).$$

$$= -19 \qquad \text{Add.}$$

So, −4(3) + (−7) is _____.

Check

Find −5(−12) + (−15).

Copyright © McGraw-Hill Education

 Talk About It!

Why was multiplication performed as the first operation in evaluating the expression?

Example 2 Order of Integer Operations

Find −4(−5)(−2) − (−8).

$$-4(-5)(-2) - (-8) = 20(-2) - (-8) \qquad \text{Multiply } -4(-5).$$

$$= -40 + 8 \qquad \text{Multiply } 20(-2). \text{ Add the additive inverse of } -8.$$

$$= -32 \qquad \text{Add } -40 + 8.$$

So, −4(−5)(−2) − (−8) is _____.

Check

Find 9(−1)(−8) − (−13).

 Talk About It!

Could you have multiplied (−5)(−2) first? Explain your reasoning.

 Go Online You can complete an Extra Example online.

Example 3 Order of Integer Operations

Evaluate $\frac{w}{xy} + y - z^3$ if $w = 36$, $x = -6$, $y = -1$, and $z = -2$.

$\frac{w}{xy} + y - z^3$	$= \frac{36}{(-6)(-1)} + (-1) - (-2)^3$	Substitute the values.
	$= \frac{36}{(-6)(-1)} + (-1) - (-8)$	$(-2)^3 = (-2)(-2)(-2)$
	$= \frac{36}{6} + (-1) - (-8)$	Multiply $(-6)(-1)$.
	$= 6 + (-1) - (-8)$	Divide $\frac{36}{6}$.
	$= 5 - (-8)$	Add $6 + (-1)$.
	$= 5 + 8$	Add the additive inverse.
	$= 13$	Add $5 + 8$.

So, the value of the expression is _____.

Check

Evaluate $\frac{q}{rs} - (r \cdot p)$ if $q = 56$, $r = -4$, $s = 2$, and $p = 1$.

Show your work here

🌐 Example 4 Order of Integer Operations

The average temperature in January in Helsinki, Finland is about $-5°C$.

Use the expression $\frac{(9C + 160)}{5}$, where C is the temperature in degrees Celsius, to find the temperature in degrees Fahrenheit. Round to the nearest degree.

$\frac{(9C + 160)}{5}$	$= \frac{9(-5) + 160}{5}$	Replace C with -5.
	$= \frac{-45 + 160}{5}$	Multiply $9(-5)$.
	$= \frac{115}{5}$	Add $-45 + 160$.
	$= 23$	Simplify.

So, the average temperature in January in Helsinki, Finland is about 23 degrees Fahrenheit.

Check

In a recent year, the average temperature during the month of June in Hall Beach, Canada was 32°F. Use the expression $\frac{5(F - 32)}{9}$, where F is the temperature in degrees Fahrenheit, to find the temperature in degrees Celsius. Round to the nearest degree.

🧭 **Go Online** You can complete an Extra Example online.

💬 **Talk About It!**

At what temperature do you think the measures in °F and °C are equal? Explain your reasoning.

Practice

🧭 **Go Online** You can complete your homework online.

Evaluate each expression. (Examples 1 and 2)

1. $-5(6) + (-9)$

2. $\dfrac{-36}{9} + (-7)$

3. $-4(-8) + (-10)$

4. $2(-5)(-6) - (-12)$

5. $10(-3)(4) - (-15)$

6. $2\left(\dfrac{-80}{4}\right) - 14$

Evaluate each expression if $a = -2$, $b = 3$, $c = -12$, and $d = -4$. (Example 3)

7. $\dfrac{bd}{a} + c$

8. $\dfrac{ac}{b} - (a + d)$

9. $\dfrac{d^3}{a^2} - (c + b)$

Evaluate each expression if $m = -32$, $n = 2$, $p = -8$, and $r = 4$. (Example 3)

10. $\dfrac{pr}{n} + m$

11. $\dfrac{p^2}{m} - (np + r)$

12. $\dfrac{p^3}{r^2} - (m + np)$

Test Practice

13. The table gives the income and expenses of a small company for one year. Use the expression $\dfrac{I - E}{12}$ where I represents the total income and where E represents the total expenses, to find the average difference between the company's income and expenses each month. (Example 4)

	Amount ($)
Income	84,000
Expenses	86,400

14. Open Response Five years ago the population at Liberty Middle School was 1,600 students. This year the population is 1,250 students. Use the expression $\dfrac{N - P}{5}$ where N represents this year's population and where P represents the previous population to find the average change in population each year.

Apply

15. The table shows the extreme temperatures for different U.S. cities in degrees Fahrenheit. Use the expression $\frac{5(F - 32)}{9}$, where F represents the temperature in degrees Fahrenheit to convert each temperature to degrees Celsius. Which city had the greatest difference in temperature extremes in degrees Celsius? Which city had the least? Round to the nearest degree. Explain.

City	Low Extreme (°F)	High Extreme (°F)
Chicago	−27	104
Nashville	−17	107
Oklahoma City	−8	110

16. The table shows the yearly initial and ending balance for each sibling in a family for their account balance with their parents. Use the expression $\frac{I - E}{12}$ where I represents the initial balance and where E represents the ending balance, to find the average difference between the initial balance and ending balance each month. Which sibling had the greatest monthly change?

Sibling	Initial Balance	Ending Balance
Alma	$226	−$50
Laurel	$200	−$64
Wes	$290	$2

17. (MP) **Find the Error** A student solved the problem shown below. Find the student's mistake and correct it.

Find $-3(-6)(-4) - (-9)$.

$$(18)(-4) - (-9) = (-72) - (-9)$$
$$= -81$$

18. **Create** Write and solve a real-world problem in which you perform more than one operation with integers.

19. When simplifying $5(-2)(9) - 3$, a student first subtracted 3 from 9. Is this the correct first step? Explain.

20. (MP) **Identify Structure** When simplifying $-7(3)(-10) + (-4)$, can you multiply 3 and −10 first and get the same result as multiplying −7 and 3 first? Explain.

Rational Numbers

I Can... divide rational numbers and convert fractions to decimal equivalents using division.

Learn Rational Numbers

A **rational number** is any number that can be written in the form $\frac{a}{b}$ where a and b are integers, and $b \neq 0$.

The table explains why each number is a rational number.

Number	Explanation
-15	You can write -15 as the ratio $\frac{-15}{1}$.
-0.8	You can write -0.8 as the ratio $\frac{-8}{10}$.
$-\frac{1}{8}$	You can write $-\frac{1}{8}$ as the ratio $\frac{-1}{8}$.
28%	You can write 28% as the ratio $\frac{28}{100}$.

Copyright © McGraw-Hill Education

Explore Rational Numbers Written as Decimals

🖱 **Online Activity** You will use Web Sketchpad to explore patterns in the decimal form of rational numbers and make a conjecture about the types of numbers that eventually repeat in zeros.

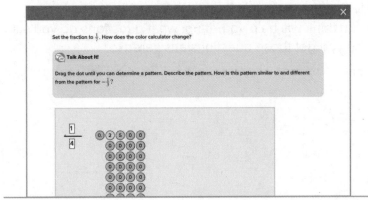

What Vocabulary Will You Learn?
bar notation
rational number
repeating decimal
terminating decimal

💬 **Talk About It!**

Is $-\frac{5}{6}$ the same as $\frac{-5}{6}$ or $\frac{5}{-6}$? Justify your response.

Learn Rational Numbers Written as Decimals

Any fraction can be expressed as a decimal by dividing the numerator by the denominator. The decimal form of a rational number either terminates in 0s or eventually repeats. **Repeating decimals** are decimals in which 1 or more digits repeat and can be represented using bar notation. In **bar notation**, a bar is drawn only over the digit(s) that repeat.

The following decimals are written in bar notation.

$$-0.44444... = -0.\overline{4}$$ The digit 4 repeats.

$$2.4343... = 2.\overline{43}$$ The digits 43 repeat.

Complete the table by writing each decimal using bar notation.

Decimal	Bar Notation
0.11111.....	
0.61111.....	
0.616161.....	
6.160000.....	
6.1611611611.....	

Every decimal can be considered a repeating decimal. Decimals with a repeating digit of zero are also called **terminating decimals**, because the repeating zeros in a terminating decimal are usually truncated, or dropped. For example, the terminating decimal $0.25\overline{0}$ is written as 0.25. The decimal $0.25\overline{0}$ can be considered repeating because the digit 0 repeats.

Pause and Reflect

Are you ready to move on to the Example? If yes, what have you learned that you think will help you? If no, what questions do you still have? How can you get those questions answered?

Record your observations here

Math History Minute

Katherine Johnson (1918-) was a research mathematician who was hired to be a "human computer" by the National Advisory Committee for Aeronautics, later known as NASA. While computers had been programmed to control the movements of astronaut John Glenn's space capsule in 1962, Glenn asked Katherine to perform the same calculations by hand to confirm that the computer's calculations were correct.

Example 1 Write Fractions as Decimals

Write $\frac{1}{40}$ as a decimal. Determine if the decimal is a terminating decimal.

Part A Write the fraction as a decimal.

Divide 1 by 40 using long division.

$$40\overline{)1.0000}$$

So, $\frac{1}{40} = 0.0250...$ or 0.025.

Part B Determine if the decimal is a terminating decimal.

$\frac{1}{40} = 0.0250...$

The decimal ends with _____ zeros.

So, this is a terminating decimal.

Check

Write $-\frac{1}{25}$ as a decimal. Determine if the decimal is a terminating decimal.

 **Think About It!**

Can you predict whether the decimal will be terminating or repeating?

Go Online You can complete an Extra Example online.

Example 2 Write Fractions as Decimals

Write $-\frac{5}{6}$ as a decimal. Determine if the decimal is a terminating decimal.

Part A Write the fraction as a decimal.

Divide 5 by 6 using long division.

$$6\overline{)5.000}$$

The remainder of 2 will repeat, so the 3 in the quotient will also repeat.

So, $-\frac{5}{6} = -0.8333....$

Part B Determine if the decimal is a terminating decimal.

The remainder is never zero, so the quotient will have a repeating 3. Because the decimal repeats, write it using bar notation.

$-\frac{5}{6} = $ [   ] or [   ]

So, this is not a terminating decimal.

Check

Write $\frac{5}{9}$ as a decimal. Determine if the decimal is a terminating decimal.

(Show your work here)

Go Online You can complete an Extra Example online.

Think About It!

How does the denominator of the fraction help you determine if the decimal will terminate?

Talk About It!

If the fraction is negative, how will this affect how it is written in decimal form?

Learn Write Repeating Decimals as Fractions

Go Online Watch the animation to learn how to use an algebraic method to write a repeating, non-terminating decimal, such as $0.\overline{4}$, as a fraction.

$N = 0.444...$	Assign a variable to the value of the decimal.
$10(N) = 10(0.444...)$	Multiply each side by a power of 10.
$10N = 4.444...$	Multiplying by 10 moves the decimal point one place to the right.
$-(N = 0.444...)$	Subtract the original equation to eliminate the repeating part.
$9N = 4$	Simplify.
$\dfrac{9N}{9} = \dfrac{4}{9}$	Divide each side by 9.
$N = \dfrac{4}{9}$	Simplify.

The decimal 0.444... is equivalent to $\dfrac{4}{9}$.

Talk About It!

How do you know that $0.\overline{4}$ cannot be written as the fraction $\dfrac{44}{100}$ nor the fraction $\dfrac{444}{1,000}$?

Example 3 Write Repeating Decimals as Fractions

Write $0.\overline{5}$ as a fraction in simplest form.

Assign a variable to the value $0.\overline{5}$. Let $N = 0.555...$. Then perform operations on N to determine its fractional value.

$N = 0.555...$	
$10(N) = 10(0.555...)$	Multiply each side by 10 because one digit repeats.
$10N = 5.555...$	Multiplying by 10 moved the decimal point one place to the right.
$-(N = 0.555...)$	Subtract $N = 0.555...$ to eliminate the repeating part.
$9N = 5$	Simplify.
$N = \dfrac{\boxed{}}{\boxed{}}$	Divide each side by 9.

So, the decimal $0.\overline{5}$ can be written as the fraction $\dfrac{5}{9}$.

Check

Write $-0.\overline{7}$ as a fraction in simplest form.

Show your work here

Talk About It!

How can you verify that you determined the correct fractional value?

Go Online
You can complete an Extra Example online.

Copyright © McGraw-Hill Education

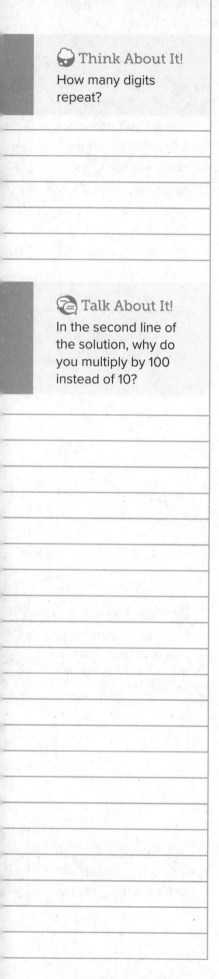

Think About It!
How many digits repeat?

Talk About It!
In the second line of the solution, why do you multiply by 100 instead of 10?

Example 4 Write Repeating Decimals as Mixed Numbers

Write $2.\overline{18}$ as a mixed number in simplest form.

Assign a variable to the value $2.\overline{18}$. Let $N = 2.181818...$. Then perform operations on N to determine its fractional value.

$N = 2.181818...$

$100(N) = 100(2.181818...)$ Multiply each side by 100 because two digits repeat.

$100N = 218.181818...$ Simplify.

$\underline{-(N = 2.181818...)}$ Subtract $N = 2.181818...$ to eliminate the repeating part.

$99N = \boxed{}$ Simplify.

$N = \dfrac{\boxed{}}{\boxed{}}$ Divide each side by 99.

$N = \boxed{}$ Write as a mixed number in simplest form.

So, the decimal $2.\overline{18}$ can be written as $2\frac{2}{11}$.

Check

Write $1.\overline{42}$ as a mixed number in simplest form.

Show your work here

Go Online You can complete an Extra Example online.

🌐 **Apply** Crafting

Two vendors at a craft show are selling signs. The lengths of signs offered by Signs & More and Wood Works are shown in the table. Each sign has the same price. Henry wants to buy the longest of each kind of sign. From which vendor should he buy each style of sign?

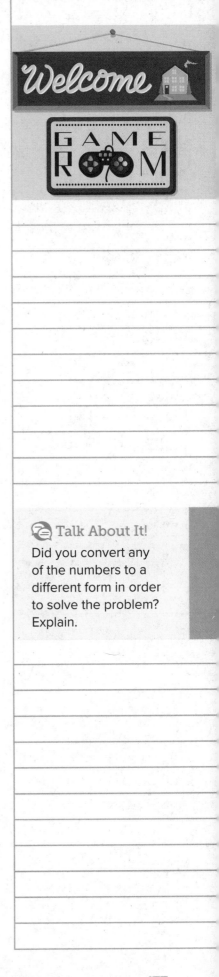

	Signs & More	Wood Works
rounded corner sign	$16\frac{1}{8}$ in.	16.25 in.
square corner sign	$16\frac{1}{3}$ in.	16.3 in.

1 What is the task?

Make sure you understand exactly what question to answer or problem to solve. You may want to read the problem three times. Discuss these questions with a partner.

First Time Describe the context of the problem, in your own words.
Second Time What mathematics do you see in the problem?
Third Time What are you wondering about?

2 How can you approach the task? What strategies can you use?

3 What is your solution?

Use your strategy to solve the problem.

4 How can you show your solution is reasonable?

🖊 **Write About It!** Write an argument that can be used to defend your solution.

💬 Talk About It!
Did you convert any of the numbers to a different form in order to solve the problem? Explain.

Check

During field day, a school had a long jump contest. The top six results of the competition are shown in the table. The teacher wants to award a ribbon to the longest jump by both a 6th grader and a 7th grader, as well as a trophy for the longest jump overall. Who receives the ribbons? Who receives the trophy?

	Grade Level	Length of Jump (ft)
Bentley	7th	$9\frac{5}{8}$
Grant	6th	$9.\overline{68}$
Luna	7th	$9\frac{2}{3}$
Miguel	6th	$9\frac{3}{5}$
Reece	7th	9.5
Trevor	6th	9.73

Show your work here

 Go Online You can complete an Extra Example online.

Pause and Reflect

Create a graphic organizer to record the steps to writing a fraction as a decimal and then determining if the decimal is a terminating decimal.

Record your observations here

Practice

Go Online You can complete your homework online.

Write each fraction as a decimal. Determine if the decimal is a terminating decimal. (Examples 1 and 2)

1. $\dfrac{5}{8}$

2. $-\dfrac{3}{4}$

3. $\dfrac{2}{9}$

4. $-\dfrac{5}{6}$

5. $-\dfrac{4}{5}$

6. $\dfrac{23}{50}$

Write each decimal as a fraction or mixed number in simplest form. (Examples 3 and 4)

7. $0.\overline{8}$

8. $-0.\overline{18}$

9. $-1.\overline{5}$

10. $4.\overline{45}$

Test Practice

11. Open Response Ms. Bradley surveyed her class about their favorite fruits. The results are shown in the table.

A. Write the fraction of students who prefer strawberries, as a decimal. Determine if the decimal is a terminating decimal.

B. Write the fraction of students who prefer kiwi, as a decimal. Determine if the decimal is a terminating decimal.

Fruit	Fraction of the Class
Apples	$\dfrac{8}{30}$
Kiwi	$\dfrac{1}{30}$
Peaches	$\dfrac{12}{30}$
Strawberries	$\dfrac{9}{30}$

Apply

12. Jessica is making matching book bags for her 4 friends. Each book bag needs $1\frac{7}{8}$ yards of fabric. Which of the fabrics shown in the table can Jessica use to make all the book bags for her friends?

Fabric	Amount of Fabric Available (yd)
Moons and Stars	$7\frac{1}{2}$
Softballs	7.4
Stripes	$7\frac{3}{5}$
Tie-Dye	7.9

13. The table shows the times of runners completing a marathon. To qualify for the next marathon, a runner's time must be less than $3\frac{1}{4}$ hours. Which runners qualify?

Runner	Time (h)
Cho	$3\frac{1}{5}$
Kevin	3.2
Ojas	$3\frac{8}{30}$
Sydney	$3\frac{13}{60}$

14. (MP) **Identify Structure** Write a fraction that is equivalent to a terminating decimal between 0.25 and 0.50.

15. (MP) **Justify Conclusions** Are there any rational numbers between $0.\overline{5}$ and $\frac{5}{9}$? Justify your answer.

16. (MP) **Use Math Tools** Eve is making pizza that calls for $\frac{2}{5}$-pound of feta cheese. The store only has packages that contain 0.375- and 0.5- pound of feta cheese. Which of the following strategies might Eve use to determine which package to buy? Use the strategy to solve the problem.

mental math, number sense, estimation

17. (MP) **Make a Conjecture** Write the following fractions as decimals: $\frac{2}{9}$, $\frac{50}{99}$, and $\frac{98}{99}$. Make a conjecture about how to express these kinds of fractions as decimals.

Add and Subtract Rational Numbers

I Can... find the additive inverse of a rational number and add and subtract rational numbers.

Learn Rational Numbers and Additive Inverses

Two rational numbers are opposites if they are represented on a number line by points that are the same distance but on opposite sides from zero.

Two points, $\frac{3}{4}$ and $-\frac{3}{4}$, are graphed. They are opposites because they are both $\frac{3}{4}$-unit from zero.

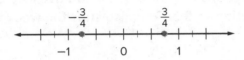

The sum of a number and its opposite, or additive inverse, is zero. The number line shows $-\frac{3}{4} + \frac{3}{4} = 0$.

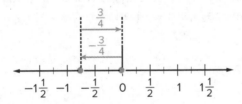

Example 1 Find Additive Inverses

Find the additive inverse of $-\frac{7}{8}$.

Graph and label a point that is the same distance from zero as $-\frac{7}{8}$.

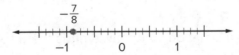

So, the additive inverse of $-\frac{7}{8}$ is ⬚.

Check

Find the additive inverse of $-\frac{1}{8}$.

 Show your work here

 Go Online You can complete an Extra Example online.

🌐 Example 2 Find Additive Inverses

Annalise earned $36.82 at her part time job, and she earned $18.50 babysitting.

Find the total amount she earned, the additive inverse, and describe a situation so that Annalise ends the week with zero dollars.

Part A Find the total amount she earned.

If p is the amount of money Annalise earned at her part time job, b is the amount of money Annalise earned while babysitting, and t is the total amount of money earned, then

$t = p + b$

$ = \$36.82 + \$18.50$

$ = \$\boxed{}$

Part B Find the additive inverse.

Annalise ended the week with $0. What number could you add to $55.32 that would result in a sum of $0?

$\boxed{}$

Part C Describe a situation so Annalise ends the week with zero dollars.

Circle the situations that represent −$55.32.

losing $55.32	getting a gift of $55.32
finding $55.32	earning $55.32
donating $55.32	spending $55.32

🌑 Think About It!

Should Annalise's babysitting and part-time job money be represented with positive or negative numbers?

💬 Talk About It!

How do you know that losing $55.32, donating $55.32, and spending $55.32 all represent a negative value?

Check

Zoey spent $12 on a video and $25.82 on a poster at the music store. Find the total amount she spent, the additive inverse, and describe a situation so that Zoey ends the week with zero dollars.

Part A What was the total amount?

Part B What is the additive inverse of the total?

Part C Which describes a situation so that Zoey ends the week with zero dollars?

Ⓐ Zoey earned $37.82 at her lemonade stand.

Ⓑ Zoey spent $37.82 on dinner and a movie with friends.

Ⓒ Zoey and a friend split the cost of a video game that cost $37.82.

Ⓓ Zoey got a $35 gift card for her birthday.

🔵 **Go Online** You can complete an Extra Example online.

Learn Add Rational Numbers

The rules that apply to adding fractions and decimals also apply to rational numbers. The rules for adding integers also apply to positive and negative rational numbers.

Use the chart to see some strategies for how to add rational numbers written in different forms.

Words	Example
Terminating Decimals $\left(-\frac{1}{4}, \frac{3}{8}, \frac{7}{10}\cdots\right)$	
If the fractions are decimals that terminate, use decimals or fractions to add.	$-\frac{1}{5} + 0.8 = -\frac{1}{5} + \frac{4}{5}$ or $-\frac{1}{5} + 0.8 = -0.2 + 0.8$
Non-Terminating Decimals $\left(-\frac{1}{9}, \frac{2}{3}, \frac{11}{15}\cdots\right)$	
If the fractions are decimals that repeat nonzero digits, use fractions to add.	$\frac{1}{3} + (-0.25) = \frac{1}{3} + \left(-\frac{1}{4}\right)$

When a fraction is negative, the sign may be applied to the fraction, the numerator, or the denominator.

$$-\frac{2}{3} = \frac{-2}{3} = \frac{2}{-3}$$

When you are adding two fractions with negative signs, the sign is usually applied to the numerator.

Talk About It!

Why would it be difficult to add -2.5 and $3.1\overline{6}$?

Example 3 Add Rational Numbers

Find $-3\frac{5}{9} + 1\frac{2}{9}$. Write in simplest form.

$$-3\frac{5}{9} + 1\frac{2}{9} = -\frac{32}{9} + \frac{11}{9}$$ Rewrite the mixed numbers as improper fractions.

$$= \frac{-32 + 11}{9}$$ Add the numerators. Assign any negative signs to the numerator.

$$= \frac{-21}{9}$$ Add.

$$= \frac{-7}{3} \text{ or } -2\frac{1}{3}$$ Simplify and rename as a mixed number.

So, the sum of $-3\frac{5}{9} + 1\frac{2}{9}$ is _____ or _____.

Check

Find $-3\frac{1}{6} + \left(-2\frac{5}{6}\right)$.

Show your work here

🅡 **Go Online** You can complete an Extra Example online.

Example 4 Add Rational Numbers

Find $-\frac{2}{5} + 2.3$.

Because the addends are written in different forms, you can first write them in the same form. The decimal form of $-\frac{2}{5}$ is a terminating decimal. So you can either write the addends as fractions or as decimals.

Method 1 Write both addends as decimals.

$$-\frac{2}{5} + 2.3 = \boxed{} + 2.3$$ Rewrite $-\frac{2}{5}$ as a decimal.

$$= \boxed{}$$ Simplify.

(continued on next page)

🍩 **Think About It!**

Because the signs are different, what do you need to do to the mixed numbers before adding them?

💬 **Talk About It!**

Why is it important when adding rational numbers to rewrite mixed numbers as improper fractions?

Method 2 Write both addends as fractions.

 Go Online Watch the animation to see how to add the two numbers by writing them both as fractions.

$$-\frac{2}{5} + 2.3 = -\frac{2}{5} + 2\frac{3}{10}$$

Rewrite the decimal as a mixed number.

$$= -\frac{4}{10} + \frac{23}{10}$$

The LCD of 5 and 10 is 10.

$$= \frac{-4 + 23}{10}$$

Add the numerators. Assign the negative sign to the numerator.

$$= \boxed{} \text{ or } \boxed{}$$

Add and rename as a mixed number.

So, the sum of $-\frac{2}{5} + 2.3$ is $\frac{19}{10}$ or $1\frac{9}{10}$.

Check

Find $\frac{2}{5} + (-0.75)$.

(Show your work here)

 Go Online You can complete an Extra Example online.

Learn Add Rational Numbers

When adding three or more rational numbers, use the Commutative Property to group numbers by signs or forms.

Grouping numbers with the same form can help simplify the expression.

$$\frac{1}{2} + 0.75 + \left(-2\frac{1}{3}\right) + (-3.7)$$

Write the expression.

$$\frac{1}{2} + 0.75 + \left(-2\frac{1}{3}\right) + (-3.7)$$

Identify each number's form as either a fraction or a **decimal**.

$$\frac{1}{2} + \left(-2\frac{1}{3}\right) + 0.75 + (-3.7)$$

Rewrite the expression, grouping the numbers by their form.

Talk About It!

In Method 1, you wrote both addends as decimals before adding. Give an example of when Method 1 is not the best method to use. Explain your reasoning.

Talk About It!

How does grouping numbers by like forms help in simplifying an addition expression?

Example 5 Add Rational Numbers

The lowest recorded elevation in the contiguous United States, 282 feet below sea level, is in Death Valley. Suppose you began a hike in the parking area near Badwater Basin at 266 feet below sea level. The stopping points of the hike and the elevation traveled are shown in the table.

What is the elevation of your stopping point?

Stop	Elevation Traveled (ft)
1	−12.5
2	$26\frac{3}{5}$
3	$-3\frac{1}{2}$
4	397.3

Because the denominators of the mixed numbers are 2 and 5, it is easy to convert those to decimals. Write all of the numbers as decimals.

$-266 + (-12.5) + 26\frac{3}{5} + \left(-3\frac{1}{2}\right) + 397.3$ Write the expression.

$= -266 + (-12.5) + 26.6 + (-3.5) + 397.3$ Write fractions as decimals.

$= -266 + (-12.5) + (-3.5) + 26.6 + 397.3$ Commutative Property

$= [-266 + (-12.5) + (-3.5)] + [26.6 + 397.3]$ Associative Property

$= -282 + 423.9$ Simplify.

$= 141.9$ Add.

So, your elevation at the end of the hike is _____ feet above sea level.

Copyright © McGraw-Hill Education

Think About It!

How will you represent the starting point of the hike in the expression?

Talk About It!

If the starting point was a rational number, such as $-266\frac{1}{9}$, would it have changed how you found the sum?

Check

During the annual Hot Air Balloon Rally, hot air balloon pilots need to track their altitude as they travel. During a flight that began at 98.1 meters above sea level, one pilot tracked and recorded her altitude every half hour.

Time (h)	Altitude Change (m)
$\frac{1}{2}$	226.86
1	$-66\frac{4}{8}$
$1\frac{1}{2}$	-15.32
2	$172\frac{3}{4}$

What is her altitude, in meters, after two hours?

Show your work here

Go Online You can complete an Extra Example online.

Learn Subtract Rational Numbers

To subtract rational numbers in different forms, write the numbers in the same form.

Go Online Watch the animation to see how to subtract a mixed number and a decimal.

The animation shows how to rewrite a problem using decimals. This method works if the mixed number can be rewritten as a terminating decimal.

Rewrite Using Decimals	
Steps	**Example**
1. Write the mixed number or fraction as a decimal.	$2\frac{2}{5} - 6.55 = 2\frac{4}{10} - 6.55$
	$= 2.4 - 6.55$
2. Subtract the decimals.	$= 2.4 + (-6.55)$
	$= -4.15$

(continued on next page)

The animation also shows how to rewrite a problem using fractions or mixed numbers. This method works when the fraction or mixed number cannot be written as a terminating decimal.

Rewrite Using Fractions or Mixed Numbers	
Steps	**Example**
1. Write the decimal as a fraction or mixed number.	$4.6 - 2\frac{2}{3} = 4\frac{6}{10} - 2\frac{2}{3}$
2. Rewrite fractions with a common denominator.	$= 4\frac{3}{5} - 2\frac{2}{3}$ $= 4\frac{9}{15} - 2\frac{10}{15}$
3. Subtract the fractions or mixed numbers.	$= 3\frac{24}{15} + \left(-2\frac{10}{15}\right)$
4. Simplify if necessary.	$= 1\frac{14}{15}$

Use the chart below to find out how to subtract rational numbers written in different forms.

Words	Example
Terminating Decimals $\left(-\frac{1}{4}, \frac{3}{8}, \frac{7}{8}\cdots\right)$	
If the fractions are decimals that terminate, use decimals or fractions to subtract.	$-0.90 - \frac{1}{10} = -\frac{9}{10} - \frac{1}{10}$ or $= -0.9 - 0.1$
Non-Terminating Decimals $\left(-\frac{1}{9}, \frac{2}{3}, \frac{11}{15}\cdots\right)$	
If the fractions are decimals that repeat nonzero digits, use fractions to subtract.	$-\frac{1}{6} - 0.125 = -\frac{1}{6} - \frac{1}{8}$

Talk About It!

How does knowing how to add rational numbers help you to subtract rational numbers?

Pause and Reflect

When adding or subtracting rational numbers, how can you use inverse operations to check your work?

Record your observations here

Example 6 Subtract Rational Numbers

Find −3.27 − (−6.7).

Use the same rules to subtract positive and negative decimals as subtracting integers.

Integers		Rational Numbers
−3 − (−6)	Write the expression.	−3.27 − (−6.7)
−3 + 6	Add the additive inverse.	−3.27 + 6.7
3	Add.	3.43

So, because $|6.7| > |−3.27|$, the sum will have the same sign as 6.7, positive.

So, −3.27 − (−6.7) is _____.

Check

Find −4.2 − 3.57.

Show your work here

 Go Online You can complete an Extra Example online.

Example 7 Subtract Rational Numbers

Find $5\frac{1}{3} - \left(-4\frac{5}{9}\right)$. Write in simplest form.

$$5\frac{1}{3} - \left(-4\frac{5}{9}\right) = \frac{16}{3} - \left(-\frac{41}{9}\right)$$ Write the mixed numbers as improper fractions.

$$= \boxed{} - \left(\boxed{}\right)$$ The LCD of 3 and 9 is 9.

$$= \frac{48}{9} + \frac{41}{9}$$ Add using the additive inverse.

$$= \boxed{} \text{ or } \boxed{}$$ Simplify.

So, the difference of $5\frac{1}{3} - \left(-4\frac{5}{9}\right)$ is $\frac{89}{9}$ or $9\frac{8}{9}$.

Check

Find $3\frac{1}{2} - \left(-1\frac{3}{10}\right)$. Write in simplest form.

Show your work here

 Go Online You can complete an Extra Example online.

Example 8 Evaluate Expressions

Evaluate $x - y$ if $x = -2\frac{4}{5}$ and $y = 1.4$.

Determine if the mixed number is a decimal that terminates. Because $-2\frac{4}{5}$ terminates, you can use either decimals or fractions to subtract.

Method 1 Evaluate using decimals.

$$x - y = -2\frac{4}{5} - 1.4 \qquad \text{Replace } x \text{ with } -2\frac{4}{5} \text{ and } y \text{ with } 1.4.$$
$$= -2.8 - 1.4 \qquad \text{Rewrite } -2\frac{4}{5} \text{ as a decimal.}$$
$$= -2.8 + (-1.4) \qquad \text{Add the additive inverse of } 1.4.$$
$$= -4.2 \qquad \text{Simplify.}$$

Method 2 Evaluate using fractions.

$$x - y = -2\frac{4}{5} - 1.4 \qquad \text{Replace } x \text{ with } -2\frac{4}{5} \text{ and } y \text{ with } 1.4.$$
$$= -2\frac{4}{5} - 1\frac{2}{5} \qquad \text{Rewrite } 1.4 \text{ as a mixed number.}$$
$$= -\frac{14}{5} - \frac{7}{5} \qquad \text{Write mixed numbers as improper fractions.}$$
$$= -\frac{14}{5} + \left(-\frac{7}{5}\right) \qquad \text{Add the additive inverse of } \frac{7}{5}.$$
$$= -\frac{21}{5} \text{ or } -4\frac{1}{5} \qquad \text{Add.}$$

So, when $x = -2\frac{4}{5}$ and $y = 1.4$, $x - y =$ ☐ or ☐ .

Check

Evaluate $x - y$ if $x = 3\frac{3}{4}$ and $y = -4.2$. Write in simplest form.

Show your work here

Go Online You can complete an Extra Example online.

📖 **Foldables** It's time to update your Foldable, located in the Module Review, based on what you learned in this lesson. If you haven't already assembled your Foldable, you can find the instructions on page FL1.

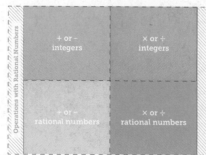

Talk About It!

Generate two different expressions that involve the subtraction of rational numbers. The first expression should be best simplified by converting any decimals to fractions. The second expression should be best simplified by converting fractions to decimals.

🌐 Apply Animal Care

A veterinarian measures the changes in a cat's weight over four months. If the cat weighed 17.25 pounds at its first visit, what is the cat's weight after its last visit?

Month	Change from Previous Month (lb)
1	-0.5
2	$-2\frac{1}{5}$
3	$-\frac{3}{10}$
4	0.35

1 What is the task?

Make sure you understand exactly what question to answer or problem to solve. You may want to read the problem three times. Discuss these questions with a partner.

First Time Describe the context of the problem, in your own words.
Second Time What mathematics do you see in the problem?
Third Time What are you wondering about?

2 How can you approach the task? What strategies can you use?

Record your observations here

3 What is your solution?

Use your strategy to solve the problem.

Show your work here

4 How can you show your solution is reasonable?

🔄 **Write About It!** Write an argument that can be used to defend your solution.

💬 Talk About It!

Why is it more efficient in this problem to write each of the values as a decimal before adding?

Check

A local forest that covers $472\frac{7}{10}$ acres has gone through periods of growth and loss over the past four years. In some years, firefighters started a controlled burn that burns part of the forest in order to encourage new growth. Each year they track the changes in the area of the forest. Find the area of the forest after the four years.

Year	Change in Area (acres)
1	$3\frac{4}{5}$
2	$-1\frac{1}{2}$
3	-7.9
4	$2\frac{2}{5}$

Go Online You can complete an Extra Example online.

Pause and Reflect

Think about scenarios, when adding or subtracting rational numbers, where it would be beneficial to change all the numbers to decimals, or all to fractions. Give examples.

Practice

Go Online You can complete your homework online.

Find the additive inverse of each rational number. (Example 1)

1. $-\dfrac{1}{2}$

2. 0.25

3. $\dfrac{9}{10}$

4. -0.4

5. Quinn earned $24.50 dog-sitting and $12.70 for recycling cans. Find the total amount he earned and describe a situation in which Quinn ends the week with zero dollars. (Example 2)

Add or Subtract. Write in simplest form. (Examples 3, 4, 6, 7)

6. $3\dfrac{5}{6} + \left(-1\dfrac{1}{6}\right)$

7. $-\dfrac{2}{3} + 2\dfrac{3}{8}$

8. $-3.7 + \dfrac{1}{4}$

9. $\dfrac{1}{3} + 4.1$

10. $-1\dfrac{1}{4} + 0.75 + 0.45$

11. $-2.45 - (-3.9)$

12. $5.47 - (-2.8)$

13. $7\dfrac{5}{12} - \left(-3\dfrac{3}{4}\right)$

14. $-\dfrac{7}{8} - 2\dfrac{1}{6}$

15. Marlee is making jewelry for a class craft show. She began with 115 inches of wire. She used 25.75 inches for rings. Then her teacher gave her $30\dfrac{1}{4}$ inches of wire to make more jewelry. She then used $38\dfrac{1}{2}$ inches for the bracelets and 60.2 inches for necklaces. How much wire does Marlee have left? (Example 5)

Test Practice

16. Equation Editor Evaluate $a - b$ if $a = -2.5$ and $b = \dfrac{2}{5}$. Write your answer in simplest form.

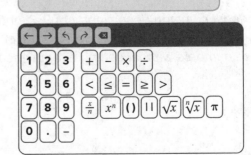

Apply

17. Jada measures the changes in her dog's weight over the entire year. She weighs her dog every 3 months and records the results. If Jada's dog weighed 55.75 pounds at the beginning of the year, what is the dog's weight at the end of the year?

Months	Difference from Previous Weight (lb)
January–March	$+2.125$
April–June	$-3\frac{1}{4}$
July–September	$-\frac{1}{2}$
October–December	$+0.875$

18. A local petting zoo allows visitors to feed the goats food pellets. The petting zoo starts the month with 525.25 pounds of food pellets. Each week, the workers track the amount of food pellets given out and any food pellet deliveries. Use the table to determine the number of pounds of food pellets the petting zoo will have at the end of the month.

Week	Change in Food Pellets (lb)
1	$-164\frac{1}{2}$
2	-189.75
3	$355\frac{7}{8}$
4	$-200\frac{3}{16}$

19. Write an addition problem with unlike mixed numbers and a least common denominator of 16. Find the sum in simplest form.

20. Suppose you use 8 instead of 4 as a common denominator when finding $7\frac{1}{2} - \left(-3\frac{1}{4}\right)$. How will that change the process for finding the difference?

21. MP **Find the Error** A student is adding $1\frac{2}{9}$, $3\frac{1}{3}$, and $-4\frac{5}{6}$. The first step the student performs is to find the common denominator of 9, 3, and 6. Find the student's mistake and correct it.

The least common denominator of 9, 3, and 6 is 36 because you can divide 36 by all of these numbers without getting a remainder.

22. MP **Use a Counterexample** Is the following statement *true* or *false*? If false, provide a counterexample.

The difference between a positive mixed number and negative mixed number is never positive.

Multiply and Divide Rational Numbers

I Can... use the rules for multiplying and dividing integers to multiply and divide rational numbers.

Learn Multiply Rational Numbers

The rules for integers also apply to positive and negative rational numbers.

State whether the product of each expression will be *positive* or *negative* using your knowledge of multiplying integers.

Multiply Rational Numbers	
Expression	**Sign of Product**
$\frac{1}{7}\left(-\frac{2}{3}\right)$	
$2\frac{3}{5} \cdot \frac{5}{7}$	
$-\frac{1}{3} \cdot \frac{7}{9}$	
$-1\frac{1}{2}\left(-5\frac{1}{4}\right)$	

Example 1 Multiply Rational Numbers

Find $-\frac{3}{4}\left(-\frac{7}{9}\right)$. Write in simplest form.

$$-\frac{3}{4}\left(-\frac{7}{9}\right) = -\frac{\cancel{3}^{1}}{4}\left(-\frac{\cancel{7}}{\cancel{9}_{3}}\right) \qquad \text{Divide by common factors.}$$

$$= \frac{-1(-7)}{4 \cdot 3} \qquad \text{Multiply the numerators and denominators.}$$

$$= \boxed{} \qquad \text{Simplify.}$$

So, $-\frac{3}{4}\left(-\frac{7}{9}\right)$ is $\frac{7}{12}$.

What Vocabulary Will You Learn?
multiplicative inverses

Copyright © McGraw-Hill Education

Check

Find $-\frac{7}{8}\left(\frac{4}{14}\right)$.

Show your work here

 Go Online You can complete an Extra Example online.

Think About It!

Before you begin to multiply the rational numbers, what do you need to do?

Talk About It!

How would the steps be different if you did not simplify the rational numbers before multiplying?

Example 2 Multiply Rational Numbers

Find $-3\frac{1}{5}\left(1\frac{1}{14}\right)$. Write in simplest form.

$-3\frac{1}{5}\left(1\frac{1}{14}\right) = -\frac{16}{5}\left(\frac{15}{14}\right)$ Write the mixed numbers as improper fractions.

$= -\frac{\overset{8}{16}}{\underset{1}{5}}\left(\frac{\overset{3}{15}}{\underset{7}{14}}\right)$ Divide by common factors.

$= \frac{-8 \cdot 3}{1 \cdot 7}$ Multiply the numerators and denominators.

$= \boxed{}$ or $\boxed{}$ Simplify.

So, the product of $-3\frac{1}{5}\left(1\frac{1}{14}\right)$ is $-\frac{24}{7}$ or $-3\frac{3}{7}$.

Check

Find $-1\frac{7}{8}\left(-2\frac{2}{5}\right)$. Write in simplest form.

Show your work here

 Go Online You can complete an Extra Example online.

Learn Multiply Rational Numbers

When multiplying rational numbers written in different forms, write the factors in the same form.

Words	Example
Terminating Decimals $\left(-\frac{1}{4}, \frac{3}{8}, \frac{7}{8}\cdots\right)$	
If the fractions or mixed numbers are decimals that terminate, use decimals or fractions to multiply.	$-3.7\left(1\frac{1}{5}\right) = -3.7 \cdot (1.2)$ or $= -3\frac{7}{10}\left(1\frac{1}{5}\right)$
Non-Terminating Decimals $\left(-\frac{1}{9}, \frac{2}{3}, \frac{11}{15}\cdots\right)$	
If the fractions or mixed numbers are decimals that repeat nonzero digits, use fractions to multiply.	$0.75\left(-\frac{2}{3}\right) = \frac{3}{4}\left(-\frac{2}{3}\right)$

Pause and Reflect

Create a graphic organizer that will help you understand how to determine if a fraction or mixed number is a terminating or non-terminating decimal.

Record your observations here

Example 3 Multiply Rational Numbers

Find $\frac{1}{3}(-2.75)$. Write in simplest form.

Because the factors are written in different forms, you first need to rewrite them in the same form.

Because $\frac{1}{3}$ repeats non-zero digits, write the second factor as a mixed number.

$\frac{1}{3}(-2.75) = \frac{1}{3}\left(-2\frac{3}{4}\right)$ Write −2.75 as a mixed number.

$= \frac{1}{3}\left(-\frac{11}{4}\right)$ Write $-2\frac{3}{4}$ as an improper fraction.

$= \frac{1(-11)}{(3 \cdot 4)}$ Multiply the numerators and denominators.

$=$ ☐   Simplify.

So, the product of $\frac{1}{3}(-2.75)$ is $-\frac{11}{12}$.

Check

Find $-0.64\left(-\frac{1}{4}\right)$.

 Go Online You can complete an Extra Example online.

Pause and Reflect

How do you know, without calculating, that $\frac{1}{3}$ repeats non-zero digits?

 Talk About It!

Is it possible to find the product without writing the numbers in the same form? Explain.

Example 4 Multiply Rational Numbers

Evaluate $\frac{1}{2}ab$ if $a = 1\frac{3}{7}$ and $b = -\frac{4}{9}$.

$\frac{1}{2}ab = \frac{1}{2}\left(1\frac{3}{7}\right)\left(-\frac{4}{9}\right)$ Replace a with $1\frac{3}{7}$ and b with $-\frac{4}{9}$.

$= \frac{1}{2}\left(\frac{10}{7}\right)\left(-\frac{4}{9}\right)$ Write $1\frac{3}{7}$ as an improper fraction.

$= \frac{1}{\underset{1}{2}}\left(\frac{\overset{5}{10}}{7}\right)\left(-\frac{4}{9}\right)$ Divide by common factors.

$= \frac{1(5)(-4)}{1(7)(9)}$ Multiply the numerators and denominators.

$=$ ☐   Simplify.

So, the value of $\frac{1}{2}ab$ when $a = 1\frac{3}{7}$ and $b = -\frac{4}{9}$ is $-\frac{20}{63}$.

Check

Evaluate $\frac{2}{3}xy$ if $x = -3\frac{1}{6}$ and $y = -5\frac{1}{4}$.

🅡 **Go Online** You can complete an Extra Example online.

Pause and Reflect

When multiplying rational numbers, why is it beneficial to divide by common factors?

Example 5 Multiply Rational Numbers

Evaluate $\frac{1}{4}xyz$ if $x = 9.5$, $y = -0.8$, and $z = 2\frac{1}{5}$.

Determine if the fraction and mixed number are terminating decimals. Because $\frac{1}{4}$ and $2\frac{1}{5}$ both terminate, you can use either decimals or fractions to multiply.

$\frac{1}{4}xyz = \frac{1}{4}\left(\boxed{}\right)\left(\boxed{}\right)\left(\boxed{}\right)$

Replace x with 9.5, y with -0.8, and z with $2\frac{1}{5}$.

$= 0.25(9.5)(-0.8)(2.2)$

Replace $\frac{1}{4}$ with 0.25 and $2\frac{1}{5}$ with 2.2.

$= \boxed{}$

Simplify.

So, the value of $\frac{1}{4}xyz$ when $x = 9.5$, $y = -0.8$, and $z = 2\frac{1}{5}$ is -4.18.

Check

Evaluate abc if $a = -2.8$, $b = -2\frac{1}{7}$, and $c = -\frac{3}{13}$.

Show your work here

Go Online You can complete an Extra Example online.

Pause and Reflect

How can you use estimation to check your work?

Record your observations here

Talk About It!

How would your answer compare if you had multiplied using fractions rather than decimals?

Learn Divide Rational Numbers

The rules for integers also apply to positive and negative rational numbers.

Two numbers whose product is 1 are called **multiplicative inverses**, or reciprocals. For example, $-\frac{1}{4}$ and -4 are multiplicative inverses because $-\frac{1}{4}(-4) = 1$.

Go Online Watch the animation to see how to divide rational numbers.

Dividing Rational Numbers	
Steps	Example
1. Rewrite the division as multiplication by the reciprocal.	$3 \div \left(-\frac{1}{4}\right) = -12$
2. Multiply.	reciprocals ↕ ↕ same result
3. Identify the quotient.	$3 \cdot (-4) = -12$

Talk About It!
Why is the reciprocal of a fraction with 1 in the numerator an integer?

Example 6 Divide Rational Numbers

Find $-\frac{2}{3} \div \frac{1}{9}$.

$-\frac{2}{3} \div \frac{1}{9} = -\frac{2}{3} \cdot \frac{9}{1}$ Multiply by the multiplicative inverse.

$= -\frac{2}{3_1}\left(\frac{9^3}{1}\right)$ Divide by common factors.

$= -\frac{6}{1}$ or -6 Simplify.

So, $-\frac{2}{3} \div \frac{1}{9}$ is _____.

Check

Find $-\frac{6}{7} \div 12$.

Go Online You can complete an Extra Example online.

Copyright © McGraw-Hill Education

Example 7 Divide Rational Numbers

Find $-4\frac{2}{5} \div \left(-2\frac{14}{15}\right)$.

$$-4\frac{2}{5} \div \left(-2\frac{14}{15}\right) = -\frac{22}{5} \div \left(-\frac{44}{15}\right) \qquad \text{Write the mixed numbers as improper fractions.}$$

$$= -\frac{22}{5}\left(-\frac{15}{44}\right) \qquad \text{Multiply by the multiplicative inverse.}$$

$$= -\frac{\overset{1}{22}}{\underset{1}{5}}\left(-\frac{\overset{3}{15}}{\underset{2}{44}}\right) \qquad \text{Divide by common factors.}$$

$$= \frac{3}{2} \text{ or } 1\frac{1}{2} \qquad \text{Simplify.}$$

So, $-4\frac{2}{5} \div \left(-2\frac{14}{15}\right)$ is ☐ or ☐ .

Check

Find $-1\frac{1}{3} \div 2\frac{2}{15}$.

Show your work here

Pause and Reflect

How is dividing rational numbers similar to multiplying rational numbers? How is it different?

Record your observations here

Think About It!

Before you divide the two rational numbers, what do you need to do?

Talk About It!

How do you know the quotient will be a positive number before you divide?

Learn Divide Rational Numbers

When dividing rational numbers written in different forms, write the factors in the same form.

Words	Example
Terminating Decimals $\left(-\dfrac{1}{4}, \dfrac{3}{8}, \dfrac{7}{8}\cdots\right)$	
If the fractions or mixed numbers are decimals that terminate, use decimals or fractions to divide.	$6.3 \div \left(-\dfrac{7}{8}\right) = 6.3 \div -0.875$ or $6.3 \div \left(-\dfrac{7}{8}\right) = 6\dfrac{3}{10} \div \left(-\dfrac{7}{8}\right)$
Non-Terminating Decimals $\left(-\dfrac{1}{9}, \dfrac{2}{3}, \dfrac{11}{15}\cdots\right)$	
If the fractions or mixed numbers are decimals that repeat nonzero digits, use fractions to divide.	$0.75 \div \left(-\dfrac{2}{3}\right) = \dfrac{3}{4} \div \left(-\dfrac{2}{3}\right)$

Pause and Reflect

Are you ready to move on to the next Example? If yes, what have you learned that you think will help you? If no, what questions do you still have? How can you get those questions answered?

Record your observations here

Example 8 Divide Rational Numbers

Evaluate $\frac{a}{b}$ if $a = \frac{3}{4}$ and $b = -0.05$.

Method 1 Evaluate with numbers as decimals.

$\frac{a}{b} = \dfrac{\frac{3}{4}}{-0.05}$ Replace a with $\frac{3}{4}$ and b with -0.05.

$= \dfrac{\boxed{}}{-0.05}$ Write $\frac{3}{4}$ as a decimal.

$= -15$ Divide.

Method 2 Evaluate with numbers as fractions.

$\frac{a}{b} = \dfrac{\frac{3}{4}}{-0.05}$ Replace a with $\frac{3}{4}$ and b with -0.05.

$= \dfrac{\frac{3}{4}}{\left(-\frac{1}{20}\right)}$ Write -0.05 as a fraction.

$= \frac{3}{4} \div \left(\boxed{}\right)$ Write the complex fraction as a division problem.

$= \frac{3}{4}\left(\boxed{}\right)$ Multiply by the multiplicative inverse.

$= \frac{3}{\underset{1}{4}}\left(-\frac{\overset{5}{20}}{1}\right)$ Divide by common factors.

$= \frac{3(-5)}{1 \cdot 1}$ Multiply the numerators and denominators.

$= \boxed{}$ Simplify.

So, the solution is -15.

Check

Evaluate $\frac{x}{y}$ if $x = -6.85$ and $y = -2\frac{2}{5}$. If your answer is in decimal form, round to the nearest hundredth.

Show your work here

Go Online You can complete an Extra Example online.

Think About It!

How will you decide whether to evaluate the expression with decimals or fractions?

Talk About It!

Is it more advantageous to use one method over the other? Explain.

🌐 Apply Temperature

A sign shows the temperature in Badger, Minnesota, at 10 P.M. is
−11.5°F. At 4 A.M., the temperature changed by $\frac{2}{3}$ of the current value.
What is the final temperature?

1 What is the task?

Make sure you understand exactly what question to answer or
problem to solve. You may want to read the problem three times.
Discuss these questions with a partner.

First Time Describe the context of the problem, in your own words.
Second Time What mathematics do you see in the problem?
Third Time What are you wondering about?

2 How can you approach the task? What strategies can you use?

3 What is your solution?

Use your strategy to solve the problem.

4 How can you show your solution is reasonable?

✏️ **Write About It!** Write an argument that can be used to defend
your solution.

💬 Talk About It!

Is the final temperature
warmer or colder than
the original
temperature? Explain.

Check

Lena was practicing freediving. On her first attempt, she dove to a depth of −63.4 feet. Her second attempt changed by $\frac{2}{5}$ of the original depth. How far did Lena dive on her second attempt?

Show your work here

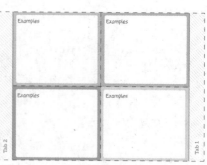

Go Online You can complete an Extra Example online.

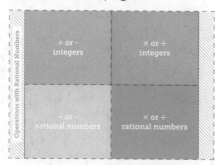

Foldables It's time to update your Foldable, located in the Module Review, based on what you learned in this lesson. If you haven't already assembled your Foldable, you can find the instructions on page FL1.

Practice

🢖 Go Online You can complete your homework online.

Multiply or divide. Write the product or quotient in simplest form. (Examples 1-3, 6-7)

1. $-\frac{1}{2}\left(-\frac{4}{5}\right)$

2. $1\frac{4}{9}\left(-2\frac{4}{7}\right)$

3. $1\frac{1}{10}\left(-6\frac{7}{8}\right)$

4. $-\frac{1}{6}(2.4)$

5. $-\frac{6}{7} \div \frac{3}{14}$

6. $\frac{2}{3} \div \left(-\frac{4}{9}\right)$

7. $7\frac{1}{2} \div \left(-2\frac{5}{6}\right)$

8. $-3\frac{4}{9} \div \left(-2\frac{1}{3}\right)$

9. $-5\frac{1}{4} \div \frac{7}{8}$

Evaluate each expression if $x = -\frac{2}{3}$, $y = 0.6$, and $z = -1\frac{7}{8}$. Write the product in simplest form. (Examples 4 and 5)

10. $\frac{1}{4}xy$

11. $-\frac{4}{5}xz$

12. $-xyz$

Test Practice

13. Evaluate $\frac{x}{y}$ if $x = \frac{4}{5}$ and $y = -0.1$. Write your answer in simplest form. (Example 3)

15. Equation Editor Evaluate $\frac{1}{2}xyz$ if $x = -8.4$, $y = 0.25$, and $z = 3\frac{4}{5}$. Write your answer in simplest form.

14. Evaluate $\frac{c}{d}$ if $c = -4.75$ and $d = -1\frac{1}{4}$. Write your answer in simplest form.
(Example 3)

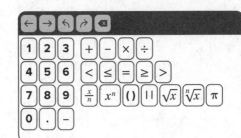

Apply

16. The table shows the change in the value of Rudo's stocks one day. The next day, the value of the All-Plus stock dropped $\frac{1}{4}$ of the amount it changed from the previous day. What was the total change in the All-Plus stock? Round to the nearest cent.

Stock	Change
All-Plus	−$2.50
True-Fit	−$3.75

17. The photography club is selling hot chocolate at soccer games to raise money for new cameras. The table shows their profit per game for the first five games. Based on the average profit per game, how much total money can the club expect to earn by the end of the 10-game season?

Game	Profit ($)
1	−12.50
2	−10.15
3	18.65
4	25.90
5	45.75

18. **MP Find the Error** A student is finding $-\frac{6}{7} \div \left(-\frac{5}{6}\right)$. Find the student's mistake and correct it.

$$-\frac{6}{7} \div \left(-\frac{5}{6}\right) = -\frac{7}{6} \cdot \left(-\frac{5}{6}\right)$$
$$= \frac{35}{36}$$

19. **MP Use a Counterexample** Is the following statement *true* or *false*? If false, provide a counterexample.

The product of a fraction between 0 and 1 and a whole number or mixed number is never less than the whole number or mixed number.

20. **MP Persevere with Problems** Find the missing fraction for each problem.

a. $-\frac{6}{11} \cdot \left(\frac{x}{y}\right) = -\frac{1}{11}$

b. $\frac{a}{b} \div \left(-\frac{2}{3}\right) = 1$

21. **MP Make an Argument** Which is greater, $20 \cdot \frac{3}{4}$ or $20 \div \frac{3}{4}$? Explain.

Copyright © McGraw-Hill Education

Apply Rational Number Operations

I Can... add, subtract, multiply, and divide rational numbers, including using those four operations to solve real-world problems.

Learn Apply Rational Number Operations

The order of operations used to simplify numerical expressions with whole numbers also applies to simplifying numerical expressions with rational numbers.

If the expression contains fractions and decimals, use the Commutative Properties or the Associative Properties to group like forms together. Simplify as much as possible, and then rename the numbers in the same form.

Example 1 Apply Rational Number Operations

Evaluate $ab + c - d$ **if** $a = \frac{5}{6}$, $b = -\frac{4}{5}$, $c = 0.75$, **and** $d = \frac{1}{3}$.

Substitute the values into the equation.

$$\frac{5}{6} \cdot \left(-\frac{4}{5}\right) + 0.75 - \frac{1}{3} = -\frac{2}{3} + 0.75 - \frac{1}{3} \qquad \text{Multiply the fractions.}$$

$$= -\frac{2}{3} + 0.75 + \left(-\frac{1}{3}\right) \qquad \text{Add the additive inverse.}$$

$$= -\frac{2}{3} + \left(-\frac{1}{3}\right) + 0.75 \qquad \text{Commutative Property}$$

$$= -1 + 0.75 \qquad \text{Add the fractions.}$$

$$= -0.25 \qquad \text{Add.}$$

So, the value of the expression is _____ .

 Think About It!

How will you use the order of operations to evaluate this expression?

 Talk About It!

Was it helpful to use the Commutative Property to change the order? Explain.

Copyright © McGraw-Hill Education

Check

Evaluate $a + bc + d$ if $a = -3.6$, $b = 2\frac{4}{5}$, $c = -\frac{1}{2}$, and $d = 6\frac{3}{4}$.

Show your work here

Example 2 Apply Rational Number Operations

Evaluate $\frac{2}{5}(x + y) + z$ if $x = \frac{7}{8}$, $y = -\frac{1}{4}$, and $z = -\frac{2}{3}$.

Substitute the values into the equation.

$$\boxed{}\left[\boxed{} + \left(\boxed{}\right)\right] + \boxed{}$$

Simplify the expression.

$$\frac{2}{5} \times \left[\frac{7}{8} + \left(-\frac{1}{4}\right)\right] + \left(-\frac{2}{3}\right)$$

$$= \frac{2}{5} \times \left[\frac{7}{8} + \left(-\frac{2}{8}\right)\right] + \left(-\frac{2}{3}\right) \qquad \text{Write } -\frac{1}{4} \text{ as } -\frac{2}{8}.$$

$$= \frac{2}{5}\left(\boxed{}\right) + \left(-\frac{2}{3}\right) \qquad \text{Add.}$$

$$= \left(\boxed{}\right) + \left(-\frac{2}{3}\right) \qquad \text{Multiply.}$$

$$= \frac{3}{12} + \left(-\frac{8}{12}\right) \qquad \text{Find a common denominator.}$$

$$= \left(\boxed{}\right) \qquad \text{Add.}$$

So, the value of the expression is $-\frac{5}{12}$.

Check

Evaluate $x + y(0.4 + z)$ if $x = -2.6$, $y = -\frac{1}{2}$, and $z = -\frac{5}{8}$.

Show your work here

🌐 **Go Online** You can complete an Extra Example online.

Think About It!
How would you begin evaluating this expression?

Talk About It!
Why would it be easier to do the operation in the parentheses rather than using the Distributive Property to solve the example?

 Apply Food

Aarya has a recipe for blonde brownies. She doesn't like a lot of walnuts, but loves chocolate, so she is using $\frac{1}{2}$ of the amount of walnuts called for and increasing the chocolate chips by $\frac{1}{3}$. She then wants to double her new recipe. How many cups of walnuts will Aarya need? How many cups of chocolate chips?

Go Online Watch the animation.

Quantity	Ingredient
$\frac{1}{2}$ cup	chopped walnuts
$\frac{3}{4}$ cup	chocolate chips

1 What is the task?

Make sure you understand exactly what question to answer or problem to solve. You may want to read the problem three times. Discuss these questions with a partner.

First Time Describe the context of the problem, in your own words.
Second Time What mathematics do you see in the problem?
Third Time What are you wondering about?

2 How can you approach the task? What strategies can you use?

Record your observations here

3 What is your solution?

Use your strategy to solve the problem.

Show your work here

4 How can you show your solution is reasonable?

Write About It! Write an argument that can be used to defend your solution.

Talk About It!
How is finding the amount of walnuts for the recipe different than finding the amount of chocolate chips?

Check

Adel City is enclosing a skate park with fencing. The skate park is $31\frac{1}{3}$ yards wide and $24\frac{2}{3}$ yards long. Fencing is sold in 8-foot sections and costs $67.99 per section. How much will it cost to fence in the entire skate park?

 Go Online You can complete an Extra Example online.

Pause and Reflect

Why is it helpful to use parentheses when substituting values for variables?

Record your observations here

Practice

🜨 **Go Online** You can complete your homework online.

Evaluate each expression if $a = \frac{7}{8}$**,** $b = -\frac{7}{16}$**,** $c = 0.8$**,** $d = \frac{1}{4}$**.** (Example 1)

1. $\frac{a}{b} - c + d$

2. $a + b + cd$

3. $c - a + \frac{b}{d}$

4. $\frac{b}{a} + \frac{c}{d}$

5. $d^2 - b$

6. $a + \frac{b}{c} - d$

Evaluate each expression if $a = \frac{7}{10}$**,** $b = \frac{3}{5}$**,** $c = -1.9$**, and** $d = -\frac{1}{5}$**.**
Write your answer in simplest form. (Example 2)

7. $\frac{1}{2}(a - b) + c$

8. $0.25(a + b) - d$

9. $b + 0.2(c - d)$

10. $(b + d) - a$

11. $b^2 + \frac{1}{10}(d - b)$

12. $b + \frac{c}{d}$

Test Practice

13. Equation Editor Evaluate $4.5(xy) - \frac{x}{5}$ if $x = 12$ and $y = \frac{2}{3}$.

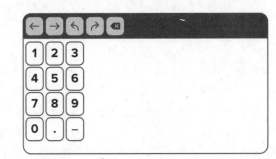

Apply

14. Jake is enclosing his vegetable garden with fencing. The table shows the dimensions of his rectangular garden. Fencing is sold in 2.5-foot sections and costs $25.99 per section. How much will it cost to fence in the entire garden?

Garden Dimension	Measurement (yards)
Length	$14\frac{2}{3}$
Width	$10\frac{1}{3}$

15. Maya has a recipe for blueberry chocolate chip muffins. She doesn't like a lot of oats, but loves blueberries, so she is using $\frac{3}{4}$ of the amount of oats called for and increasing the amount of blueberries by $\frac{1}{4}$ of the original amount. She then wants to double her new recipe. How many cups of oats will Maya need? How many cups of blueberries?

Ingredient	Amount (cups)
Blueberries	$\frac{3}{4}$
Chocolate Chips	1
Oats	$\frac{1}{2}$
White Sugar	1

16. (MP) **Justify Conclusions** A student says that the Commutative Property and Associative Property cannot be used to simplify expressions with rational numbers. Is the student correct? Justify your answer.

17. (MP) **Persevere with Problems** There are 120 bouncy balls in a display bin. Of the bouncy balls in the bin, one-half are priced at $0.25 each, one-fifth are priced at $0.75, and the remaining bouncy balls are priced at $1.50. What is the total value of the bouncy balls in the bin? Write a numerical expression that can be used to solve the problem. Then solve it.

18. A preschool teacher has 17.5 feet of yarn for her students to make necklaces to raise money for a local animal shelter. Each necklace requires 14 inches of yarn. If the necklaces sell for $2.50, how much money will they raise if they sell all the necklaces? Explain how you solved.

19. Create Write and solve a multi-step real-world problem where you apply rational number operations.

📖 **Foldables** Use your Foldable to help review the module.

Operations With Rational Numbers

Tab 1

Rule	Rule
Rule	Rule

Tab 2

Rate Yourself!

Complete the chart at the beginning of the module by placing a checkmark in each row that corresponds with how much you know about each topic after completing this module.

Write about one thing you learned.

_____

_____

_____

_____

_____

_____

Write about a question you still have.

_____

_____

_____

_____

_____

_____

Reflect on the Module

Use what you learned about operations with Integers and rational numbers to complete the graphic organizer.

℮ Essential Question

How are operations with rational numbers related to operations with integers?

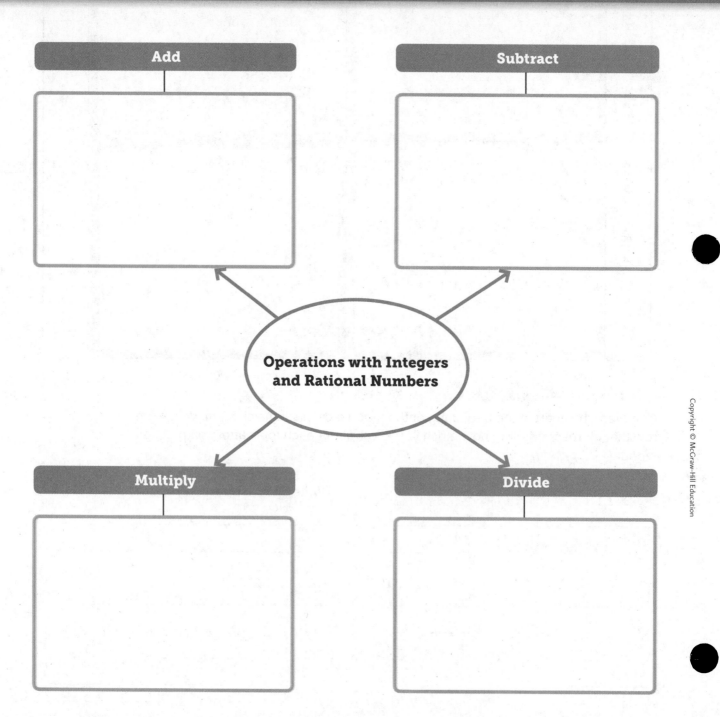

Add

Subtract

Operations with Integers and Rational Numbers

Multiply

Divide

Test Practice

1. Open Response Christine is finding the sum $28 + (-15) + 22$. (Lesson 1)

A. What property of addition could she use to write the integers in a different order? Explain why she might want to do this.

B. Find the sum of $28 + (-15) + 22$.

2. Equation Editor The table shows the high temperature on the moon during the day and the overnight low temperature. (Lesson 2)

Time of Day	Temperature
Day	253°F
Night	-387°F

What is the range between the moon's minimum and maximum temperature in degrees Fahrenheit?

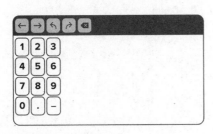

3. Equation Editor Find ab^2c if $a = -3$, $b = -2$, and $c = 6$. (Lesson 3)

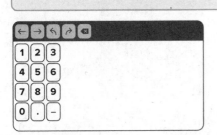

4. Table Item Determine whether the quotient of each expression will be positive or negative. (Lesson 4)

	positive	negative
$-14 \div (-2)$		
$\frac{64}{-4}$		
$1,256 \div 8$		

5. Open Response The equation $C = \frac{5(F - 32)}{9}$ can be used to convert temperatures in degrees Fahrenheit to degrees Celsius. (Lesson 5)

State of Water	Temperature (°F)
Freezing Point	32
Boiling Point	212

Find the freezing point and boiling point of water in degrees Celsius.

6. Multiple Choice Which of the following is the decimal representation of the rational number $-\frac{5}{12}$? (Lesson 6)

(A) $-0.41\overline{6}$

(B) $0.41\overline{6}$

(C) $-0.\overline{416}$

(D) $0.\overline{416}$

7. **Open Response** The manager of Happy Puppy dog boarding company begins the month with 85.4 pounds of dog food and tracks the weekly use, along with the supply delivery received in the middle of the month. (Lesson 7)

Week	Weekly Dog Food Use (lb)
1	−28.4
2	$-21\frac{5}{8}$
3	$97\frac{3}{4}$
4	−30.25

How much dog food does the company have on hand at the end of the month?

[   ]

8. **Equation Editor** At the beginning of the day, Meredith has 85.4 pounds of apples to sell at a farmer's market. The table shows the hourly change in the amount of apples she has after the first several hours. (Lesson 7)

Hour	Change in Apples (lb)
1	−8.5
2	$-12\frac{2}{3}$
3	$-16\frac{1}{5}$
4	−10.2

After 4 hours, about how many pounds of apples does Meredith have left?

[   ]

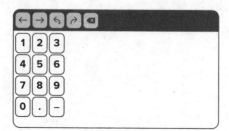

9. **Multiple Choice** What is the result when $-\frac{7}{8}$ is multiplied by −3.5? (Lesson 8)

Ⓐ $3\frac{1}{16}$

Ⓑ $2\frac{11}{16}$

Ⓒ $-2\frac{11}{16}$

Ⓓ $-3\frac{1}{16}$

10. **Equation Editor** Find $-\frac{11}{12} \div \left(-\frac{2}{3}\right)$. Express your answer as a fraction or mixed number in simplest form. (Lesson 8)

[   ]

11. **Open Response** Nathaniel has 132 DVDs in his collection. Each DVD case has the dimensions shown. He plans to buy shelves for the collection that are 31.5 inches wide and cost $15.75 each. How much will it cost Nathaniel to shelve his DVD collection? Explain how you found your answer. (Lesson 9)

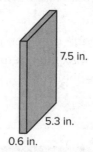

7.5 in.

5.3 in.

0.6 in.

[   ]

Exponents and Scientific Notation

e Essential Question

Why are exponents useful when working with very large or very small numbers?

What Will You Learn?

Place a checkmark (✓) in each row that corresponds with how much you already know about each topic **before** starting this module.

KEY ◻ — I don't know. ◈ — I've heard of it. ★ — I know it!	Before ◻	Before ◈	Before ★	After ◻	After ◈	After ★
writing numerical products as powers						
evaluating powers						
multiplying numerical and algebraic powers						
dividing numerical and algebraic powers						
finding powers of numerical and algebraic powers						
finding powers of numerical and algebraic products						
simplifying expressions with negative exponents and zero exponents						
writing numbers in scientific notation						
computing with numbers in scientific notation						

▦ Foldables Cut out the Foldable and tape it to the Module Review at the end of the module. You can use the Foldable throughout the module as you learn about exponents and scientific notation.

What Vocabulary Will You Learn?

Check the box next to each vocabulary term that you may already know.

☐ base

☐ evaluate

☐ exponent

☐ monomial

☐ negative exponent

☐ Product of Powers Property

☐ power

☐ Power of a Power Property

☐ Power of a Product Property

☐ Quotient of Powers Property

☐ scientific notation

☐ standard form

☐ term

☐ Zero Exponent Rule

Are You Ready?

Study the Quick Review to see if you are ready to start this module. Then complete the Quick Check.

Quick Review

Example 1

Multiply integers.

Find $3 \cdot 2 \cdot 3 \cdot 2 \cdot 2$.

$3 \cdot 2 \cdot 3 \cdot 2 \cdot 2 = 3 \cdot 3 \cdot 2 \cdot 2 \cdot 2$

$\qquad\qquad = (3 \cdot 3) \cdot (2 \cdot 2 \cdot 2)$

$\qquad\qquad = 9 \cdot 8$

$\qquad\qquad = 72$

Example 2

Multiply rational numbers.

Find $2.8 \cdot 2.8$.

$$\begin{array}{r} 2.8 \\ \times\ 2.8 \\ \hline 7.84 \end{array}$$

2.8 ←—— one decimal place

× 2.8 ←—— one decimal place

7.84 ←—— two decimal places

Quick Check

1. A coach watched game film for $4 \cdot 2 \cdot 4 \cdot 4 \cdot 2$ hours last season. How many hours did the coach watch game film?

2. Find $(-1.3)(-1.3)(-1.3)$.

How Did You Do?

Which exercises did you answer correctly in the Quick Check? Shade those exercise numbers at the right.

 ① ②

Powers and Exponents

I Can... use integer exponents to show repeated multiplication of rational numbers.

Explore Exponents

Online Activity You will explore how to write repeated multiplication using exponents.

Emily is doubling the amount she saves each week, then adds this to her total savings. Complete the table to find these values. The first two columns have been completed for you.

Week	0	1	2	3	4	5
Weekly Savings	1¢	2¢	☐¢	☐¢	☐¢	☐¢
Total Savings	1¢	3¢	☐¢	☐¢	☐¢	☐¢

Clear All   Check Answer

Talk About It!

Learn Write Products as Powers

A product of repeated factors can be expressed as a **power**, that is, using an **exponent** and a **base**.

Go Online Watch the animation to learn how to write an expression as a power.

The animation shows that to write $3 \cdot 3 \cdot 3 \cdot 3$ as a power, express the number of times 3 is used as a factor as an exponent.

$$\overbrace{3 \cdot 3 \cdot 3 \cdot 3}^{\square \text{ factors}} = 3^4$$

The base, 3, is the common factor that is being multiplied.

The exponent, 4, tells how many times the base is used as a factor.

(continued on next page)

What Vocabulary Will You Learn?
base
evaluate
exponent
power

Your Notes

The expression $2 \cdot 2 \cdot 2 \cdot 2 \cdot (-4) \cdot (-4) \cdot (-4)$ has two different bases. To write this expression using exponents, express the number of times the base, 2, is used as a factor. Then express the number of times −4 is used as a factor.

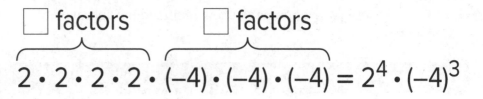

☐ factors   ☐ factors

$$2 \cdot 2 \cdot 2 \cdot 2 \cdot (-4) \cdot (-4) \cdot (-4) = 2^4 \cdot (-4)^3$$

Label each part of the expression with the correct term.

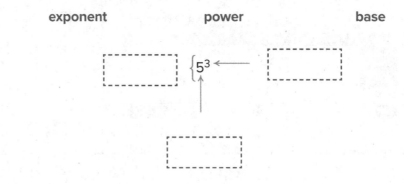

exponent   power   base

5^3

Complete the following statements about how powers are read.

$3^1 = 3$ 3 to the _____ power

$3^2 = 3 \cdot 3$ 3 to the _____ power or 3 _____

$3^3 = 3 \cdot 3 \cdot 3$ 3 to the _____ power or 3 _____

$3^4 = 3 \cdot 3 \cdot 3 \cdot 3$ 3 to the _____ power or 3 _____

n factors

$3^n = 3 \cdot 3 \cdot 3 \cdot ... \cdot 3$ 3 to the _____ power or 3 to the _____

Example 1 Write Numerical Products as Powers

Write the expression $(-2) \cdot (-2) \cdot (-2) \cdot \frac{3}{4} \cdot \frac{3}{4} \cdot \frac{3}{4} \cdot \frac{3}{4}$ using exponents.

The base -2 is used as a factor _____ times, and the base $\frac{3}{4}$ is

used as a factor _____ times. So, the exponents are _____

and _____ .

So, $(-2) \cdot (-2) \cdot (-2) \cdot \frac{3}{4} \cdot \frac{3}{4} \cdot \frac{3}{4} \cdot \frac{3}{4} = (-2)^3 \cdot \left(\frac{3}{4}\right)^4$.

Check

Write the expression $\frac{2}{3} \cdot \frac{2}{3} \cdot \frac{2}{3} \cdot \frac{2}{3} \cdot (-5) \cdot (-5) \cdot (-5) \cdot (-5) \cdot (-5)$ using exponents.

Show your work here

 **Go Online** You can complete an Extra Example online.

Pause and Reflect

Compare and contrast a product of repeated factors with its equivalent power.

Record your observations here

 Think About It!
What are the repeated factors?

Talk About It!
Describe an advantage of writing a product of repeated factors as a power.

Example 2 Write Algebraic Products as Powers

Write the expression $a \cdot b \cdot b \cdot a \cdot b$ using exponents.

$$a \cdot b \cdot b \cdot a \cdot b = a \cdot a \cdot b \cdot b \cdot b \qquad \text{Commutative Property}$$
$$= a^2 \cdot b^3 \qquad\qquad \text{The base } a \text{ is a factor 2 times,}$$
and base b is a factor 3 times.

So, $a \cdot b \cdot b \cdot a \cdot b =$ _____.

Check

Write the expression $a \cdot a \cdot b \cdot a \cdot b \cdot a \cdot a \cdot b \cdot b$ using exponents.

Go Online You can complete an Extra Example online.

Learn Negative Bases and Parentheses

For expressions that contain negative signs and/or parentheses, the inclusion and placement of parentheses can result in distinct expressions that have different values. For example, do you think that $(-a)^b$ and $-a^b$ have the same value?

Complete the following which compares and contrasts the expressions $(-a)^b$ and $-a^b$.

	$(-a)^b$	$-a^b$
Words	The expression $(-a)^b$ indicates that $-a$ is used as a factor b times.	The expression $-a^b$ means the opposite of a^b.
Variables	$(-a)^b = \underbrace{(-a) \cdot (-a) \cdot \ldots \cdot (-a)}_{b \text{ times}}$	$-a^b = -\underbrace{(a \cdot a \cdot \ldots \cdot a)}_{b \text{ times}}$
Numbers	$(-3)^4 = (-3) \cdot (-3) \cdot (-3) \cdot (-3)$ $= \boxed{}$	$-3^4 = -(3 \cdot 3 \cdot 3 \cdot 3)$ $= \boxed{}$

Talk About It!

In the first step, why was the Commutative Property used?

Talk About It!

When you evaluate $(-3)^4$ and -3^4, their results are opposites. When you evaluate $(-5)^3$ and -5^3, are the results opposites? Explain why or why not.

Learn Evaluate Powers

To **evaluate** an expression with a power, write the power as a product and then multiply.

When evaluating expressions with powers or more than one operation, it is important to remember to use the order of operations.

Order of Operations

1. Simplify the expression inside the grouping symbols.

2. Evaluate all powers.

3. Perform multiplication and division in order from left to right.

4. Perform addition and subtraction in order from left to right.

Example 3 Evaluate Numerical Expressions

Evaluate $(-2)^3 + (3.5)^2$.

$(-2)^3 + (3.5)^2 = (-2) \cdot (-2) \cdot (-2) + (3.5) \cdot (3.5)$ Write powers as products.

$= \boxed{} + \boxed{}$ Multiply.

$= \boxed{}$ Add.

So, $(-2)^3 + (3.5)^2 = 4.25$.

Check

Evaluate $8^3 + (-3)^3$.

 Think About It!
According to the order of operations, what are the first evaluations you will need to do?

 Talk About It!
Are the expressions $(-2)^3 + 5^2$ and $-2^3 + 5^2$ equivalent? Justify your response.

Copyright © McGraw-Hill Education

Go Online You can complete an Extra Example online.

Example 4 Evaluate Algebraic Expressions

Evaluate $a^2 + b^4$ if $a = 3$ and $b = \frac{1}{2}$.

$$a^2 + b^4 = 3^2 + \left(\frac{1}{2}\right)^4 \qquad \text{Replace } a \text{ with 3 and } b \text{ with } \frac{1}{2}.$$

$$= (3 \cdot 3) + \left(\frac{1}{2} \cdot \frac{1}{2} \cdot \frac{1}{2} \cdot \frac{1}{2}\right) \qquad \text{Write the powers as products.}$$

$$= \boxed{} + \frac{\boxed{}}{\boxed{}} \qquad \text{Multiply.}$$

$$= \boxed{} \qquad \text{Add.}$$

So, $a^2 + b^4$ when $a = 3$ and $b = \frac{1}{2}$, is $9\frac{1}{16}$.

Check

Evaluate $a^4 + b^2$ if $a = -3$ and $b = 6$.

Example 5 Evaluate Algebraic Expressions

Evaluate $d^3 + (c^2 - 2)$ if $c = -4$ and $d = \frac{2}{5}$.

$$d^3 + (c^2 - 2) = \left(\frac{2}{5}\right)^3 + [(-4)^2 - 2] \qquad \text{Replace } c \text{ with } -4 \text{ and } d \text{ with } \frac{2}{5}.$$

$$= \left(\frac{2}{5}\right)^3 + \boxed{} \qquad \text{Perform the operations in grouping symbols.}$$

$$= \boxed{} \qquad \text{Simplify.}$$

So, $d^3 + (c^2 - 2)$ when $c = -4$ and $d = \frac{2}{5}$, is $14\frac{8}{125}$.

Check

Evaluate $x^2 + (y^3 - 85)$ if $x = -\frac{1}{3}$ and $y = 5$.

Go Online You can complete an Extra Example online.

 Think About It!

Before you can evaluate the expressions, what do you need to do?

 Talk About It!

A classmate evaluated the expression $\left(\frac{1}{2}\right)^4$ by writing $\frac{1^4}{2}$ and simplifying it to $\frac{1}{2}$. Find and correct the error.

 Talk About It!

When you replace c with -4, why should you evaluate $(-4)^2$ and not -4^2?

⊕ Apply Mammals

The table shows the average weights of two endangered mammals. How much more does the brown bear weigh than the panther?

Animal	Weight (lb)
Panther	$2^3 \cdot 3 \cdot 5$
Brown Bear	$2 \cdot 5^2 \cdot 7$

1 What is the task?

Make sure you understand exactly what question to answer or problem to solve. You may want to read the problem three times. Discuss these questions with a partner.

First Time Describe the context of the problem, in your own words.
Second Time What mathematics do you see in the problem?
Third Time What are you wondering about?

2 How can you approach the task? What strategies can you use?

3 What is your solution?

Use your strategy to solve the problem.

4 How can you show your solution is reasonable?

✏ **Write About It!** Write an argument that can be used to defend your solution.

> 💬 Talk About It!
> A female gorilla weighs $2^3 \cdot 5^2$ pounds. How does this compare to the panther's and brown bear's average weight?

Check

Interstate 90 stretches almost $2^4 \cdot 7^2 \cdot 4$ miles across the United States. Interstate 70 stretches almost $2^3 \cdot 5^2 \cdot 11$ miles across the United States. How much longer is Interstate 90 than Interstate 70?

Show your work here

Go Online You can complete an Extra Example online.

Pause and Reflect

How well do you understand the concepts from today's lesson? What questions do you still have? How can you get those questions answered?

Record your observations here

Practice

🧭 **Go Online** You can complete your homework online.

Write each expression using exponents. (Examples 1 and 2)

1. $(-7) \cdot (-7) \cdot 5 \cdot 5 \cdot 5 \cdot 5 =$ _____

2. $n \cdot n \cdot p \cdot p \cdot r \cdot r \cdot r =$ _____

Evaluate each numerical expression. (Example 3)

3. $3^4 - (-4)^2 =$ _____

4. $6 + 2^6 =$ _____

5. Evaluate $x^3 - y^2$ if $x = 2$ and $y = \frac{3}{4}$.
   (Example 4)

6. Evaluate $(g + h)^3$ if $g = 2$ and $h = -3$.
   (Example 5)

7. Replace ☐ with $<$, $>$, or $=$ to make a true statement: $(-3)^4$ ☐ $(-4)^3$.

Test Practice

8. A scientist estimates that, after a certain amount of time, there would be $2^5 \cdot 3^3 \cdot 10^5$ bacteria in a Petri dish. How many bacteria is this?

9. Multiselect Select all of the expressions that evaluate to negative rational numbers.

☐ $(-9)^4$

☐ $\left(-\frac{4}{5}\right)^3$

☐ $3^5 - 10^4$

☐ $(9.8)^2 - 10^2$

☐ $\left(-\frac{3}{8}\right)^2$

Apply

10. The table shows the approximate number of species of each type of tree. How many more species of palm tree are there than maple tree?

Tree Type	Number of Species
Palm	$2^3 \cdot 3 \cdot 5^3$
Maple	2^7

11. The table shows the approximate number of lakes in two different states. How many more lakes does Florida have than Minnesota?

State	Number of Lakes
Florida	$3 \cdot 10^4$
Minnesota	$2^3 \cdot 5^2 \cdot 109$

12. (MP) **Identify Structure** Without evaluating, explain why $(-8.4)^5$ is less than 2^2.

13. (MP) **Find the Error** A student evaluated the expression $(x^2 - y)^3$ if $x = -4$ and $y = 7$. Find his mistake and correct it.

$$(x^2 - y)^3 = (-4^2 - 7)^3$$
$$= (-16 - 7)^3$$
$$= (-23)^3$$
$$= -12{,}167$$

14. Determine if the statement is *true* or *false*. Justify your response.

The expressions $(-x)^y$ and $-x^y$ have the same value.

15. A student finds that $\left(\frac{1}{4}\right)^2 = \left(\frac{1}{2}\right)^4$ and concludes, then, that $\left(\frac{1}{3}\right)^2 = \left(\frac{1}{2}\right)^3$. Is this reasoning correct? Explain.

Multiply and Divide Monomials

I Can... use the Laws of Exponents to multiply and divide monomials with common bases.

What Vocabulary Will You Learn?

monomial

Product of Powers Property

Quotient of Powers Property

term

Explore Product of Powers

Online Activity You will use Web Sketchpad to explore how to simplify a product of powers with like bases.

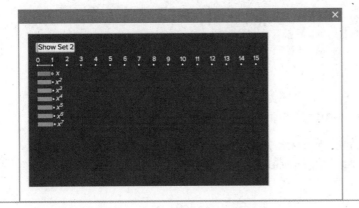

Learn Monomials

A **monomial** is a number, a variable, or a product of a number and one or more variables. For example, $6y^2$ is a monomial because it is a product of 6 and y^2. The expression $x + 3$ is not a monomial since it is a sum of two monomials.

When addition or subtraction signs separate an algebraic expression into parts, each part is called a **term**. A monomial only has one term. The expression $x + 3$ is a sum of two monomials and therefore not a monomial itself because it has two terms.

Write each expression in the appropriate bin. An example of each type is given.

$$x \qquad 80 \qquad x^2 - y^2 \qquad 8x \qquad x + 5$$

Monomials	Not Monomials
$6y^2$	$x + 3$

Learn Product of Powers

🅡 **Go Online** Watch the animation to simplify the product of powers with the same base.

For example, to simplify $4^2 \cdot 4^3$, the animation shows to:

Step 1 Write 4^2 as a product of 2 factors.

Step 2 Write 4^3 as a product of 3 factors.

$$\underset{\underbrace{\hspace{5cm}}_{\text{5 factors}}}{4^2 \cdot 4^3 = \overbrace{(4 \cdot 4)}^{\text{2 factors}} \cdot \overbrace{(4 \cdot 4 \cdot 4)}^{\text{3 factors}}}$$

There are 5 common factors all together, so the product is 4^5.

Notice that the sum of the exponents of the original powers is the exponent in the final product.

$$4^2 \cdot 4^3 = 4^{2+3}$$
$$= 4^5$$

You can multiply powers with the same base by adding the exponents.

You can simplify a product of powers with like bases using the **Product of Powers Property**. The group of integer exponent properties, including the Product of Powers Property, is called the Laws of Exponents.

Words	Algebra
To multiply powers with the same base, add their exponents.	$a^m \cdot a^n = a^{m+n}$
	Numbers
	$2^4 \cdot 2^3 = 2^{4+3}$ or ☐

Copyright © McGraw-Hill Education

💬 Talk About It!

When simplifying a product of powers using the Product of Powers Property, why do the bases have to be the same? For example, why can't you use the Product of Powers Property to simplify $x^3 \cdot y^2$?

Example 1 Multiply Numerical Powers

Simplify $5^4 \cdot 5$.

$5^4 \cdot 5 = 5^4 \cdot 5^1$ $5 = 5^1$

$ = 5^{4+1}$ Product of Powers Property

$ = \boxed{}$ or $\boxed{}$ Add the exponents. Simplify.

So, $5^4 \cdot 5 = 5^5$ or 3,125.

Check

Simplify $4^5 \cdot 4^3$.

 Show your work here

Example 2 Multiply Algebraic Powers

Simplify $c^3 \cdot c^5$.

$c^3 \cdot c^5 = c^{3+5}$ Product of Powers Property

$ = \boxed{}$ Simplify.

So, $c^3 \cdot c^5 = c^8$.

Check

Simplify $x^4 \cdot x^6$.

 Show your work here

 Go Online You can complete an Extra Example online.

Think About It!

How will the Product of Powers Property help you simplify the expression?

Talk About It!

Describe another method you can use to simplify the expression.

Talk About It!

Explain why you were able to add the exponents to simplify the expression $c^3 \cdot c^5$.

Copyright © McGraw-Hill Education

Think About It!

When you multiply the two monomials, what will you do with the coefficients?

Talk About It!

When simplifying the expression, why were the coefficients multiplied but the exponents added?

Example 3 Multiply Monomials

Simplify $-3x^2 \cdot 4x^5$.

$$-3x^2 \cdot 4x^5 = -3 \cdot x^2 \cdot 4 \cdot x^5 \qquad \text{Definition of coefficient}$$
$$= (-3 \cdot 4)(x^2 \cdot x^5) \qquad \text{Commutative and Associative Property}$$
$$= \boxed{}(x^2 \cdot x^5) \qquad \text{Multiply the coefficients.}$$
$$= -12(x^{2+5}) \qquad \text{Product of Powers Property}$$
$$= \boxed{} \qquad \text{Simplify.}$$

So, $-3x^2 \cdot 4x^5 = -12x^7$.

Check

Simplify $-2a(3a^4)$.

Show your work here

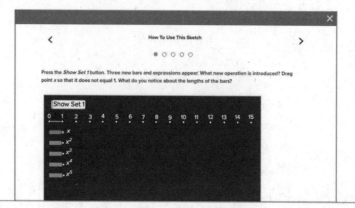
Go Online You can complete an Extra Example online.

Explore Quotient of Powers

Online Activity You will use Web Sketchpad to explore how to simplify a quotient of powers with like bases.

Learn Quotient of Powers

Go Online Watch the animation to learn how to simplify the quotient of powers with the same base.

For example, to simplify $\frac{3^6}{3^2}$, the animation shows to:

Step 1 Write 3^6 as a product of 6 factors.

Step 2 Write 3^2 as a product of 2 factors.

Step 3 Divide out common factors and simplify.

$$\frac{3^6}{3^2} = \frac{\overbrace{3 \cdot 3 \cdot 3 \cdot 3 \cdot \cancel{3} \cdot \cancel{3}}^{\text{6 factors}}}{\underbrace{\cancel{3} \cdot \cancel{3}}_{\text{2 factors}}} = \frac{3 \cdot 3 \cdot 3 \cdot 3}{1} \text{ or } 3^4$$

Four factors of 3 remain in the numerator and 1 in the denominator, so the quotient is 3^4.

Notice that the difference of the exponents of the original powers is the exponent in the final quotient.

$$\frac{3^6}{3^2} = 3^{6-2}$$
$$= 3^4$$

You can divide powers with the same base by subtracting the exponents.

You can simplify a quotient of powers with like bases using the **Quotient of Powers Property**.

Words	Algebra
To divide powers with the same base, subtract their exponents.	$\frac{a^m}{a^n} = a^{m-n}$, where $a \neq 0$.
	Numbers
	$\frac{3^7}{3^3} = 3^{7-3}$ or ☐

> **Talk About It!**
>
> When simplifying a quotient of powers using the Quotient of Powers Property, why do the bases have to be the same? For example, why can't you use the Quotient of Powers Property to simplify $\frac{x^8}{y^3}$?

Example 4 Divide Algebraic Powers

Simplify $\dfrac{x^8}{x^2}$.

$\dfrac{x^8}{x^2} = x^{8-2}$ Quotient of Powers Property

$= \boxed{}$ Subtract the exponents.

So, $\dfrac{x^8}{x^2} = x^6$.

Check

Simplify $\dfrac{z^{12}}{z^4}$.

Show your work here

🧭 **Go Online** You can complete an Extra Example online.

🌐 Example 5 Divide Powers

Hawaii's total shoreline is about 2^{10} miles long. New Hampshire's shoreline is about 2^7 miles long.

About how many times longer is Hawaii's shoreline than New Hampshire's?

To determine how many times longer Hawaii's shoreline is than New Hampshire's, divide their lengths.

$\dfrac{2^{10}}{2^7} = 2^{10-7}$ Quotient of Powers Property

$= \boxed{}$ Subtract the exponents.

$= \boxed{}$ Simplify.

So, Hawaii's shoreline is about 8 times longer than New Hampshire's shoreline.

💬 Talk About It!

Describe another method you could use to simplify the expression.

🍎 Think About It!

What operation will you perform to find how many times longer one shoreline is than the other?

💬 Talk About It!

Alaska's shoreline is about 3^8 miles long. Explain why you could not use the Quotient of Powers Property to simplify the expression $\dfrac{3^8}{2^7}$ to find out how many times longer Alaska's shoreline is than New Hampshire's.

Check

The table shows the seating capacity of two different facilities. About how many times as great is the capacity of Madison Square Garden in New York City than a typical movie theater?

Place	Seating Capacity
Movie Theater	3^5
Madison Square Garden	3^9

Show your work here

🔽 **Go Online** You can complete an Extra Example online.

Example 6 Divide Numerical Powers

Simplify $\dfrac{2^5 \cdot 3^5 \cdot 5^2}{2^2 \cdot 3^4 \cdot 5}$.

$\dfrac{2^5 \cdot 3^5 \cdot 5^2}{2^2 \cdot 3^4 \cdot 5} = \left(\dfrac{2^5}{2^2}\right)\left(\dfrac{3^5}{3^4}\right)\left(\dfrac{5^2}{5}\right)$ Group by common base.

$= 2^{5-2} \cdot 3^{5-4} \cdot 5^{2-1}$ Quotient of Powers Property

$= \boxed{} \cdot \boxed{} \cdot \boxed{}$ Subtract the exponents.

$= \boxed{} \cdot \boxed{} \cdot \boxed{}$ Evaluate the powers.

$= \boxed{}$ Simplify.

So, $\dfrac{2^5 \cdot 3^5 \cdot 5^2}{2^2 \cdot 3^4 \cdot 5} = 120$.

💬 **Talk About It!**

Why is it important to group terms by their common base(s)?

Check

Simplify $\dfrac{2^2 \cdot 3^3 \cdot 4^5}{2 \cdot 3 \cdot 4^4}$.

💭 **Think About It!**

How can you begin solving the problem?

Example 7 Divide Monomials

Simplify $\dfrac{12w^5}{2w}$.

$$\dfrac{12w^5}{2w} = \left(\dfrac{12}{2}\right)\left(\dfrac{w^5}{w}\right) \qquad \text{Associative Property}$$

$$= \boxed{}\left(\dfrac{w^5}{w}\right) \qquad \text{Divide the coefficients.}$$

$$= 6(w^{5-1}) \qquad \text{Quotient of Powers Property}$$

$$= \boxed{} \qquad \text{Subtract the exponents.}$$

So, $\dfrac{12w^5}{2w} = 6w^4$.

💬 **Talk About It!**

Why were the coefficients divided, but the exponents subtracted?

Check

Simplify $\dfrac{24k^9}{6k^5}$.

🔵 **Go Online** You can complete an Extra Example online.

🌐 **Apply** Computer Science

The processing speed of a certain computer is 10^{12} instructions per second. A second computer has a processing speed that is 10^5 times as fast as the first computer. A third computer has a processing speed of 10^{15} instructions per second. Which computer has the fastest processing speed?

1 What is the task?

Make sure you understand exactly what question to answer or problem to solve. You may want to read the problem three times. Discuss these questions with a partner.

First Time Describe the context of the problem, in your own words.
Second Time What mathematics do you see in the problem?
Third Time What are you wondering about?

2 How can you approach the task? What strategies can you use?

3 What is your solution?

Use your strategy to solve the problem.

 Talk About It!

How did understanding the Laws of Exponents help you solve the problem?

4 How can you show your solution is reasonable?

✏️ **Write About It!** Write an argument that can be used to defend your solution.

Check

Bolivia has a population of 10^7 people, while the population of Aruba is 10^5 people. The Philippines has a population that is 10^3 times greater than Aruba's population. Which statement is true about the population of the three countries?

Ⓐ Aruba has the least population.

Ⓑ Boliva has the greatest population.

Ⓒ The Phillipines has the least population.

Ⓓ Bolivia and the Phillipines have the same population.

Show your work here

🔾 **Go Online** You can complete an Extra Example online.

📖 **Foldables** It's time to update your Foldable, located in the Module Review, based on what you learned in this lesson. If you haven't already assembled your Foldable, you can find the instructions on page FL1.

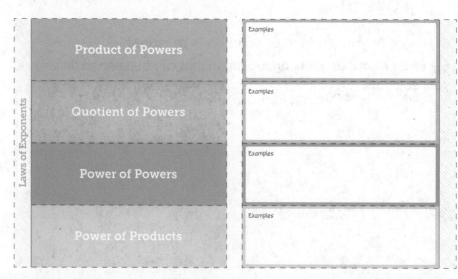

Practice

Go Online You can complete your homework online.

Simplify each expression. (Examples 1–3)

1. $3^8 \cdot 3 =$ _____

2. $m^5 \cdot m^2 =$ _____

3. $3m^3n^2 \cdot 8mn^3 =$ _____

4. $9p^4 \cdot (-8p^2) =$ _____

5. Simplify $\dfrac{b^{12}}{b^5}$. (Example 4)

6. Simplify $\dfrac{5^5 \cdot 6^3 \cdot 8^{10}}{5^3 \cdot 6 \cdot 8^9}$. (Example 6)

7. A publisher sells 10^6 copies of a new science fiction book and 10^3 copies of a new mystery book. How many times as many science fiction books were sold than mystery books? (Example 5)

Test Practice

8. Simplify $\dfrac{45x^{15}}{9x^{10}}$. (Example 7)

9. Equation Editor Simplify $\dfrac{a^4c^6}{a^2c}$.

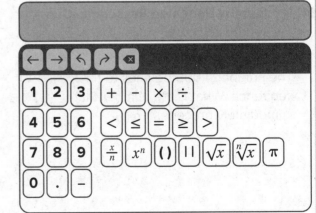

10. Fouster's Farms has 5^3 fruit trees on their land. Myrna's Farms has five times as many fruit trees as Fouster's. A commercial farm has 500 fruit trees. Which farm has the most fruit trees?

11. The table shows the number of bacteria in each Petri dish. Dish C has 8^2 times as many bacteria has Dish A. Which Petri dish holds the most number of bacteria?

Petri Dish	Number of Bacteria
A	8^6
B	8^9

12. Write a multiplication expression with a product of 8^{13}.

13. (MP) **Persevere with Problems** What is four times 4^{15}? Write using exponents and explain your reasoning.

14. (MP) **Identify Repeated Reasoning** Consider the sequence below:

   2, 4, 8, 16, 32, 64, ...

   The number 4,096 belongs to this sequence. What is the number that immediately precedes it?

15. What value of n makes a true statement?

$$4^n \cdot 4^2 = 16{,}384$$

Powers of Monomials

I Can... use the Power of a Power Property and the Power of a Product Property to simplify expressions with integer exponents.

What Vocabulary Will You Learn?
Power of a Power Property
Power of a Product Property

Explore Power of a Power

Online Activity You will explore how to simplify a power raised to another power.

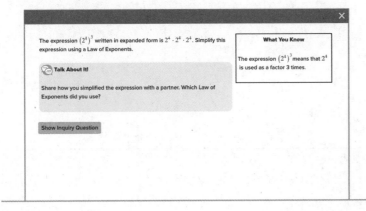

The expression $\left(2^4\right)^3$ written in expanded form is $2^4 \cdot 2^4 \cdot 2^4$. Simplify this expression using a Law of Exponents.

What You Know

The expression $\left(2^4\right)^3$ means that 2^4 is used as a factor 3 times.

Talk About It!

Share how you simplified the expression with a partner. Which Law of Exponents did you use?

Show Inquiry Question

Learn Power of a Power

You can use the rule for finding the product of powers to illustrate how to find the power of a power.

$(6^4)^3 = (6^4)(6^4)(6^4)$ Expand the 3 factors.

$= 6^{4\,+\,4\,+\,4}$ Product of Powers Property

$= \boxed{}$ Add the exponents.

Notice, that the product of the original exponents, 4 and 3, is the final power, 12. You can simplify a power raised to another power using the **Power of a Power Property.**

Words	Algebra
	$(a^m)^n = a^{m \cdot n}$
	Numbers
To find the power of a power, multiply the exponents.	$(5^2)^3 = 5^{2 \cdot 3}$ or $\boxed{}$

How can the Power of a Power Property help you simplify the expression?

 Talk About It!

Describe another method you could use to check your solution.

Example 1 Power of a Power

Simplify $(8^6)^3$.

$(8^6)^3 = 8^{6 \cdot 3}$ Power of a Power Property

$= \boxed{}$ Simplify.

So, $(8^6)^3 = 8^{18}$.

Check

Simplify the expression $(6^3)^5$.

Show your work here

Example 2 Power of a Power

Simplify $(k^7)^5$.

$(k^7)^5 = k^{7 \cdot 5}$ Power of a Power Property

$= \boxed{}$ Simplify.

So, $(k^7)^5 = k^{35}$.

Check

Simplify the expression $(x^4)^7$.

Show your work here

 Go Online You can complete an Extra Example online.

Learn Power of a Product

The following example demonstrates how the Power of a Power Property can be extended to find the power of a product.

$(4a^2)^3 = (4a^2)(4a^2)(4a^2)$ Expand the 3 factors.

$\quad\quad = 4 \cdot 4 \cdot 4 \cdot a^2 \cdot a^2 \cdot a^2$ Commutative Property

$\quad\quad = 4^3 \cdot (a^2)^3$ Definition of power

$\quad\quad = 64 \cdot a^6 \text{ or } 64a^6$ Power of a Power Property

Notice, that each factor inside the parentheses is raised to the

_____ power.

You can simplify a product raised to a power using the **Power of a Product Property.**

Words	Algebra
To find the power of a product, find the power of each factor and multiply.	$(ab)^m = a^m b^m$
	Numbers
	$(2x^3)^4 = (2)^4 \cdot (x^3)^4 \text{ or } \boxed{}$

Example 3 Power of a Product

Simplify $(2p^3)^4$.

$(2p^3)^4 = 2^4 \cdot (p^3)^4$ Power of a Product Property

$\quad\quad = 2^4 \cdot p^{3 \cdot 4}$ Power of a Power Property

$\quad\quad = \boxed{}$ Simplify.

So, $(2p^3)^4 = 16p^{12}$.

Check

Simplify $(7w^7)^3$.

 Think About It!

What parts of the monomial, $2p^3$, are raised to the fourth power?

Talk About It!

Why was the exponent 4 applied to both of the factors in the first step?

Go Online You can complete an Extra Example online.

Think About It!

What parts of the monomial, $(-2m^7n^6)$, are raised to the fifth power?

 Talk About It!

Why were the exponents multiplied when simplifying the expression?

Example 4 Power of a Product

Simplify $(-2m^7n^6)^5$.

$(-2m^7n^6)^5 = (-2)^5 \cdot (m^7)^5 \cdot (n^6)^5$ Power of a Product Property

$= \boxed{}$ Simplify.

So, $(-2m^7n^6)^5 = -32m^{35}n^{30}$.

Check

Simplify $(-5w^2z^8)^3$.

🔵 **Go Online** You can complete an Extra Example online.

Pause and Reflect

Did you ask questions about the properties used in this lesson? Why or why not?

Record your observations here

🌐 **Apply** Geometry

A square floor has a side length of $8x^3y^2$ units. A square tile has a side length of xy units. How many tiles will it take to cover the floor?

🚀 Go Online
Watch the animation.

1 What is the task?

Make sure you understand exactly what question to answer or problem to solve. You may want to read the problem three times. Discuss these questions with a partner.

First Time Describe the context of the problem, in your own words.
Second Time What mathematics do you see in the problem?
Third Time What are you wondering about?

2 How can you approach the task? What strategies can you use?

3 What is your solution?

Use your strategy to solve the problem.

💬 Talk About It!

To find the number of tiles needed to cover the floor, why do you need to divide the areas?

4 How can you show your solution is reasonable?

✏️ **Write About It!** Write an argument that can be used to defend your solution.

Check

The floor of the commons room at King Middle School is in the shape of a square with side lengths of $3x^2y^3$ feet. New tile is going to be put on the floor of the room. The square tile has side lengths of xy feet. How many tiles will it take to cover the floor?

Show your work here

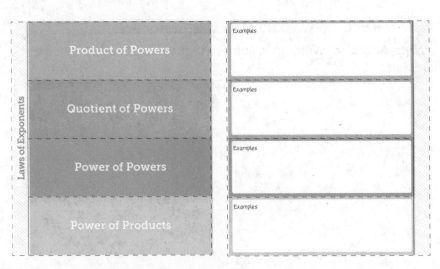

 Go Online You can complete an Extra Example online.

 Foldables It's time to update your Foldable, located in the Module Review, based on what you learned in this lesson. If you haven't already assembled your Foldable, you can find the instructions on page FL1.

Practice

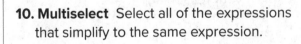 Go Online You can complete your homework online.

Simplify each expression. (Examples 1–4)

1. $(7^2)^3 = $ _____

2. $(8^3)^3 = $ _____

3. $(d^7)^6 = $ _____

4. $(z^7)^3 = $ _____

5. $(2m^5)^6 = $ _____

6. $(7a^5b^6)^4 = $ _____

7. $(-3w^3z^8)^5 = $ _____

8. $(-5r^4s^{10})^4 = $ _____

9. Which is greater: 1,000 or $(6^2)^3$? Explain.

Test Practice

10. Multiselect Select all of the expressions that simplify to the same expression.

☐ $(x^3y^4)^2$

☐ $(x^2y)^2$

☐ $(x^3)^2y^6$

☐ $x^6(y^4)^2$

☐ $(x^3)^2(y^2)^4$

Apply

11. A square floor has a side length of $7x^5y^6$ units. A square tile has a side length of x^2y units. How many tiles will it take to cover the floor?

12. A cube has a side length of $3x^6$ units. A smaller cube has a side length of x^2 units. How many smaller cubes will fit in the larger cube?

13. Make an argument for why $(4^2)^4 = (4^4)^2$.

14. **MP** **Persevere with Problems** Describe all the positive integers that would make $\left[\left(\frac{1}{2}\right)^4\right]^n$ less than $\left(\frac{1}{2}\right)^7$. Explain your reasoning.

15. Without computing, determine which number is greater, $[(-4)^4]^5$ or $-[(4^{12})^3]$. Explain your reasoning.

16. **MP** **Identify Structure** Charlotte states that $(4^3)^3$ can be rewritten as 2^{18}. Explain how she is correct.

Zero and Negative Exponents

I Can... use the Zero Exponent Rule and the Quotient of Powers Property to simplify expressions with zero and negative integer exponents.

What Vocabulary Will You Learn?
negative exponent
Zero Exponent Rule

Explore Exponents of Zero

Online Activity You will explore how to simplify expressions with exponents of zero and make a conjecture about the value of expressions with exponents of zero.

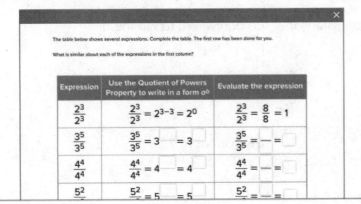

Learn Exponents of Zero

Use the **Zero Exponent Rule** to simplify expressions containing exponents of zero.

Words
Any nonzero number to the zero power is equivalent to 1.
Algebra
$x^0 = 1, x \neq 0$
Numbers
$5^0 = \boxed{}$

(continued on next page)

Complete the pattern in the table to demonstrate that any nonzero number, n, to the zero power is equivalent to 1.

Power	Equivalent Form
n^5	$n \cdot n \cdot n \cdot n \cdot n$
n^4	
n^3	
n^2	
n^1	
n^0	

n^{5-1} … $\div n$
n^{4-1} … $\div n$
n^{3-1} … $\div n$
n^{2-1} … $\div n$
n^{1-1} … $\div n$

The pattern in the table shows that as you decrease the exponent by 1, the value of the power is divided by n each time. By extending the pattern, $n \div n = 1$. So, $n^0 = 1$.

Write the expressions that are equivalent to 1 in the bin.

$$2^1 \qquad 1^2 \qquad 2^0 \qquad 1^0 \qquad 0^1 \qquad \left(\frac{1}{2}\right)^0$$

Equivalent to 1

Pause and Reflect

Do all powers with any rational number base and an exponent of zero equal 1? Explain.

Record your observations here

Example 1 Exponents of Zero

Simplify 12^0.

Any nonzero number to the zero power is _____.

So, $12^0 = 1$.

Check

Simplify m^0, where $m \neq 0$.

 Go Online You can complete an Extra Example online.

Explore Negative Exponents

 Online Activity You will use Web Sketchpad to explore how to simplify expressions with negative exponents.

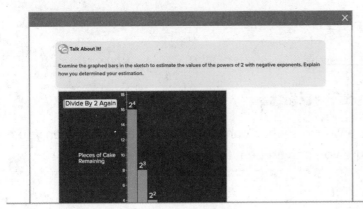

Learn Negative Exponents

A **negative exponent** is the result of repeated division. You can use negative exponents to represent very small numbers.

Complete the table below. Every time you divide by 10, the exponent decreases by one. The pattern in the table shows that 10^{-2} can be defined as $\frac{1}{100}$ or $\frac{1}{10^2}$.

Exponential Form	Standard Form
$10^3 = 10 \cdot 10 \cdot 10$	
$10^2 = 10 \cdot 10$	
10^1	
10^0	
10^{-1}	
10^{-2}	
10^{-3}	

$\div 10$
$\div 10$
$\div 10$
$\div 10$
$\div 10$
$\div 10$

Words
Any nonzero number to the negative *n* power is the multiplicative inverse of its *n*th power.
Algebra
$x^{-n} = \frac{1}{x^n}, x \neq 0$
Numbers
$7^{-3} = \frac{1}{7} \cdot \frac{1}{7} \cdot \frac{1}{7}$ or $\frac{1}{7^3}$

Pause and Reflect

Did you make any errors when completing the Learn? What can you do to make sure you don't repeat that error in the future?

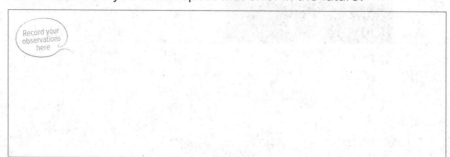

Record your observations here

Example 2 Negative Exponents

Express 6^{-3} using a positive exponent.

$$6^{-3} = \frac{1}{\boxed{}} \qquad \text{Definition of negative exponent}$$

So, 6^{-3} expressed using a positive exponent is $\frac{1}{6^3}$.

Check

Express a^{-5} using a positive exponent.

Example 3 Negative Exponents

Express the fraction $\frac{1}{c^5}$ using a negative exponent.

$$\frac{1}{c^5} = c^{-5} \qquad \text{Definition of negative exponent}$$

So, $\frac{1}{c^5}$ expressed using a negative exponent is _____.

Check

Express the fraction $\frac{1}{9^7}$ using a negative exponent.

 Think About It!

What is the multiplicative inverse of 6?

 Talk About It!

Explain how to write x^{-5} using a positive exponent.

Go Online You can complete an Extra Example online.

Example 4 Negative Exponents

Simplify $5^3 \cdot 5^{-5}$.

An expression is in simplest form if it contains no like bases and no negative exponents.

$5^3 \cdot 5^{-5} = 5^{3 + (-5)}$ Product of Powers Property

$= \boxed{}$ Simplify.

$= \dfrac{1}{\boxed{}}$ Write using positive exponents.

$= \dfrac{1}{\boxed{}}$ Simplify.

So, $5^3 \cdot 5^{-5} = \dfrac{1}{25}$.

Check

Simplify $3^9 \cdot 3^{-4}$.

Show your work here

Example 5 Negative Exponents

Simplify $\dfrac{w^{-1}}{w^{-4}}$.

$\dfrac{w^{-1}}{w^{-4}} = w^{-1 - (-4)}$ Quotient of Powers Property

$= \boxed{}$ Simplify.

So, $\dfrac{w^{-1}}{w^{-4}} = w^3$.

Check

Simplify $\dfrac{b^{-7}}{b^{-4}}$.

Show your work here

 Go Online You can complete an Extra Example online.

⊕ **Apply** Measurement

One strand of human hair is about 0.001 inch in diameter. The diameter of a certain wire for jewelry is 10^{-2} inch. How many times larger is the diameter of the wire than a strand of hair? Write the decimal as a power of 10.

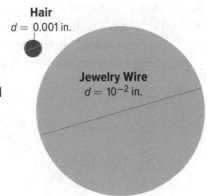

Hair
$d = 0.001$ in.

Jewelry Wire
$d = 10^{-2}$ in.

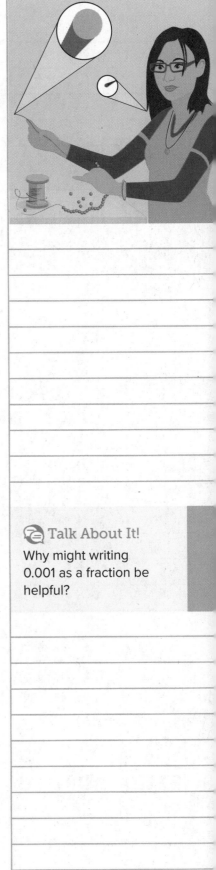

1 What is the task?

Make sure you understand exactly what question to answer or problem to solve. You may want to read the problem three times. Discuss these questions with a partner.

First Time Describe the context of the problem, in your own words.
Second Time What mathematics do you see in the problem?
Third Time What are you wondering about?

2 How can you approach the task? What strategies can you use?

Record your observations here

3 What is your solution?

Use your strategy to solve the problem.

Show your work here

4 How can you show your solution is reasonable?

✏ **Write About It!** Write an argument that can be used to defend your solution.

> 💬 **Talk About It!**
> Why might writing 0.001 as a fraction be helpful?

Check

An American green tree frog tadpole is about 0.00001 kilometer in length when it hatches. The largest indoor swimming pool is 10^{-1} kilometer wide. How many times longer is the width of the swimming pool than the length of the green tree frog tadpole?

Go Online You can complete an Extra Example online.

Pause and Reflect

Where did you encounter struggle in this lesson, and how did you deal with it? Write down any questions you still have.

Practice

Go Online You can complete your homework online.

Simplify each expression. (Example 1)

1. $46^0 =$ _____

2. w^0, where $w \neq 0$, = _____

Express each using a positive exponent. (Example 2)

3. $8^{-4} =$ _____

4. $y^{-9} =$ _____

Express each fraction using a negative exponent. (Example 3)

5. $\dfrac{1}{d^6} =$ _____

6. $\dfrac{1}{10^5} =$ _____

Simplify each expression. (Examples 4 and 5)

7. $9^4 \cdot 9^{-6} =$ _____

8. $y^{-9} \cdot y^3 =$ _____

9. $\dfrac{x^{-8}}{x^{-12}} =$ _____

10. $\dfrac{d^{-13}}{d^{-2}} =$ _____

11. Simplify $8^{-7} \cdot 8^7 \cdot 10^4 \cdot 10^{-4}$.

Test Practice

12. Multiselect Select all of the expressions that are simplified.

- ☐ n^4
- ☐ $\dfrac{1}{n^{-5}}$
- ☐ $n^6 \cdot n^{-8}$
- ☐ $n^7 \cdot p^8$
- ☐ $\dfrac{1}{n^3}$

Apply

13. A dish containing bacteria has a diameter of 0.0001 kilometer. The diameter of a bacterium is 10^{-16} kilometer. How many times larger is the diameter of the dish than the diameter of the bacterium?

14. The table shows the diameters of two objects. How many times larger is the diameter of a pinhead than the diameter of a cloud water droplet?

Object	Diameter (m)
Cloud Water Droplet	10^{-5}
Pinhead	0.001

15. Without evaluating, order 5^{-7}, 5^4, and 5^0 from least to greatest. Explain your reasoning.

16. 🆔 **Persevere with Problems** Explain how to find the value of $\left(\frac{1}{3^{-2}}\right)^{-3}$.

17. Determine if the following numerical expressions are equivalent. Explain your reasoning.

$$\left[\left(\frac{1}{10^{-2}}\right)^0\right]^3 \text{ and } [(10^{-4})^3]^0$$

18. Determine if the following statement is *true* or *false*. Explain your reasoning.

If a number between 0 and 1 is raised to a negative power, the result is a number greater than 1.

Scientific Notation

I Can... write very large and very small numbers using scientific notation.

What Vocabulary Will You Learn?

scientific notation

standard form

Explore Scientific Notation

Online Activity You will explore how to write very large and very small numbers using scientific notation.

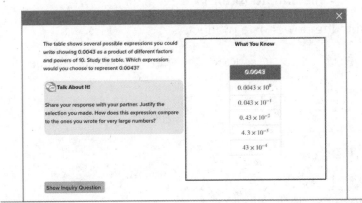

Learn Scientific Notation

Numbers that do not contain exponents are written in **standard form**. However, when you work with very large or very small numbers, it can be difficult to keep track of the place value. **Scientific notation** is a compact way of writing very large or very small numbers.

Words
Scientific notation is a way of expressing a number as the product of a factor and an integer power of 10. When the number is positive, the factor must be greater than or equal to 1 and less than 10.
Symbols
For positive numbers written in scientific notation, $a \times 10^n$, where $1 \le a < 10$ and n is an integer
Examples
$425{,}000{,}000 = 4.25 \times 10^8$ $0.00003 = 3 \times 10^{-5}$

(continued on next page)

Circle yes or no to determine whether each number is written in scientific notation.

10.6×10^{14}	Yes	No
2.019×10^{12}	Yes	No
9.5×10^{-7}	Yes	No
0.526×10^{-2}	Yes	No

Example 1 Write Numbers in Standard Form

Write 5.34×10^4 in standard form.

Method 1 Write the power as a product.

$5.34 \times 10^4 = 5.34 \times 10 \times 10 \times 10 \times 10$ Write the power as a product.

$\qquad = 5.34 \times \boxed{}$ Multiply.

$\qquad = \boxed{}$ Multiply.

So, $5.34 \times 10^4 = 53{,}400$.

Method 2 Move the decimal point.

Moving the decimal point one place to the right is a result of multiplying a number by 10. So, when you multiply a number by 10^4, the decimal point moves four places to the right.

The exponent on the power of 10 indicates the number of places, in this case, _____. A positive exponent indicates the direction, to the _____.

$5.34 \times 10^4 = 53{,}400$

Check

Write 9.931×10^5 in standard form.

Think About It!

What is standard form?

Talk About It!

Compare the two methods used to write the number in standard form.

Go Online You can complete an Extra Example online.

Example 2 Write Numbers in Standard Form

Write 3.27 × 10⁻³ in standard form.

Method 1 Write the power as a product.

$3.27 \times 10^{-3} = 3.27 \times 10^{-1} \times 10^{-1} \times 10^{-1}$ Product of Powers Property

$\qquad = 3.27 \times \dfrac{1}{10} \times \dfrac{1}{10} \times \dfrac{1}{10}$ Definition of negative exponents

$\qquad = 3.27 \times \dfrac{1}{1,000}$ Multiply the fractions.

$\qquad = 3.27 \div 1,000$ Definition of reciprocal

$\qquad = \boxed{}$ Divide.

So, $3.27 \times 10^{-3} = 0.00327$.

Method 2 Move the decimal point.

When you divide a number by 10, the decimal point moves one place to the left. Dividing by 10 is the same as multiplying by 10^{-1}. So, when you multiply a number by 10^{-3}, the decimal point moves three places to the left.

The exponent on the power of 10 indicates the number of places, in this case, _____. A negative sign on the exponent indicates the direction, to the _____.

$3.27 \times 10^{-3} = 0.00327$

Check

Write 6.02×10^{-4} in standard form.

Show your work here

🅚 **Go Online** You can complete an Extra Example online.

💭 **Think About It!**

How would you begin solving the problem?

💬 **Talk About It!**

Compare the two methods used to write the number in standard form.

Learn Scientific Notation and Technology

Go Online Watch the video to learn how to interpret scientific notation that has been generated by a calculator.

The animation shows that a calculator will convert numbers to scientific notation only if they are very large or very small. The value after the E represents the exponent on the power of 10.

8E11 means 8 × [　　]　　　　2.5E−12 means 2.5 × [　　]

You can also use a calculator to enter a number in scientific notation.

Step 1 Type the number that is multiplied by the power of 10.

Step 2 Press the 2ND button.

Step 3 Press the EE button, located above the comma button. (Note this will only display as E on the screen.)

Step 4 Type the exponent on the power of 10.

🌐 Example 3 Scientific Notation and Technology

Calculators often use the E symbol to indicate scientific notation.

The calculator screen shows the approximate wavelength, in meters, of violet light.

Write this number in standard form.

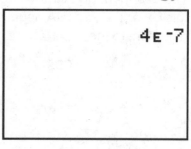

Step 1 Represent the number on the calculator screen in scientific notation.

4E−7 ⟶ [　] × [　]　　4 is the factor and −7 is the exponent.

Step 2 Write the number in standard form.

To write 4×10^{-7} in standard form, move the decimal point _____ places to the _____.

So, 4×10^{-7} is 0.0000004.

Think About It!

What do you think the symbol E represents on the calculator screen?

Check

The calculator screen shows the diameter of a grain of sand in inches. Which numbers represent the number shown on the calculator screen? Select all that apply.

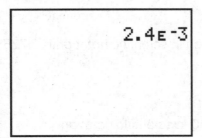

2.4ᴇ⁻3

☐ 2,400

☐ 0.0024

☐ 2.4×10^3

☐ 2.4×10^{-3}

☐ 2.4^{-3}

Show your work here

⟲ **Go Online** You can complete an Extra Example online.

Learn Write Numbers in Scientific Notation

When writing a positive number in scientific notation, the sign of the exponent can be determined by examining the number in standard form.

If the number in standard form is . . .

	greater than or equal to 10,	between 0 and 1,
Words	then the exponent in the power of 10 is *positive*.	then the exponent in the power of 10 is *negative*.
Example	$5{,}860{,}000 = 5.86 \times 10^6$	$0.000586 = 5.86 \times 10^{-4}$

💬 Talk About It!

If the number in standard form is greater than or equal to 1 and less than 10, what is the exponent on the power of 10 when the number is written in scientific notation? Explain.

Copyright © McGraw-Hill Education

Example 4 Write Numbers in Scientific Notation

Write 3,725,000 in scientific notation.

$3{,}725{,}000 = 3.725 \times 1{,}000{,}000$ Write the number as a product.

$= 3.725 \times \boxed{}$ Since 3,725,000 is >10, the exponent is positive.

So, 3,725,000 written in scientific notation is 3.725×10^{6}.

Check

Write 8,785,000,000 in scientific notation.

  Talk About It!

Why was the decimal point moved 6 places? How can you determine the value of the exponent?

Example 5 Write Numbers in Scientific Notation

Write 0.000316 in scientific notation.

$0.000316 = 3.16 \times 0.0001$ Write the number as a product.

$= 3.16 \times \dfrac{1}{\boxed{}}$ Write the decimal as a fraction with power of 10.

$= 3.16 \times \boxed{}$ Since $0 < 0.000316 < 1$, the exponent is negative.

So, 0.000316 written in scientific notation is 3.16×10^{-4}.

Check

Write 0.524 in scientific notation.

Talk About It!

Why was the decimal point moved 4 places? Why is the exponent negative?

Go Online You can complete an Extra Example online.

Learn Use Scientific Notation

When using very large or very small quantities, it is important to choose the appropriate size of measurement.

For example, it is more appropriate to represent the distance from California to New York as 2,441 miles as opposed to 1.55×10^8 inches.

Circle the units that are more appropriate for measuring the following.

the time it takes to travel from Florida to Michigan

minutes   hours

the thickness of a penny

millimeters   meters

the length of a football field

inches   yards

the weight of a paperclip

ounces   pounds

You can estimate very large or very small numbers by expressing them in the form of a single digit times an integer power of 10. Estimating very large or very small numbers makes them easier to work with.

Complete the examples to estimate the numbers.

The population of the United States in a recent year was 324,430,860.

$324{,}430{,}860 \approx 300{,}000{,}000$

$300{,}000{,}000 = \boxed{} \times \boxed{}$

The diameter of an animal cell was measured at 0.000635 centimeter.

$0.000635 \approx 0.0006$

$0.0006 = \boxed{} \times \boxed{}$

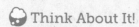
Talk About It!

Describe some situations in which it might make sense to use a unit of measure that was *not* as meaningful.

🌐 **Example 6** Choose Units of Appropriate Size

If you could walk at a rate of 2 meters per second, it would take you 1.92×10^8 seconds to walk to the Moon.

Is it more appropriate to report this time as 1.92×10^8 seconds or 6.09 years? Explain your reasoning.

Representing the time in seconds gives a large number, 1.92×10^8 seconds or 192,000,000 seconds. It would be more meaningful to report the time as 6.09 years, so choosing the larger unit of measure is more appropriate in this situation.

So, the measure _____ is the more appropriate time.

Check

A plant cell has a diameter of 1.3×10^{-8} kilometer. Is it more appropriate to report the diameter of a plant cell as 1.3×10^{-8} kilometer or 1.3×10^{-2} millimeter?

🌐 **Example 7** Estimate with Scientific Notation

The population of Missouri is 6,063,589 people.

Write an estimation in scientific notation for the population.

$6{,}063{,}589 \approx$ ⬚   Estimate.

$6{,}000{,}000 =$ ⬚ $\times$ ⬚   Write in scientific notation.

So, the population of Missouri is about 6,000,000 or 6×10^6 people.

Check

The distance from the Earth to the Moon is 238,900 miles. Write an estimation in scientific notation for the distance.

🌐 **Go Online** You can complete an Extra Example online.

🌎 Apply Travel

The table shows the approximate number of international visitors to the United States, from several countries, for July and August of a recent year. Which country had the greatest percent of change in the approximate number of visitors from July to August?

Country	Visitors (Jul.)	Visitors (Aug.)
Canada	2.09×10^6	2.43×10^6
Mexico	1.55×10^6	1.63×10^6
United Kingdom	4.2×10^5	4.9×10^5
Japan	3.05×10^5	3.9×10^5

1 What is the task?

Make sure you understand exactly what question to answer or problem to solve. You may want to read the problem three times. Discuss these questions with a partner.

First Time Describe the context of the problem, in your own words.
Second Time What mathematics do you see in the problem?
Third Time What are you wondering about?

2 How can you approach the task? What strategies can you use?

3 What is your solution?

Use your strategy to solve the problem.

4 How can you show your solution is reasonable?

✏️ **Write About It!** Write an argument that can be used to defend your solution.

Copyright © McGraw-Hill Education

💬 **Talk About It!**
What method did you use to solve the problem? Explain why you chose that method.

Check

The approximate attendance and average ticket price for four Major League baseball teams for a recent year is shown in the table.

Team	Attendance	Ticket Price ($)
Los Angeles Angels	3,020,000	32.70
Miami Marlins	1.71×10^6	28.31
Pittsburgh Pirates	2,450,000	29.96
St. Louis Cardinals	3.44×10^6	34.20

Which team had the most revenue from ticket sales?

Show your work here

Go Online You can complete an Extra Example online.

Pause and Reflect

What have you learned about scientific notation in this lesson? Include a real-world example of scientific notation in your summary.

Record your observations here

Practice

Go Online You can complete your homework online.

Write each number in standard form. (Examples 1 and 2)

1. $1.6 \times 10^3 =$ _____

2. $1.49 \times 10^{-7} =$ _____

3. A calculator screen shows a number in scientific notation as 8.3E−6. Write this number in standard form. (Example 3)

4. A calculator screen shows a number in scientific notation as 7E11. Write this number in standard form. (Example 3)

Write each number in scientific notation. (Examples 4 and 5)

5. $2,204,000,000 =$ _____

6. $0.0000000642 =$ _____

7. A common race is a 5K race, where runners travel 5 kilometers. Is it more appropriate to report the distance as 5 kilometers or 5×10^6 millimeters? Explain your reasoning. (Example 6)

8. The population of Florida was recently recorded as 20,612,439 people. Write an estimation in scientific notation for the population. (Example 7)

Test Practice

9. The diameter of a grain of sand is 0.0024 inch. Write an estimation in scientific notation for the diameter. (Example 7)

10. Equation Editor The mass of planet Earth is about 5.98×10^{24} kilograms. When this number is written in standard notation, how many zeros are in the number?

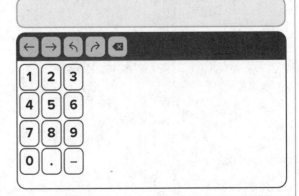

Apply

11. The table shows the approximate number of insects in each group. What is the combined population of the four insect groups?

Insect	Population
Honeybees	7.3498×10^4
Ants	6.822×10^3
Termites	9.8×10^5
Aphids	2.9502×10^4

12. The table shows the approximate population and land area for several states. Which state has the least population density?

State	Population	Land Area (mi²)
Arizona	6.83×10^6	113,594
Colorado	5.46×10^6	103,642
Kentucky	4.43×10^6	39,486
Minnesota	5.45×10^6	79,627

13. Simplify the following expression and write in scientific notation.

$$\frac{(0.00045)(500,000)}{0.0015}$$

14. (MP) **Be Precise** The amount of time spent doing homework is 2.7×10^3 seconds. Choose a more appropriate measurement to report the amount of time spent doing homework. Justify your answer.

15. (MP) **Find the Error** Katrina states that 3.5×10^4 is greater than 2.1×10^6 because $3.5 > 2.1$. Explain her mistake and correct it.

16. Give a number in scientific notation that is between the two numbers on a number line.

$$7.1 \times 10^3 \text{ and } 71,000,000$$

Compute with Scientific Notation

I Can... perform computations with numbers written in scientific notation.

Learn Multiply and Divide with Scientific Notation

To solve real-world problems involving numbers written in scientific notation, you may need to multiply or divide these numbers. Since numbers written in scientific notation contain exponents, sometimes the Laws of Exponents can be used when operating with them.

With all operations, it is often helpful to write the numbers in the same form, scientific notation or standard form, before computing.

When performing these operations with numbers written in scientific notation...	Remember to...
Multiplication	Use the Product of Powers Property.
Division	Use the Quotient of Powers Property.

Example 1 Multiply with Scientific Notation

Scientists estimate that there are over 3.5×10^6 ants per acre in the Amazon rain forest.

If the Amazon rain forest covers approximately 1 billion acres, find the total number of ants. Write in scientific notation.

Step 1 Write the number of acres in scientific notation.

1 billion = ☐ × ☐

Step 2 Multiply to find the total number of ants.

$(3.5 \times 10^6) \times (1 \times 10^9)$	Write the expression.
$= (3.5 \times 1) \times (10^6 \times 10^9)$	Commutative and Associative Properties
$= (3.5) \times (10^6 \times 10^9)$	Multiply 3.5 by 1.
$= 3.5 \times 10^{6 + 9}$	Product of Powers Property
$= 3.5 \times$ ☐	Add the exponents.

So, the total number of ants is approximately 3.5×10^{15}.

 Think About It!
How will you set up the multiplication to solve this problem?

Talk About It!
In the 2nd step, why were the Commutative and Associative Properties used?

Copyright © McGraw-Hill Education

Check

A dime is 1.35×10^{-3} meter thick. What would be the height, in meters, of a stack of 1 million dimes?

Show your work here

🅺 **Go Online** You can complete an Extra Example online.

🌐 **Example 2** Divide with Scientific Notation

Neurons are cells in the nervous system that process and transmit information. An average neuron is about 5×10^{-6} meter in diameter. A standard table tennis ball is 0.04 meter in diameter.

About how many times as great is the diameter of a ball than a neuron? Write your answer in scientific notation.

Step 1 Write the numbers in the same form.

Write the diameter of the table tennis ball in scientific notation.

$$0.04 = \boxed{} \times \boxed{}$$

Step 2 Divide the diameter of the ball by the diameter of the neuron.

$$\frac{4 \times 10^{-2}}{5 \times 10^{-6}} = \left(\frac{4}{5}\right) \times \left(\frac{10^{-2}}{10^{-6}}\right) \qquad \text{Associative Property}$$

$$= \boxed{} \times \left(\frac{10^{-2}}{10^{-6}}\right) \qquad \text{Divide 4 by 5.}$$

$$= 0.8 \times 10^{[-2 - (-6)]} \qquad \text{Quotient of Powers Property}$$

$$= 0.8 \times \boxed{} \qquad \text{Subtract the exponents.}$$

Step 3 Write in scientific notation.

Since 0.8×10^4 is not written in scientific notation, move the decimal one place to the right and subtract 1 from the exponent.

$$0.8 \times 10^4 = \boxed{} \times \boxed{}$$

So, the diameter of the table tennis ball is about 8×10^3 or _____ times larger than the diameter of the neuron.

💭 **Think About It!**

How would you begin solving the problem?

🗨 **Talk About It!**

Why is 0.8×10^4 not in scientific notation? When writing 0.8×10^4 in scientific notation, why do you need to subtract 1 from the exponent?

Check

In a recent year, the population of China was about 1.3×10^9. According to the census data, the population of the United States was about 308,745,538. About how many times greater was the population of China than the population of the United States?

Show your work here

R Go Online You can complete an Extra Example online.

Learn Add and Subtract with Scientific Notation

To add or subtract numbers written in scientific notation, it is helpful to rewrite the numbers so that the exponents on each power of 10 have the same value.

To add 3.6×10^4 to the numbers shown, each number needs to be rewritten so the exponents on each power of 10 are the same. Rewrite each number so that it has an exponent of 4.

$1.43 \times 10^3 = \boxed{} \times 10^4$

$1.43 \times 10^5 = \boxed{} \times 10^4$

$1.43 \times 10^2 = \boxed{} \times 10^4$

Once numbers written in scientific notation have the same exponent on the power of 10, you can use the Distributive Property to add or subtract.

$(4.36 \times 10^5) + (2.09 \times 10^6)$

$\qquad = (4.36 \times 10^5) + (20.9 \times 10^5)$ Rewrite with the same power of 10.

$\qquad = (4.36 + 20.9) \times 10^5$ Distributive Property

$\qquad = (25.26) \times 10^5$ Add 4.36 and 20.9.

$\qquad = 2.526 \times 10^6$ Rewrite in scientific notation.

You can also write each number in standard form before computing or to check your work.

Talk About It!

Explain why you cannot add the exponents on the powers of 10, when adding the expressions.

Think About It!

Do the powers of 10 have the same exponent?

Talk About It!

Describe another method you can use to evaluate the expression.

Example 3 Add or Subtract with Scientific Notation

Evaluate $(1.45 \times 10^9) - (2.84 \times 10^8)$. Express the result in scientific notation.

$(1.45 \times 10^9) - (2.84 \times 10^8)$ Write the expression.

$= (14.5 \times 10^8) - (2.84 \times 10^8)$ Write 1.45×10^9 as 14.5×10^8.

$= (\boxed{} - \boxed{}) \times 10^8$ Distributive Property

$= \boxed{} \times 10^8$ Subtract 2.84 from 14.5.

$= \boxed{} \times 10^9$ Rewrite in scientific notation.

So, $(1.45 \times 10^9) - (2.84 \times 10^8) = 1.166 \times 10^9$.

Check

Evaluate $(8.41 \times 10^3) + (9.71 \times 10^4)$. Express the result in scientific notation.

Show your work here

 Go Online You can complete an Extra Example online.

Pause and Reflect

How is addition of numbers written in scientific notation similar to subtraction of numbers written in scientific notation? How is it different?

Record your observations here

🌐 Apply Population

In a recent year, the world population was about
7,289,332,000. The population of the United States was
about 3×10^8. About how many times larger was the world
population than the population of the United States?

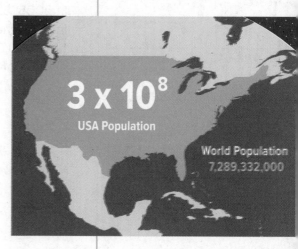

3 x 10⁸
USA Population

World Population
7,289,332,000

1 What is the task?

Make sure you understand exactly what question to answer
or problem to solve. You may want to read the problem
three times. Discuss these questions with a partner.

First Time Describe the context of the problem, in your
own words.
Second Time What mathematics do you see in the problem?
Third Time What are you wondering about?

2 How can you approach the task? What strategies can you use?

Record your
observations
here

3 What is your solution?

Use your strategy to solve the problem.

Show
your work
here

💬 Talk About It!
Why might it be helpful
to round the world
population?

4 How can you show your solution is reasonable?

✏️ **Write About It!** Write an argument that can be used to defend
your solution.

Check

In 2005, 8.1×10^{10} text messages were sent in the United States. In 2010, the number of annual text messages had risen to 1,810,000,000,000. About how many times as great was the number of text messages in 2010 than 2005?

Go Online You can complete an Extra Example online.

Pause and Reflect

Have you ever wondered when you might use the concepts you learn in math class? What are some everyday scenarios in which you might use what you learned today?

Practice

Go Online You can complete your homework online.

1. There are about 3×10^{11} stars in our galaxy and about 100 billion galaxies in the observable universe. Suppose every galaxy has as many stars as ours. How many stars are in the observable universe? Write in scientific notation. **(Example 1)**

2. Humpback whales are known to weigh as much as 80,000 pounds. The tiny krill they eat weigh only 2.1875×10^{-3} pound. About how many times greater is the weight of a humpback whale? **(Example 2)**

Evaluate. Express each result in scientific notation. (Example 3)

3. $(1.28 \times 10^5) + (1.13 \times 10^3) =$

_____

4. $(7.26 \times 10^6) - (1.3 \times 10^4) =$

_____

Test Practice

5. The speed of light is about 1.86×10^5 miles per second. The star Sirius is about 5.062×10^{13} miles from Earth. About how many seconds does it take light to travel from Sirius to Earth? Write in scientific notation, rounded to the nearest hundredth.

6. **Table Item** The table shows the amount of money raised by each region. The four regions raised a total of $\$(5.38 \times 10^4)$. How much did the West raise?

Region	Amount Raised ($)
East	1.46×10^4
North	2.38×10^4
South	6.75×10^3
West	

Apply

7. A bacterium was found to have a mass of 2×10^{-12} gram. After 30 hours, one bacterium was replaced by a population of 480,000,000 bacteria. What is the mass of the population of bacteria after 30 hours? Write your answer in scientific notation.

8. At the beginning of the business day, a bank's vault held $575,900. By the end of the day, (3.5×10^3) had been added to the vault. How much money did the bank's vault hold at the end of the business day? Write your answer in standard form.

9. Explain how the Product of Powers and Quotient of Powers Properties help you to multiply and divide numbers in scientific notation.

10. MP **Persevere with Problems** An Olympic-sized swimming pool holds 6.6043×10^5 gallons of water. A standard garden hose can deliver 9 gallons of water per minute. If the garden hose was filling the pool 24 hours per day, how many days would it take to fill the Olympic-sized pool? Explain your reasoning.

11. MP **Find the Error** Miguel found the quotient $\frac{5.78 \times 10^5}{0.000002}$ as 0.289. Find his mistake and correct it.

12. MP **Be Precise** A *googol* is the number 1 followed by 100 zeros. Earth's mass is 5.972×10^{24} kilograms. How many Earths are needed to have a total mass of 1 googol kilograms?

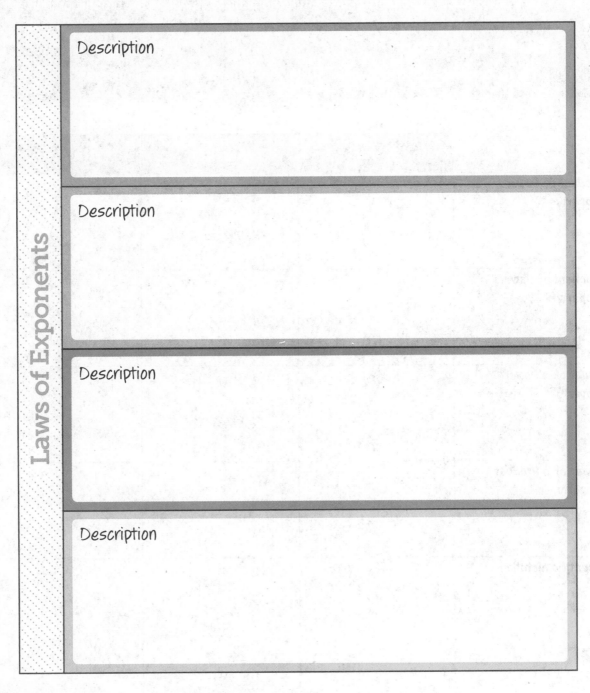

Foldables Use your Foldable to help review the module.

Laws of Exponents

Description

Description

Description

Description

Rate Yourself!

Complete the chart at the beginning of the module by placing a checkmark in each row that corresponds with how much you know about each topic after completing this module.

Reflect on the Module

Use what you learned about exponents and scientific notation to complete the graphic organizer.

e Essential Question

Why are exponents useful when working with very large or very small numbers?

	Words	Algebra	Numbers
Product of Powers Property			
Quotient of Powers Property			
Power of a Power Property			
Power of a Product Property			
Scientific Notation			

Test Practice

1. Multiple Choice Which of the following represents the expression $a \cdot b \cdot a \cdot a \cdot b \cdot a \cdot b \cdot b \cdot a \cdot b \cdot b \cdot b$ using exponents? **(Lesson 1)**

Ⓐ $a^7 \cdot b^5$

Ⓑ $a^5 \cdot b^7$

Ⓒ $a^5 \cdot b^5$

Ⓓ $a^7 \cdot b^7$

2. Equation Editor The table shows the number of admission tickets to an amusement park sold over the last 3 days. How many more tickets were sold on Sunday than on Friday?
(Lesson 1)

Day	Tickets Sold
Friday	$2^4 \cdot 3^2 \cdot 5^3$
Saturday	$2^2 \cdot 3^4 \cdot 7 \cdot 13$
Sunday	$3^6 \cdot 5 \cdot 7$

3. Open Response Louis is comparing the populations of several nearby cities. The population of Liberty Crossing can be written as 4^9 people, and the population of Harrisburg can be written as 4^7 people. The population of Glenview is 4^3 times the population of Harrisburg. Write the names of the cities from the city with the least population to the city with the greatest population. **(Lesson 2)**

4. Open Response Simplify the expression $(-3m^2)(-5m^7)$. Explain your process and justify each step by naming the appropriate property. **(Lesson 2)**

5. Table Item Indicate whether each statement is *true* or *false*. **(Lesson 3)**

	true	false
$(d^4)^3 = d^{12}$		
$(n^5)^{10} = n^{15}$		
$(k^7)^3 = k^{21}$		

6. Multiselect A square room has side lengths that can be represented by the expression $6x^3$ feet. Kyra wants to cover the floor in square tiles with side lengths that can be represented by the expression $3x$ feet.
(Lesson 3)

A. Which of the following statements are correct? Select all that apply.

☐ The area of one tile is $3x^2$ ft².

☐ The area of one tile is $9x^2$ ft².

☐ The area of the floor is $36x^3$ ft².

☐ The area of the floor is $18x^4$ ft².

☐ The area of the floor is $36x^6$ ft².

B. If $x = 3$, then what number of tiles will be needed to cover the floor?

7. Open Response Express the fraction $\frac{1}{6^4}$ using a negative exponent. (**Lesson 4**)

8. Table Item Place an X in the correct column to indicate if each statement is *sometimes*, *always*, or *never* true. (**Lesson 4**)

	sometimes	always	never
$4^0 = 1$			
$n^0 = 1$			
$z^0 = 1, z \neq 0$			

9. Multiple Choice Which expression represents $w^{-7} \cdot w^3$ written in simplest form? (**Lesson 4**)

(A) $\frac{1}{w^4}$

(C) $\frac{1}{w^{-4}}$

(B) w^{-4}

(D) w^4

10. Equation Editor Write 33,800,000 in scientific notation. (**Lesson 5**)

11. Open Response Write 7.2×10^{-3} in standard form. (**Lesson 5**)

12. Multiple Choice Which of the following is most appropriate to describe the average distance from Earth to the Sun? (**Lesson 5**)

(A) 1.5×10^{14} mm

(B) 1.5×10^{13} cm

(C) 1.5×10^{11} m

(D) 1.5×10^{8} km

13. Open Response Evaluate the expression $(7.2 \times 10^7) + (5.5 \times 10^6)$. Express the result in scientific notation. (**Lesson 6**)

14. Equation Editor Ryan downloaded a video game to his computer. The size of the file was 2.4×10^5 kilobytes. After the game was installed, he downloaded a patch file for the game that was 48,000 kilobytes in size. (**Lesson 6**)

A. What is the size of the game in standard form?

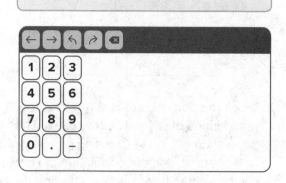

B. How many times larger was the game than the patch file?

GAS

Real Numbers

e Essential Question

Why do we classify numbers?

What Will You Learn?

Place a checkmark (✓) in each row that corresponds with how much you already know about each topic **before** starting this module.

KEY			Before			After		
▢ — I don't know. ◊ — I've heard of it. ★ — I know it!			▢	◊	★	▢	◊	★
finding square roots and cube roots								
identifying real numbers								
describing sets of real numbers								
estimating irrational numbers								
comparing and ordering real numbers								
graphing real numbers on a number line								

📖 Foldables Cut out the Foldable and tape it to the Module Review at the end of the module. You can use the Foldable throughout the module as you learn about real numbers.

What Vocabulary Will You Learn?

Check the box next to each vocabulary term that you may already know.

☐ counterexample

☐ cube root

☐ inverse operations

☐ irrational number

☐ natural numbers

☐ perfect cube

☐ perfect square

☐ principal square root

☐ radical sign

☐ real number

☐ square root

☐ truncating

Are You Ready?

Study the Quick Review to see if you are ready to start this module.
Then complete the Quick Check.

Quick Review

Example 1
Classify numbers.

Which numbers in the following list are natural numbers?

$$5, -7, \frac{1}{2}, -\frac{2}{3}, 0, 2$$

A natural number is a counting number, such as 1, 2, 3, ...

So, 5 and 2 are natural numbers.

Example 2
Compare rational numbers.

Fill in the blank with $<$, $>$, or $=$ to make $\frac{3}{5}$ _____ 0.6666... a true statement.

Since $\frac{3}{5} = 0.6$, and 0.6 is less than 0.6666..., $\frac{3}{5} < 0.6666...$.

Quick Check

1. The temperature fell 6°F, which can be represented by −6. Is −6 a natural number or integer?

2. Fill in the blank with $<$, $>$, or $=$ to make $-2\frac{1}{3}$ _____ −2.5 a true statement.

How Did You Do?
Which exercises did you answer correctly in the Quick Check?
Shade those exercise numbers at the right.

Roots

I Can... find square and cube roots, and use square and cube roots to solve equations involving perfect squares and cubes.

Explore Find Square Roots Using a Square Model

Online Activity You will use Web Sketchpad to explore how to use square models to find square roots.

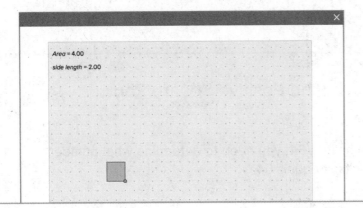

Area = 4.00

side length = 2.00

Learn Square Roots

Words
A **square root** of a number is one of its two equal factors.
Symbols
If $x^2 = y$, then x is the square root of y.
Example
$5^2 = 25$, so 5 is a square root of 25.

A **perfect square** is a rational number whose square root is a whole number. Complete the table for the following perfect squares.

Perfect Square	1	4	9	16	25	36	49	64
Square Root	1	2	3	4				

What Vocabulary Will You Learn?

cube root

inverse operations

perfect cube

perfect square

principal square root

radical sign

square root

Talk About It!

Why do you think the term *perfect square* uses the adjective *perfect*? What are some numbers that aren't perfect squares? Explain why.

(continued on next page)

💬 **Talk About It!**

In how many different ways can you represent and describe the square root of 16?

🧁 **Think About It!**

How do you know this is a positive square root?

Every positive number has *both* a positive and negative square root.

Because 3 • 3 = 9, _____ is a square root of 9.

Because (–3) • (–3) = 9, _____ is a square root of 9.

Therefore, 9 has two square roots, 3 and –3.

In most real-world situations, only the positive or **principal square root** is considered. A **radical sign**, $\sqrt{}$, is used to indicate the principal square root. When both the positive and negative square roots are asked for, the ± symbol is used before the radical sign.

$$\sqrt{25} = \boxed{} \qquad -\sqrt{25} = \boxed{} \qquad \pm\sqrt{25} = \boxed{}$$

Example 1 Find Positive Square Roots

Simplify $\sqrt{64}$.

In order to simplify $\sqrt{64}$, you need to determine what number, multiplied by itself equals 64.

Find the factors of 64. _____

Find the square root.

$\sqrt{64} = \boxed{}$ Find the positive square root of 64; $8^2 = 64$.

So, $\sqrt{64} = 8$.

Check

Simplify $\sqrt{225}$.

Go Online You can complete an Extra Example online.

Example 2 Find Both Square Roots

Simplify $\pm\sqrt{1.21}$.

Step 1 Since there is an even number of decimal places, consider the square root of 121. Determine what number, multiplied by itself, equals 121.

$$\sqrt{121} = \boxed{}$$

Step 2 Determine where to place the decimal point in 11. There are only two options for placing the decimal point in the number 11, either before each digit (0.11) or between the digits (1.1). Since 1.21 has two decimal places, then the sum of the number of decimal places in each factor must equal 2.

If you multiply 0.11 • 0.11, then the product will have _____ decimal places.

If you multiply 1.1 • 1.1, then the product will have _____ decimal places.

So, $\pm\sqrt{1.21} = \pm 1.1$.

Check

Simplify $\pm\sqrt{1.44}$.

Go Online You can complete an Extra Example online.

Example 3 Find Negative Square Roots

Simplify $-\sqrt{\dfrac{25}{36}}$.

Find the square root of $\dfrac{25}{36}$. Then add a negative sign.

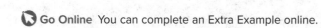

$-\sqrt{\dfrac{25}{36}} = -\dfrac{\boxed{}}{\boxed{}}$ Find the negative square root of $\dfrac{25}{36}$; $\left(\dfrac{5}{6}\right)^2 = \dfrac{25}{36}$.

So, $-\sqrt{\dfrac{25}{36}} = -\dfrac{5}{6}$.

Think About It!

What does the symbol $\pm$ before the radical sign indicate?

Check

Simplify $-\sqrt{\dfrac{49}{64}}$.

Show
your work
here

Example 4 Square Roots of Negative Numbers

Simplify $\sqrt{-16}$ using rational numbers. If the expression cannot be simplified, explain why.

In order to simplify $\sqrt{-16}$, you need to determine what number, multiplied by itself equals -16.

There is no rational number square root of -16 because

_____ times itself is equal to -16.

So, $\sqrt{-16}$ cannot be simplified and has no rational number solution.

Check

Simplify $\sqrt{-81}$ using rational numbers. If the expression cannot be simplified, explain why.

Show
your work
here

Think About It!

Can you think of a number when multiplied by itself, equals -16? Why or why not?

Talk About It!

What is the difference between $\sqrt{-16}$ and $-\sqrt{16}$?

 Go Online You can complete an Extra Example online.

Learn Use Square Roots to Solve Equations

You can solve equations by using inverse operations. **Inverse operations** undo each other. Squaring and taking a square root are inverse operations.

Square a Number

$9^2 = \boxed{}$

Take the Square Root

$\sqrt{81} = \boxed{}$

To solve an equation of the form $x^2 = p$ for x, undo the operations of squaring x by taking the square root of each side.

$x^2 = p$ Write the equation.

$\pm\sqrt{x^2} = \pm\sqrt{p}$ Take the square root of each side.

$x = \pm p$ Simplify.

There will be two solutions, a positive square root and a negative square root.

Talk About It!

Consider the equation $x^2 = 121$.

- If x is a positive number, what is x^2?
- If x is a negative number, what is x^2?
- How many solutions does the equation have? What are they?

Example 5 Use Square Roots to Solve Equations

Solve $t^2 = 169$. Check your solution.

$t^2 = 169$ Write the equation.

$\pm\sqrt{t^2} = \pm\sqrt{169}$ Take the square root of each side.

$t = \pm\boxed{}$ Definition of square root

$t = 13$ and -13 Simplify.

So, the solutions to the equation are $t = $ _____ and _____.

Check

Solve $y^2 = 256$.

Talk About It!

Why do we take the square root of each side of the equation? Why does this equation have two solutions?

 Go Online You can complete an Extra Example online.

Learn Cube Roots

Words
A **cube root** of a number is one of its three equal factors.
Symbols
If $x^3 = y$, then $x = \sqrt[3]{y}$.
Numbers
Since $2^3 = 8$, 2 is the cube root of 8.
Since $(-6)^3 = -216$, -6 is the cube root of -216.

Every integer has exactly one cube root. Complete the table that demonstrates this concept.

	Rule	Example
Cube Root of a Positive Number	The cube root of a positive number is positive.	$\sqrt[3]{27} = \boxed{}$ $\sqrt[3]{125} = \boxed{}$
Cube Root of Zero	The cube root of zero is zero.	$\sqrt[3]{0} = \boxed{}$
Cube Root of a Negative Number	The cube root of a negative number is negative.	$\sqrt[3]{-27} = \boxed{}$ $\sqrt[3]{-125} = \boxed{}$

A **perfect cube** is a number that is the cube of an integer.

Complete the table for the following perfect cubes.

Perfect Cube	1	−1	8	−8	27	−27	64	−64
Cube Root	1	−1	2	−2				

Pause and Reflect

Compare and contrast square roots and cube roots.

Record your observations here

Example 6 Cube Roots of Positive Numbers

Simplify $\sqrt[3]{125}$.

In order to simplify $\sqrt[3]{125}$, you need to determine what number, multiplied 3 times, is equal to 125.

$\sqrt[3]{125} = \boxed{}$ $5^3 = 5 \cdot 5 \cdot 5$ or 125

So, $\sqrt[3]{125} = 5$.

Check

Simplify $\sqrt[3]{216}$.

Example 7 Cube Roots of Negative Numbers

Simplify $\sqrt[3]{-27}$.

In order to simplify $\sqrt[3]{-27}$ you need to determine what number, multiplied 3 times, is equal to −27.

$\sqrt[3]{-27} = \boxed{}$ $(-3)^3 = (-3) \cdot (-3) \cdot (-3)$ or −27

So, $\sqrt[3]{-27} = -3$.

Check

Simplify $\sqrt[3]{-1,000}$.

😮 Think About It!

Will the answer be a positive or negative number?

💬 Talk About It!

What is the difference between the cube root of a negative number and the square root of a negative number?

⬤ Go Online You can complete an Extra Example online.

🌐 **Example 8** Use Cube Roots to Solve Equations

Dylan has a planter in the shape of a cube that holds 15.625, or $\frac{125}{8}$, cubic feet of potting soil.

Solve the equation $s^3 = \frac{125}{8}$ to find the side length s of the container. Check your solution.

To solve an equation of the form $x^3 = p$, take the cube root of each side of the equation.

$s^3 = \dfrac{125}{8}$ Write the equation.

$\sqrt[3]{s} = \sqrt[3]{\dfrac{125}{8}}$ Take the cube root of each side.

$s = \dfrac{5}{2}$ or $\boxed{}$ Definition of cube root

So, each side of the container is $2\frac{1}{2}$ feet.

Check the solution.

$s^3 = \dfrac{125}{8}$ Write the equation.

$\left(\dfrac{5}{2}\right)^3 \stackrel{?}{=} \dfrac{125}{8}$ Replace s with $\frac{5}{2}$.

$\dfrac{125}{8} = \dfrac{125}{8}$ Simplify. The solution, $\frac{5}{2}$, or $2\frac{1}{2}$, is correct.

Check

A box that is shaped like a cube has a volume of $\frac{512}{27}$ cubic inches.

Solve $s^3 = \frac{512}{27}$ to find the length s of one side of the box.

Show your work here

🌐 **Go Online** You can complete an Extra Example online.

🌐 **Apply** Bulletin Boards

A bulletin board consists of four equal-sized cork squares arranged in a row to form a rectangle. If the total area of all four cork squares is 36 square feet, what is the length in feet of the bulletin board?

1 What is the task?

Make sure you understand exactly what question to answer or problem to solve. You may want to read the problem three times. Discuss these questions with a partner.

First Time Describe the context of the problem, in your own words.
Second Time What mathematics do you see in the problem?
Third Time What are you wondering about?

2 How can you approach the task? What strategies can you use?

Record your observations here

3 What is your solution?

Use your strategy to solve the problem.

Show your work here

4 How can you show your solution is reasonable?

 Write About It! Write an argument that can be used to defend your solution.

💬 Talk About It!

How would the length and width of the bulletin board change if the four cork squares were arranged in a square? How would the area be affected?

Check

A set of windows consists of three equal-sized squares arranged in a row to form a rectangle. If the total area of all three windows is 108 square feet, what is the length, in feet, of the windows?

Show your work here

Go Online You can complete an Extra Example online.

Pause and Reflect

Create a graphic organizer that will help you study the vocabulary and concepts from this lesson.

Record your observations here

Practice

🔵 **Go Online** You can complete your homework online.

Simplify using rational numbers. If the expression cannot be simplified, explain why. (Examples 1–4)

1. $\sqrt{361} =$ _____

2. $\pm\sqrt{1.96} =$ _____

3. $-\sqrt{\dfrac{9}{16}} =$ _____

4. $\sqrt{-441} =$ _____

5. Solve $m^2 = 0.04$. (Example 5) _____

Simplify using rational numbers. (Examples 6 and 7)

6. $\sqrt[3]{343} =$ _____

7. $\sqrt[3]{-512} =$ _____

8. A basin of a water fountain is cube shaped and has a volume of 91.125 cubic feet. Solve $s^3 = 91.125$ to find the length s of one side of the basin. (Example 8)

Test Practice

9. Moesha has 196 pepper plants that she wants to plant in a square formation. How many pepper plants should she plant in each row?

10. **Equation Editor** What is the value of p in the equation shown?

$p^3 = -0.027$

Apply

11. A cement path consists of six equal-sized cement squares arranged in a row to form a rectangle. If the total area of the path is 96 square feet, what is the length in feet of the path?

12. A photo collage consists of seven equal-sized square photos arranged in a row to form a rectangle. If the total area of the collage is 567 square inches, what is the length of the collage?

13. (MP) **Reason Inductively** Explain why $\sqrt[3]{8}$ is a rational number, but $\sqrt{8}$ is not a rational number.

14. Give an example of when the decimal equivalent of a square root would be rounded to an approximate value. Explain why it is appropriate to round.

15. Write a number that completes the analogy:

x^2 is to 441 as x^3 is to _____.

16. (MP) **Identify Repeated Reasoning** Simplify each expression. Then write a rule for the pattern.

a. $(\sqrt{81})^2 =$ _____

b. $\left(\sqrt{\dfrac{9}{16}}\right)^2 =$ _____

c. $(\sqrt{0.04})^2 =$ _____

d. $(\sqrt{t})^2 =$ _____

Real Numbers

I Can... identify irrational numbers and name the set(s) of real numbers to which a given real number belongs.

What Vocabulary Will You Learn?

counterexample

irrational number

natural numbers

real numbers

Explore Real Numbers

Online Activity You will use number lines to explore the set of real numbers.

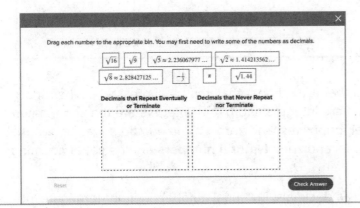

Learn Real Numbers

Real numbers are numbers that can be found on the number line.

Go Online Watch the animation to plot the numbers -4.5, -1, $\frac{3}{4}$, π, and $\sqrt{21}$ on the number line.

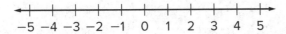

Real numbers are either rational, with a decimal expansion that terminates or repeats, or irrational. An **irrational number** is a number that cannot be expressed as the ratio $\frac{a}{b}$, where a and b are integers and $b \neq 0$. Irrational numbers have decimal expansions that are non-terminating and non-repeating.

The square root of any number that is not a perfect square is irrational.

$\sqrt{3} \approx 1.732050808...$ $\sqrt{5} \approx 2.2360679775...$

(continued on next page)

Copyright © McGraw-Hill Education

Write the following numbers in the Venn diagram.

$-8, -3, -\frac{3}{4}, 0, 18\%, 0.\overline{4}, 0.45, \frac{1}{2}, \pi, 1.21231234..., \sqrt{3}, 4, 9, 10$

Real Numbers

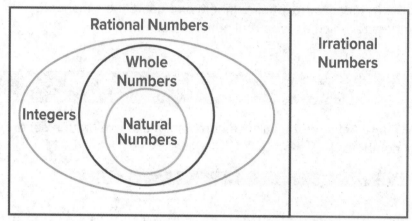

A *rational number* is a number that can be expressed as the ratio $\frac{a}{b}$, where *a* and *b* are integers and $b \neq 0$. *Integers* are the set of natural numbers, their opposites, and zero. *Whole numbers* are the set of natural numbers and zero. **Natural numbers** are the set of counting numbers.

When written as decimals, irrational numbers neither terminate, nor repeat eventually.

Example 1 Identify Real Numbers

Determine whether −25 is *rational* or *irrational*.

Can −25 be expressed as a ratio in the form $\frac{a}{b}$?

  Yes   No

When written in the form $\frac{a}{b}$, are *a* and *b* integers and $b \neq 0$?

  Yes   No

A rational number is a number that can be expressed as the ratio of two integers.

Since $-25 = \dfrac{\boxed{}}{\boxed{}}$, −25 is a rational number.

Talk About It!

What sets of numbers do the rational numbers include? Why is $\sqrt{2}$ an irrational number? π?

Talk About It!

The number −25 belongs to what other set(s) of numbers?

Check

Determine whether $-\frac{2}{9}$ is *rational* or *irrational*.

Example 2 Classify Real Numbers

Name all the sets of numbers to which the real number 0.2525... belongs.

The set of real numbers include natural numbers, whole numbers, integers, rational numbers, and irrational numbers.

Can 0.2525... be expressed as a ratio in the form $\frac{a}{b}$, where a and b are integers and $b \neq 0$?

     Yes   No

Is 0.2525... from the set of integers {..., −3, −2, −1, 0, 1, 2, 3, ...}?

     Yes   No

So, 0.2525... is a rational number because it is equivalent to .

Talk About It!
What is another way to write 0.2525...?

Check

Name all the sets of numbers to which the real number $\frac{21}{\sqrt{4}}$ belongs.

Go Online You can complete an Extra Example online.

Think About It!

Can the square root be simplified?

Example 3 Classify Real Numbers

Name all the sets of numbers to which the real number $\sqrt{36}$ belongs.

Can $\sqrt{36}$ be expressed as a ratio in the form $\frac{a}{b}$, where a and b are integers and $b \neq 0$?

     Yes   No

Is $\sqrt{36}$ from the set of integers {..., −3, −2, −1, 0, 1, 2, 3, ...}?

     Yes   No

Is $\sqrt{36}$ from the set of whole numbers {0, 1, 2, 3, ...}?

     Yes   No

Is $\sqrt{36}$ from the set of natural numbers {1, 2, 3, ...}?

     Yes   No

Since $\sqrt{36} = $ _____, it is a natural number, a whole number, an integer, and a rational number.

Check

Name all the sets of numbers to which the real number $-\sqrt{64}$ belongs.

Go Online You can complete an Extra Example online.

Example 4 Classify Real Numbers

**Name all the sets of numbers to which the real number
$-\sqrt{7}$ belongs.**

$-\sqrt{7} \approx -2.645751311...$

Does the decimal terminate? repeat eventually?

> Yes   No

Can $-\sqrt{7} \approx -2.645751311...$ be expressed as a ratio in the form $\frac{a}{b}$, where a and b are integers and $b \neq 0$?

> Yes   No

The decimal value of $-\sqrt{7}$ neither terminates nor repeats eventually, so it is an irrational number.

Check

Name all the sets of numbers to which the real number π belongs.

Pause and Reflect

Use your own words to describe how to name all the sets of numbers to which a real number belongs.

🔵 **Go Online** You can complete an Extra Example online.

💬 **Talk About It!**
How can you tell, just by studying the expression, that $-\sqrt{7}$ is irrational?

Learn Describe Sets of Real Numbers

Some sets of numbers are subsets of other sets of numbers. For example, rational numbers and irrational numbers are subsets of real numbers.

A Venn diagram can be used to describe the relationship between sets of real numbers.

Real Numbers

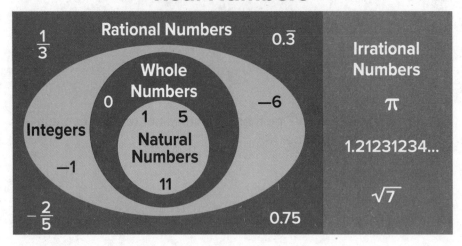

Go Online Watch the animation, or use the Venn diagram, to complete the following sentences.

_____ numbers are a subset of whole numbers.

_____ numbers are subsets of integers.

Integers are a subset of _____ numbers.

Rational numbers and irrational numbers are subsets of _____ numbers.

If a given statement about real numbers is false, you can provide a **counterexample**, which is a statement or example that shows a conjecture is false.

Pause and Reflect

Describe the decimal form of irrational numbers. How are they represented in the Venn diagram?

💬 **Talk About It!**

Natural numbers are a subset of whole numbers. What other subsets of numbers are shown in the Venn diagram?

Example 5 Describe Sets of Real Numbers

Use the Venn diagram to determine whether the statement is *true* or *false*. If the statement is *true*, explain your reasoning. If the statement is *false*, provide a counterexample.

All rational numbers are integers.

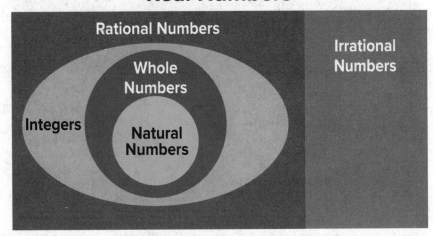

Part A Determine whether the statement is *true* or *false*.
Integers are a subset of _____ numbers. So, the statement is false.

Part B Provide a counterexample.
One possible counterexample is 0.6. The decimal 0.6 is a rational number, but not an integer.

Check

Use the Venn diagram above to determine whether the statement is *true* or *false*. If the statement is *true*, explain your reasoning. If the statement is *false*, provide a counterexample.

All whole numbers are natural numbers.

Part A

Determine whether the statement is *true* or *false*.

Part B

If the statement is *true*, explain your reasoning. If the statement is *false*, provide a counterexample.

(Show your work here)

Go Online You can complete an Extra Example online.

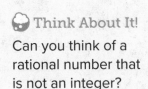
Think About It!
Can you think of a rational number that is not an integer?

Talk About It!
What are some other possible counterexamples for the statement *all rational numbers are integers*?

Example 6 Describe Sets of Real Numbers

Use the Venn diagram to determine whether the statement is *true* or *false*. If the statement is *true*, explain your reasoning. If the statement is *false*, provide a counterexample.

All irrational numbers are real numbers.

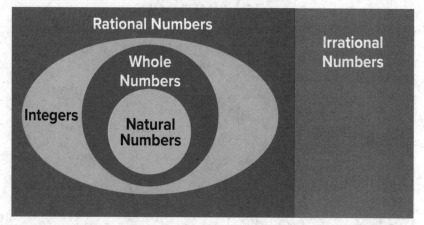

Real Numbers

By definition, irrational numbers are a subset of _____ numbers. So, irrational numbers are real numbers. So, the statement is true.

Check

Use the Venn diagram above to determine whether the statement is *true* or *false*.

All natural numbers are whole numbers.

Part A
Determine whether the statement is *true* or *false*.

Part B
If the statement is *true*, explain your reasoning. If the statement is *false*, provide a counterexample.

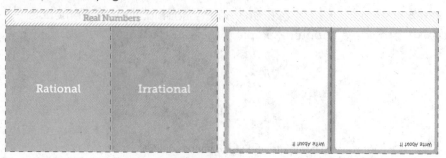
Foldables It's time to update your Foldable, located in the Module Review, based on what you learned in this lesson. If you haven't already assembled your Foldable, you can find the instructions on page FL1.

Practice

Go Online You can complete your homework online.

Identify whether each number is *rational* or *irrational*. (Example 1)

1. $-\sqrt{10}$

2. $-\dfrac{3}{11}$

3. $0.\overline{3}$

4. $\sqrt{81}$

5. 0

6. $-\dfrac{\sqrt{2}}{2}$

7. $\sqrt{7}$

8. $\dfrac{\sqrt{2}}{\sqrt{2}}$

Select all the sets of numbers to which each real number belongs. (Examples 2–4)

9. $\sqrt[3]{343}$

Ⓐ Rational
Ⓑ Irrational
Ⓒ Integer
Ⓓ Whole
Ⓔ Natural

10. $\dfrac{7}{\sqrt{2}}$

Ⓐ Rational
Ⓑ Irrational
Ⓒ Integer
Ⓓ Whole
Ⓔ Natural

11. $-\dfrac{7}{1}$

Ⓐ Rational
Ⓑ Irrational
Ⓒ Integer
Ⓓ Whole
Ⓔ Natural

Determine whether each statement is *true* or *false*. If the statement is *true*, explain your reasoning. If the statement is *false*, provide a counterexample. (Examples 5 and 6)

12. A number cannot be irrational and an integer.

13. All integers are rational.

14. **Multiselect** Select the numbers that are part of the set of rational numbers.

☐ $-\dfrac{11}{\sqrt{9}}$

☐ $\dfrac{1}{\sqrt{2}}$

☐ $\sqrt{-16}$

☐ $\dfrac{\sqrt[3]{-4,096}}{\sqrt{16}}$

☐ $\sqrt[3]{4}$

☐ $0.333\ldots$

15. **Use Math Tools** Explain how you could use a calculator to determine if $\sqrt{8}$ expressed as a decimal ever terminates.

 Justify Conclusions Determine whether each statement is *true* or *false*. If the statement is false, give a counterexample or explain your reasoning.

16. The product of a non-zero rational number and an irrational number is rational.

17. Expressing $\sqrt{2}$ as the ratio $\dfrac{\sqrt{2}}{1}$ means that $\sqrt{2}$ is a rational number.

18. The product of two irrational numbers is irrational.

Estimate Irrational Numbers

I Can... estimate irrational numbers by approximating their locations on a number line or by truncating their decimal expansions.

What Vocabulary Will You Learn?
truncating

Explore Roots of Non-Perfect Squares

Online Activity You will use Web Sketchpad to explore how to find the square root of a non-perfect square.

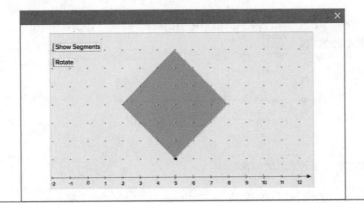

Learn Estimate Irrational Numbers Using a Number Line

The decimal expansion of an irrational number never repeats nor terminates. To write the decimal expansion of an irrational number, you can denote an approximation using the ≈ symbol. Use more place values to give more accurate approximations.

Talk About It!
How do you know that $\sqrt{8}$ is between 2 and 3? How do you know that $\sqrt{8}$ is closer to 3 than 2?

Go Online Watch the animation to see how to estimate $\sqrt{8}$ using a number line using the following steps:

Step 1 Since 8 is between two perfect squares, 4 and 9, $\sqrt{4} < \sqrt{8} < \sqrt{9}$.

Step 2 Graph $\sqrt{4} = 2$ and $\sqrt{9} = 3$ on the number line.

Step 3 Since 8 is closer to 9 than to 4, graph $\sqrt{8}$ closer to $\sqrt{9}$ than to $\sqrt{4}$.

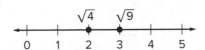

Think About It!

How would you begin estimating the square root? What are some perfect squares that are close to 83?

Example 1 Estimate Square Roots to the Nearest Integer

Estimate $\sqrt{83}$ to the nearest integer.

Step 1 Find two perfect squares between which 83 lies. Find their square roots.

The greatest perfect square less than 83 is _____, and $\sqrt{81} =$ _____.

The least perfect square greater than 83 is _____, and $\sqrt{100} =$ _____.

Step 2 Plot $\sqrt{81}$, $\sqrt{83}$, and $\sqrt{100}$ on the number line. Approximate the location of $\sqrt{83}$.

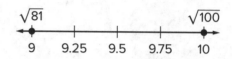

$\sqrt{81}$ $\sqrt{100}$

9   9.25   9.5   9.75   10

Step 3 Estimate the square root.

$81 < 83 < 100$ Write an inequality.

$9^2 < 83 < 10^2$ $81 = 9^2$ and $100 = 10^2$

$\sqrt{9^2} < \sqrt{83} < \sqrt{10^2}$ Find the square root of each number.

$\boxed{} < \sqrt{83} < \boxed{}$ Simplify.

So, $\sqrt{83}$ is between 9 and 10. Since $\sqrt{83}$ is closer to _____ on a number line, the best integer estimate for $\sqrt{83}$ is 9.

Check

Estimate $\sqrt{135}$ to the nearest integer.

Show your work here

Go Online You can complete an Extra Example online.

Example 2 Estimate Square Roots to the Nearest Tenth –

Estimate $\sqrt{83}$ to the nearest tenth.

Step 1 Determine the integer closest to $\sqrt{83}$.

$\sqrt{83}$ is much closer to $\sqrt{81}$, or _____, than it is to $\sqrt{100}$, or

_____.

Step 2 Test intervals close to 9. Start with the interval between 9 and 9.1.

$9 < \sqrt{83} < 9.1$ Write an inequality.

$9^2 < (\sqrt{83})^2 < 9.1^2$ Square the values.

$\boxed{} \overset{?}{<} 83 \overset{?}{<} \boxed{}$ Simplify. Is the inequality true?

The inequality is not true because 83 is not between 81 and 82.81.

Step 3 Test the next interval, 9.1 to 9.2.

$9.1 < \sqrt{83} < 9.2$ Write an inequality.

$9.1^2 < (\sqrt{83})^2 < 9.2^2$ Square the values.

$\boxed{} \overset{?}{<} 83 \overset{?}{<} \boxed{}$ Simplify. Is the inequality true?

The inequality is true because 83 is between 82.81 and 84.64.

So, $\sqrt{83}$ is between 9.1 and 9.2. Since 83 is closer to _____ than it is to 84.64, $\sqrt{83} \approx 9.1$.

Check

Estimate $\sqrt{106}$ to the nearest tenth.

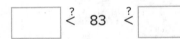 **Go Online** You can complete an Extra Example online.

 Think About It!

Between what two integers does $\sqrt{83}$ lie on a number line?

Talk About It!

How can you use a graph to verify that $\sqrt{83}$ is closer to 9.1 than 9.2? How could you continue on to get a better approximation of $\sqrt{83}$?

Example 3 Estimate Cube Roots to the Nearest Integer

Estimate $\sqrt[3]{320}$ to the nearest integer.

Step 1 Find two perfect cubes between which 320 lies. Find their cube roots.

The greatest perfect cube less than 320 is _____, and $\sqrt[3]{216} = $ _____.

The least perfect cube greater than 320 is _____, and $\sqrt[3]{343} = $ _____.

Step 2 Plot $\sqrt[3]{216}$, $\sqrt[3]{320}$, and $\sqrt[3]{343}$ on the number line.

Approximate the location of $\sqrt[3]{320}$.

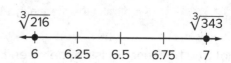

Step 3 Estimate the cube root.

$216 < \quad 320 \quad < 343$ Write an inequality.

$6^3 < \quad 320 \quad < 7^3$ $216 = 6^3$ and $343 = 7^3$

$\sqrt[3]{6^3} < \sqrt[3]{320} < \sqrt[3]{7^3}$ Find the cube root of each number.

$\boxed{} < \sqrt[3]{320} < \boxed{}$ Simplify.

So, $\sqrt[3]{320}$ is between 6 and 7. Since 320 is closer to _____, the best integer estimate for $\sqrt[3]{320}$ is 7.

Check

Estimate $\sqrt[3]{51}$ to the nearest integer.

Go Online You can complete an Extra Example online.

Learn Estimate Irrational Numbers by Truncating

You can estimate irrational numbers using a calculator by truncating the decimal expansion. **Truncating** is a process of approximating a decimal number by eliminating all decimal places past a certain point without rounding.

Truncate $\sqrt{12} \approx 3.464101615\ldots$ to the specified decimal places.

Tenths

$\sqrt{12} \approx 3.4\cancel{64101615}\ldots$

Truncate, or drop, the digits after the tenths place.

Hundredths

$\sqrt{12} \approx$ _____

Truncate, or drop, the digits after the hundredths place.

Thousandths

$\sqrt{12} \approx$ _____

Truncate, or drop, the digits after the thousandths place.

Talk About It!

How could you continue truncating $\sqrt{12}$ to get better approximations?

Pause and Reflect

Explain how truncating can be used to estimate the value of π.

> Record your observations here

Talk About It!

If you *rounded* $\sqrt{12}$ to the nearest tenth, what would be the approximation?

If you *truncated* $\sqrt{12}$ to the nearest tenth, what would be the approximation?

What is the difference between truncating and rounding?

Think About It!

Between which two whole numbers is $\sqrt{2}$?

Talk About It!

Why will Wyatt need more than 5.6 meters, but less than 6.0 meters of fencing?

Talk About It!

Could you truncate the decimal expansion differently? Explain how it would affect the answer.

Example 4 Estimate by Truncating

Wyatt wants to fence in a square portion of the yard to make a play area for his new puppy. The area covered is 2 square meters.

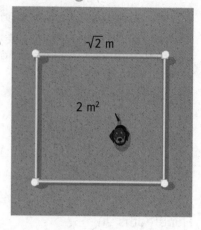

How much fencing should he buy?

The amount of fencing he will need is the _____ of the square, $4 \cdot \sqrt{2}$ meters.

Step 1 Approximate $4\sqrt{2}$ by first truncating the decimal expansion of $\sqrt{2}$ to the tenths place.

$\sqrt{2} \approx$ _____   Use a calculator.

$\sqrt{2} \approx$ _____   Truncate the digits after the tenths place.

So, $\sqrt{2}$ is between 1.4 and 1.5.

Step 2 Find the amount of fencing.

Since $\sqrt{2}$ is between 1.4 and 1.5, the perimeter, $4(\sqrt{2})$, is between $4(1.4)$ and $4(1.5)$.

$4(1.4) < 4(\sqrt{2}) < 4(1.5)$ Write the inequality.

$\boxed{} < 4(\sqrt{2}) < \boxed{}$ Multiply.

So, Wyatt will need between 5.6 and 6.0 meters of fencing. Therefore, he should buy 6 meters of fencing.

Check

Tobias dropped a tennis ball from a height of 60 feet. The time in seconds it takes for the ball to fall 60 feet is found using the expression $0.25 \cdot \sqrt{60}$. Determine the number of seconds it takes for the ball to fall 60 feet. Truncate the value of $\sqrt{60}$ to the tenths place.

Show your work here

🌐 **Go Online** You can complete an Extra Example online.

 **Apply** Golden Rectangle

The *golden rectangle* can be seen in the structure of a nautilus shell. The ratio of the longer side length to the shorter side is equal to $\dfrac{1+\sqrt{5}}{2}$. Estimate this value.

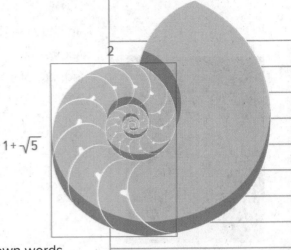

1 What is the task?

Make sure you understand exactly what question to answer or problem to solve. You may want to read the problem three times. Discuss these questions with a partner.

First Time Describe the context of the problem, in your own words.
Second Time What mathematics do you see in the problem?
Third Time What are you wondering about?

2 How can you approach the task? What strategies can you use?

3 What is your solution?

Use your strategy to solve the problem.

Talk About It!
What is another estimate for the ratio?

4 How can you show your solution is reasonable?

Write About It! Write an argument that can be used to defend your solution.

Check

In Little League, the bases are squares with sides of 14 inches. The expression $\sqrt{s^2 + s^2}$ represents the distance *diagonally* across a square of side length s. Estimate the diagonal distance across a base to the nearest inch.

Show your work here

 **Go Online** You can complete an Extra Example online.

Pause and Reflect

Compare and contrast estimating rational numbers on a number line with estimating irrational numbers on a number line.

Record your observations here

Math History Minute

Hindu mathematician **Bhāskara II (1114–1185)** is considered by many to have been the leading mathematician of the 12th century. One of his many accomplishments was to produce several approximations of the number π, including $\frac{22}{7}$, which is still used today.

Practice

🔘 **Go Online** You can complete your homework online.

Estimate each square root or cube root to the nearest integer. (Examples 1 and 3)

1. $\sqrt{125} \approx$ _____

2. $\sqrt{55} \approx$ _____

3. $\sqrt[3]{70} \approx$ _____

4. $\sqrt[3]{923} \approx$ _____

Estimate each square root to the nearest tenth. (Example 2)

5. $\sqrt{296} \approx$ _____

6. $\sqrt{5} \approx$ _____

7. $\sqrt{11} \approx$ _____

8. $\sqrt{62} \approx$ _____

9. The formula $s = \sqrt{18d}$ can be used to find the speed s of a car in miles per hour when the car needs d feet to come to a complete stop after stepping on the brakes. If it took a car 25 feet to come to a complete stop after stepping on the brakes, estimate the speed of the car. Truncate the value of $\sqrt{18d}$, when $d = 25$, to the tenths place. (Example 4)

Test Practice

10. If the area of a square is 32 square feet, estimate the length of each side of the square to the nearest whole number.

11. Equation Editor Estimate the square root to the nearest tenth.

$\sqrt{489}$

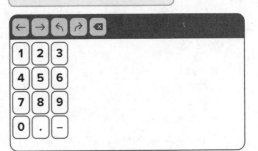

Apply

12. The formula $t = \frac{\sqrt{h}}{4}$ represents the time t in seconds that it takes an object to fall from a height of h feet. If a rock falls from 125 feet, estimate how long it will take the rock to hit the ground. Estimate the square root to the nearest integer.

13. The radius of a circle with area A can be approximated using the formula $r = \sqrt{\frac{A}{3}}$. Estimate the radius of a wrestling mat circle with an area of 452 square feet.

14. (MP) **Find the Error** A classmate estimated $\sqrt{397}$ to be about 200. Explain the mistake and correct it.

15. (MP) **Be Precise** Explain how to write the exact value for the square root of a non-perfect square. Give an example.

16. Carrie is packing her clothes in moving boxes that are in the shape of a cube. Each box has a volume of 3 cubic feet. The moving truck has shelves that are 12 inches in height. Will the moving boxes fit? Explain.

17. (MP) **Make an Argument** Explain how you could estimate $\sqrt[4]{20}$ to the nearest integer.

Compare and Order Real Numbers

I Can... use rational approximations to compare and order real numbers, including irrational numbers.

Learn Compare and Order Real Numbers

You can compare and order real numbers by writing them in the same form. One way to do this is to use or approximate the decimal expansion of each number in order to compare or order a set of numbers.

Complete the following to compare and order the set of numbers shown.

$\frac{18}{5}$, π, $\sqrt{11}$

Write each number in decimal notation.

$\frac{18}{5} = \boxed{}$ π ≈ $\boxed{}$ $\sqrt{11}$ ≈ $\boxed{}$

Compare each set of numbers using <, >, or =.

3.14 $\boxed{}$ 3.32 $\boxed{}$ 3.6

π $\boxed{}$ $\sqrt{11}$ $\boxed{}$ $\frac{18}{5}$

Graph each number on the number line.

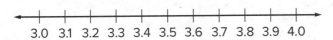

Order the set of numbers from least to greatest.

_____ , _____ , _____

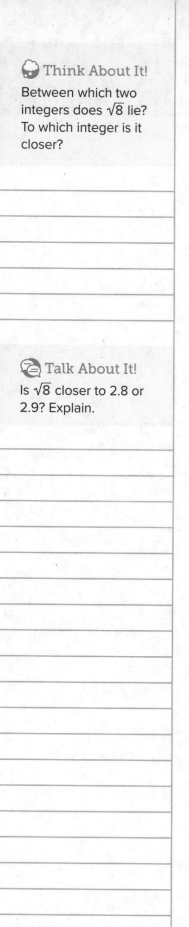

Think About It!

Between which two integers does $\sqrt{8}$ lie? To which integer is it closer?

Talk About It!

Is $\sqrt{8}$ closer to 2.8 or 2.9? Explain.

Example 1 Compare Real Numbers

Which symbol, <, >, or =, would complete the statement $\sqrt{8}$ _____ $2\frac{2}{3}$ to make a true statement? Then graph the numbers on a number line.

Part A Compare the numbers.

Approximate the decimal expansion of each number.

$\sqrt{8} \approx$ ☐   Estimate to the nearest tenth.

$2\frac{2}{3} =$ ☐   Write using bar notation.

Since 2.8 is greater than $2.\overline{6}$, $\sqrt{8} > 2\frac{2}{3}$.

Part B Graph the numbers on the number line.

Approximate the location of each number.

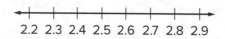

Pause and Reflect

When you first saw this Example, what was your reaction? Did you think you could solve the problem? Did what you already know help you solve the problem?

Record your observations here

Check

Which symbol, $<$, $>$, or $=$, would complete the statement

$\dfrac{\sqrt{25}}{2}$ _____ $\sqrt{6.25}$ to make a true statement? Then graph the

numbers on a number line.

Part A

Write the symbol, $<$, $>$, or $=$, that makes $\dfrac{\sqrt{25}}{2}$ _____ $\sqrt{6.25}$ a true

statement.

Part B

Which of the following is the correct graph of $\dfrac{\sqrt{25}}{2}$ and $\sqrt{6.25}$ on a

number line?

Ⓐ
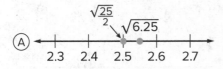

Ⓑ

Ⓒ

Ⓓ

🐦 **Go Online** You can complete an Extra Example online.

Example 2 Compare Real Numbers

Which symbol, <, >, or =, would complete the statement

$-\sqrt{6}$ _____ $-\dfrac{\pi}{2}$ **to make a true statement? Then graph the numbers on a number line.**

Part A Compare the numbers.

Approximate the decimal expansion of each number.

$-\sqrt{6} \approx$ [ ]   Estimate to the nearest tenth.

$-\dfrac{\pi}{2} \approx -\dfrac{3.14}{2}$ or [ ]   Estimate to the nearest hundredth.

Since -2.4 is less than -1.57, $-\sqrt{6} < -\dfrac{\pi}{2}$.

Part B Graph the numbers on the number line.

Approximate the location of each number. When graphing numbers on the number line, greater numbers are graphed farther to the right.

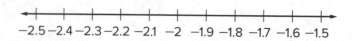

-2.5 -2.4 -2.3 -2.2 -2.1 -2 -1.9 -1.8 -1.7 -1.6 -1.5

Pause and Reflect

How does graphing numbers on a number line help you know whether to use <, >, or = when comparing them?

Record your observations here

Check

Which symbol, $<$, $>$, or $=$, would complete the statement $-\sqrt{5}$ _____ -210% to make a true statement? Then graph the numbers on a number line.

Part A

Write the symbol, $<$, $>$, or $=$, that makes $-\sqrt{5}$ _____ -210% a true statement.

Part B

Which of the following is the correct graph of $-\sqrt{5}$ and -210% on a number line?

(A)

$-\sqrt{5}$ -210%

-2.4 -2.3 -2.2 -2.1 -2.0

(B)

$-\sqrt{5}$ -210%

-2.4 -2.3 -2.2 -2.1 -2.0

(C)

$-\sqrt{5}$ -210%

-2.4 -2.3 -2.2 -2.1 -2.0

(D)

$-\sqrt{5}$ -210%

-2.4 -2.3 -2.2 -2.1 -2.0

Go Online You can complete an Extra Example online.

Think About It!

How would you begin ordering the numbers?

Example 3 Order Real Numbers

Order the set $\left\{\sqrt{30}, 6, 5\frac{4}{5}, 5.3\overline{6}\right\}$ from least to greatest. Then graph the set on the number line.

Part A Order the set of numbers.

Approximate the decimal expansion of each number.

$\sqrt{30} \approx$ [　]　　　　　Estimate to the nearest tenth.

$6 = 6.00$　　　　　Write as a decimal.

$5\frac{4}{5} =$ [　]　　　　　Write as a decimal.

$5.3\overline{6} \approx$ [　]　　　　　Write as a decimal to the nearest hundredth.

Write the decimals from least to greatest.

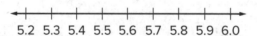

———————, ———————, ———————, ———————

So, from least to greatest, the order is $5.3\overline{6}$, $\sqrt{30}$, $5\frac{4}{5}$, and 6.

Part B Graph the numbers on the number line.

Approximate the location of $\sqrt{30}$ and $5.3\overline{6}$.

Pause and Reflect

How are comparing real numbers and ordering real numbers related to each other?

> Record your observations here

Check

Order the set $\left\{ \sqrt{3}, \dfrac{\pi}{2}, 1.60\%, 1.3\overline{5} \right\}$ from least to greatest. Then graph the set on the number line.

Part A

Order of the set from least to greatest.

Show your work here

Part B

Which of the following is the correct graph of the set of numbers?

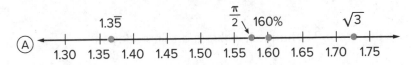

(A)

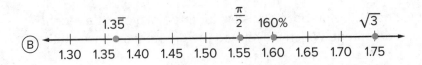

(B)

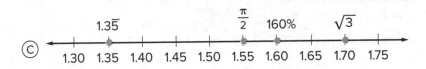

(C)

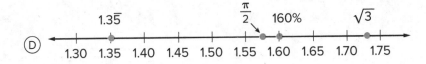

(D)

Go Online You can complete an Extra Example online.

On Wednesday, there is an $83\frac{1}{3}\%$ chance of rain. On Thursday, there is a $\frac{9}{10}$ chance of rain. On Friday, there is a 6 out of 7 chance that it will rain.

On which day is there the greatest chance of rain?

Step 1 Write each number in decimal notation. Round to the nearest hundredth.

$83\frac{1}{3}\% \approx$ ☐ $\frac{9}{10} =$ ☐   6 out of 7 $\approx$ ☐

Step 2 Order the decimals.

Since $0.9 > 0.86 > 0.83$, then $\frac{9}{10}$ ☐ $\frac{6}{7}$ ☐ $83\frac{1}{3}\%$.

So, there is the greatest chance it will rain on Thursday.

Check

The table shows the on-base statistics for three players at a recent baseball tournament. Which player had the greatest on-base statistic?

Player	On-Base Statistic
1	15 out of 21
2	$\frac{14}{19}$
3	72.5%

Show your work here

Copyright © McGraw-Hill Education

🐦 **Go Online** You can complete an Extra Example online.

🌐 Apply Line of Sight

On a clear day, the number of miles a person can see to the horizon is about $1.23\sqrt{h}$, where h is the person's height from the ground in feet. Suppose Frida is at the Empire State Building observation deck at 1,050 feet and Logan is at the Freedom Tower observation deck at 1,254 feet. How much farther can Logan see than Frida from the observation deck?

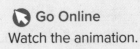 **Go Online**
Watch the animation.

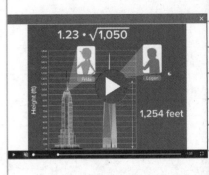

1 What is the task?

Make sure you understand exactly what question to answer or problem to solve. You may want to read the problem three times. Discuss these questions with a partner.

First Time Describe the context of the problem, in your own words.
Second Time What mathematics do you see in the problem?
Third Time What are you wondering about?

2 How can you approach the task? What strategies can you use?

Record your observations here

3 What is your solution?

Use your strategy to solve the problem.

Show your work here

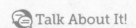 **Talk About It!**

How could you use the nearest perfect squares to check for reasonableness?

4 How can you show your solution is reasonable?

✍️ **Write About It!** Write an argument that can be used to defend your solution.

Check

The time in seconds that it takes an object to fall d feet can be found using the expression $\frac{\sqrt{d}}{4}$. Suppose Aiden drops a tennis ball from a height of 50 feet at the same time Mason drops a similar tennis ball from a height of 20 feet. How much longer will it take Aiden's tennis ball to reach the ground than Mason's tennis ball? Round to the nearest hundredth.

Show your work here

Go Online You can complete an Extra Example online.

Pause and Reflect

Review the Examples from this module. Which one did you find most challenging? What are the steps you would take to solve a problem of this type?

Record your observations here

Practice

Go Online You can complete your homework online.

Complete each statement using $<$, $>$, or $=$. Then graph the numbers on the number line. (Examples 1 and 2)

1. $\sqrt{11}$ _____ $3\frac{2}{3}$

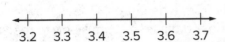

2. $\sqrt{3}$ _____ $\frac{\sqrt{10}}{2}$

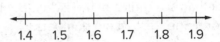

3. $-\pi^2$ _____ $-\sqrt{93}$

4. $-\sqrt{12}$ _____ -320%

5. Order the set $\left\{3\frac{1}{2}, \frac{10}{3}, \pi, \sqrt{13}\right\}$ from least to greatest. Then graph the set on the number line. (Example 3)

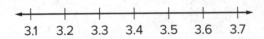

6. The table shows the foul-shot statistics for three players in a recent basketball game. Which player had the greatest foul-shot statistic? (Example 4)

Player	Foul-Shot Statistic
1	$\frac{7}{9}$
2	72%
3	8 out of 10

Test Practice

7. **Open Response** Is $\sqrt{27}$ *less than, greater than,* or *equal to* $\frac{\sqrt{95}}{2}$?

Apply

8. The radius of a circle can be approximated using the expression $\sqrt{\frac{A}{3}}$. A circular kiddie swimming pool has an area of about 28 square feet. An inflatable full-size circular pool has an area of about 113 square feet. How much greater is the radius of the full-size pool than the radius of the kiddie pool? Round to the nearest whole number.

9. The time in seconds that it takes an object to fall d feet can found using the expression $\frac{\sqrt{d}}{4}$. In an egg drop contest, Clara successfully dropped her egg container from a height of 35 feet, while Vladimir successfully dropped his egg container from a height of 23 feet. How much longer did it take Clara's egg to reach the floor than Vladimir's egg? Round to the nearest tenth.

10. Which One Doesn't Belong? Identify the number that does not belong in the group. Explain your reasoning.

$\boxed{-23.\overline{2}}$ $\boxed{-23\frac{1}{5}}$ $\boxed{-\sqrt{23}}$ $\boxed{-23.2}$

11. 🅜🅟 **Justify Conclusions** Which number is greater, 3.14 or π? Justify your answer.

12. 🅜🅟 **Find the Error** Kendra states that $\sqrt{3} > 2$ because 3 is greater than 2. Explain Kendra's mistake and correct it.

13. Identify two numbers, one rational and one irrational, that are between 1.6 and 1.8. Write an inequality to compare the two numbers.

📖 **Foldables** Use your Foldable to help review the module.

Real Numbers

Examples	Examples

Rate Yourself!

Complete the chart at the beginning of the module by placing a checkmark in each row that corresponds with how much you know about each topic after completing this module.

Write about one thing you learned.

_____

_____

_____

_____

_____

_____

Write about a question you still have.

_____

_____

_____

_____

_____

Reflect on the Module

Use what you learned about real numbers to complete the graphic organizer.

ⓔ Essential Question

Why do we classify numbers?

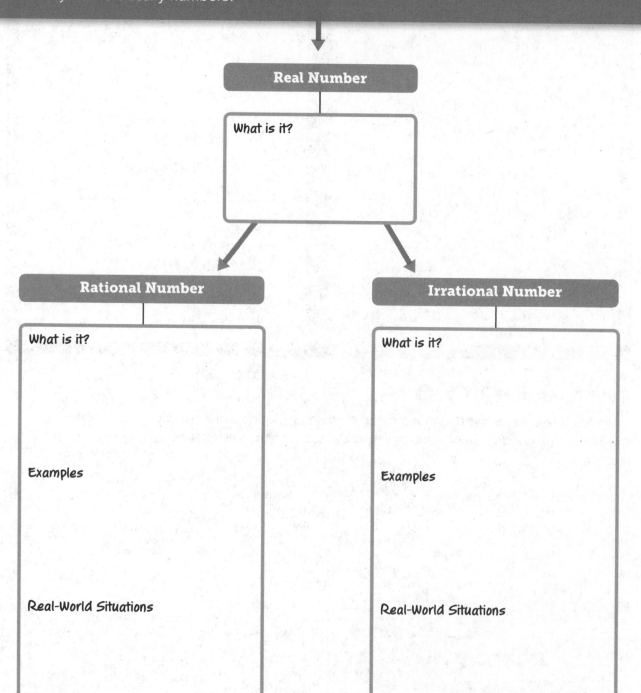

Real Number

What is it?

Rational Number

What is it?

Examples

Real-World Situations

Irrational Number

What is it?

Examples

Real-World Situations

Test Practice

1. Multiselect Simplify $\pm\sqrt{1.69}$. Select all that apply. **(Lesson 1)**

☐ −0.13

☐ 0.13

☐ −1.3

☐ 1.3

☐ 13

☐ −13

2. Open Response A vertical shelving unit consists of five equal-sized square shelves arranged in a column to form a rectangle. The total area of all five shelves is 500 square inches. **(Lesson 1)**

A. What is the height, in inches, of the shelving unit?

B. Explain how to find the height of the shelving unit.

3. Multiselect Simplify $-\sqrt{\frac{81}{121}}$. Select all that apply. **(Lesson 1)**

☐ −0.66942...

☐ $\frac{9}{11}$

☐ $-\frac{9}{11}$

☐ −0.8181...

☐ 0.8181...

4. Table Item Indicate whether each real number is rational or irrational. **(Lesson 2)**

	Rational	Irrational
−6		
$\sqrt{7}$		
$3\frac{1}{2}$		

5. Open Response Use the Venn diagram. **(Lesson 2)**

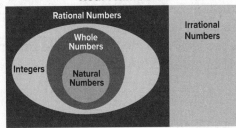

A. Determine whether the statement is *true* or *false*.
   All integers are natural numbers.

B. If the statement is *true*, explain your reasoning. If the statement is *false*, provide a counterexample.

6. Multiselect To which sets of numbers does the real number −25 belong? Select all that apply. **(Lesson 2)**

☐ rational

☐ irrational

☐ integer

☐ whole

☐ natural

7. Table Item Indicate the integer to which each square root is closest on a number line. (Lesson 3)

	7	8	9
$\sqrt{70}$			
$\sqrt{79}$			
$\sqrt{88}$			
$\sqrt{52}$			
$\sqrt{60}$			
$\sqrt{47}$			
$\sqrt{65}$			

8. Equation Editor A shipping box, in the shape of a cube, has a volume of 2,300 cubic inches. Estimate the length of the side of the shipping box to the nearest integer. (Lesson 3)

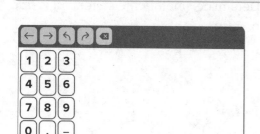

9. Open Response Winston wants to put trim board around his square shaped windows. Each window has an area of 3 square feet. Estimate the perimeter of each window. Approximate your answer by truncating the decimal expansion to the hundredths place.

10. Table Item Indicate which symbol makes each statement correct.

	<	>	=
$\sqrt[3]{24}$ ____ $\dfrac{\sqrt{400}}{6}$			
$-7\dfrac{5}{8}$ ____ $-\sqrt{55}$			
$-\sqrt{0.25}$ ____ $-\dfrac{1}{2}$			

11. Open Response Consider the real numbers $-2\dfrac{1}{5}$ and $-\sqrt{7}$. (Lesson 4)

A. Compare the numbers. Use <, >, or =.
$-2\dfrac{1}{5}$ ____ $-\sqrt{7}$

B. Graph $-\sqrt{7}$ and $-2\dfrac{1}{5}$ on the number line.

12. Open Response The table shows the number of aces per serving attempts for three players at a recent volleyball tournament. Order the players from least aces per serving attempt to greatest aces per serving attempt. (Lesson 4)

Player	Number of Aces
Angela	9 out of 22
Jaylin	$22\dfrac{2}{9}\%$
Mya	$\dfrac{3}{10}$

Algebraic Expressions

e Essential Question

Why is it beneficial to rewrite expressions in different forms?

What Will You Learn?

Place a checkmark (✓) in each row that corresponds with how much you already know about each topic **before** starting this module.

KEY ⬛ — I don't know. ◗ — I've heard of it. ⭐ — I know it!	Before ⬛	◗	⭐	After ⬛	◗	⭐
simplifying algebraic expressions by combining like terms						
using the Distributive Property to expand linear expressions						
adding linear expressions						
subtracting linear expressions						
finding the greatest common factors of monomials						
factoring linear expressions						
simplifying linear expressions						

📖 Foldables Cut out the Foldable and tape it to the Module Review at the end of the module. You can use the Foldable throughout the module as you learn about simplifying algebraic expressions.

What Vocabulary Will You Learn?

Check the box next to each vocabulary term that you may already know.

☐ coefficient   ☐ greatest common factor

☐ constant   ☐ like terms

☐ factor   ☐ linear expression

☐ factored form   ☐ simplest form

Are You Ready?

Study the Quick Review to see if you are ready to start this module.
Then complete the Quick Check.

Quick Review

Example 1
Subtract integers.

Simplify $-15 - (-3)$.

$-15 - (-3)$

$\quad = -15 + 3$ Add the additive inverse.

$\quad = -12$ Find the difference of the absolute values. The sign of the sum is negative because -15 has a greater absolute value than 3.

Example 2
Multiply integers.

Simplify $6(-7)$.

$6(-7) = -42$ The product is negative because the signs of the factors are different.

Quick Check

1. Simplify $24 - 81$.

2. Simplify $37 - (-16)$.

3. Simplify $-5(-8)$.

4. Simplify $-4(11)$.

How Did You Do?
Which exercises did you answer correctly in the Quick Check?
Shade those exercise numbers at the right.

  ③

Simplify Algebraic Expressions

I Can... simplify algebraic expressions by identifying and combining like terms.

Copyright © McGraw-Hill Education

What Vocabulary
Will You Learn?
coefficient
constant
like terms
simplest form

Explore Use Algebra Tiles to Add Integers

Online Activity You will use algebra tiles to explore how to simplify algebraic expressions.

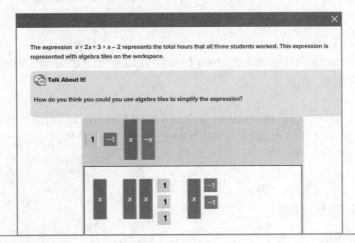

The expression $x + 2x + 3 + x - 2$ represents the total hours that all three students worked. This expression is represented with algebra tiles on the workspace.

Talk About It!

How do you think you could you use algebra tiles to simplify the expression?

Learn Like Terms

When addition or subtraction signs separate an algebraic expression into parts, each part is called a *term*. **Like terms** contain the same variables to the same powers. For example, $3x^2$ and $-7x^2$ are like terms because the variables and their exponents are the same. But, $5y^3$ and $9y^4$ are not like terms because the exponents are different.

The numerical factor of a term that contains a variable is called the **coefficient** of the variable. For example, in the term $3x^2$, 3 is the coefficient. A term without a variable is called a **constant**. Constant terms are also like terms.

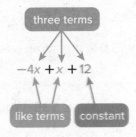

three terms

$$-4x + x + 12$$

like terms   constant

(continued on next page)

Talk About It!

Why are the expressions $3x$ and $3x^2$ *not* like terms?

 Talk About It!

Without using the Distributive Property, what is another way you could add $6n$ and $-8n$?

 Talk About It!

When might it be more advantageous to simplify the expression then evaluate versus evaluating first, then simplifying?

Sort the terms by writing like terms in the appropriate bins.

$$5 \qquad 8x \qquad -9x^2 \qquad 6x \qquad -12 \qquad 30x^2 \qquad -1.5x \qquad \frac{1}{2}$$

$4x^2$	$-2x$	3

Learn Combine Like Terms

An algebraic expression is in **simplest form** if it has no like terms and no parentheses.

Go Online Watch the animation to learn how the Distributive Property can be used to combine like terms.

$6n - 1 - 8n + 9$

$= 6n + (-1) + (-8n) + 9$	Rewrite each subtraction as addition.
$= 6n + (-8n) + (-1) + 9$	Apply the Commutative Property.
$= [6 + (-8)]n + (-1) + 9$	Apply the Distributive Property.
$= -2n + 8$	Simplify.

Example 1 Combine Like Terms

The cost of a jacket j after a 5% markup can be represented by the expression $j + 0.05j$.

Simplify the expression.

$j + 0.05j = 1j + 0.05j$	Identity Property: $j = 1j$
$= 1.05j$	Combine like terms.

Increasing the jacket's price by 5% is the same as multiplying the price by _____ .

Suppose the original cost of the jacket is $35. What is the cost of the jacket after the 5% markup? $ [ ]

Check

The cost of a new pair of shoes after a 3% markup can be represented by the expression $c + 0.03c$. Simplify the expression.

Example 2 Combine Like Terms

Simplify $-5x + y + 6 - 5y - 3$.

$-5x + y + 6 - 5y - 3$ Write the expression.

$= -5x + y + 6 + (-5y)\ \boxed{}\ \left(\boxed{}\right)$ Rewrite subtraction as addition.

$= -5x + y + \left(\boxed{}\right) + \boxed{} + (-3)$ Commutative Property

$= -5x + \left(\boxed{}\right) + \boxed{}$ Combine like terms.

$= -5x - 4y + 3$ Because parentheses are in the expression, it is not simplified. Rewrite using subtraction.

So, $-5x + y + 6 - 5y - 3 = -5x - 4y + 3$.

Check

Simplify $-5 - 3w + 9 - 6z + 8w - 4z$.

🖰 **Go Online** You can complete an Extra Example online.

> 💬 **Talk About It!**
> In Steps 2 and 3, why must subtraction be written as addition in order to use the Commutative Property? Why is the Commutative Property used?

Example 3 Combine Like Terms

Simplify $\frac{3}{4}a - \frac{2}{3} - \frac{1}{2}a + \frac{5}{6}$.

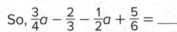

$\frac{3}{4}a - \frac{2}{3} - \frac{1}{2}a + \frac{5}{6}$　　　　　Write the expression.

$= \frac{3}{4}a + \left(-\frac{2}{3}\right) + \left(-\frac{1}{2}a\right) + \frac{5}{6}$　　Rewrite subtraction as addition.

$= \frac{3}{4}a + \left(-\frac{1}{2}a\right) + \left(-\frac{2}{3}\right) + \frac{5}{6}$　　Commutative Property

$= \frac{3}{4}a + \left(-\frac{2}{4}a\right) + \left(-\frac{4}{6}\right) + \frac{5}{6}$　　Rewrite fractions with common denominators.

$= \frac{1}{4}a + \frac{1}{6}$　　　　　　　Combine like terms.

So, $\frac{3}{4}a - \frac{2}{3} - \frac{1}{2}a + \frac{5}{6} =$ _____ .

Check

Simplify $-\frac{1}{4}m + \frac{5}{6} + \frac{3}{8}m - \frac{2}{3}$.

 Go Online You can complete an Extra Example online.

Pause and Reflect

Reflect on the process of simplifying algebraic expressions containing fractions. What concepts did you use? How are they used to simplify an expression?

Talk About It!

In Step 4, why were two different common denominators found?

Learn Expand Linear Expressions

The Distributive Property can be used to expand linear expressions.

You learned about this property in an earlier grade.

Words
The Distributive Property states that to multiply a sum or difference by a number, multiply each term inside the parentheses by the number outside the parentheses.

Symbols	Examples
$a(b + c) = ab + ac$	$4(x + 2) = 4 \cdot x + 4 \cdot 2$
	$\quad\quad\quad\quad = 4x + 8$
$a(b - c) = ab - ac$	$3(x - 5) = 3 \cdot x - 3 \cdot 5$
	$\quad\quad\quad\quad = 3x - 15$

Fill in the boxes to model the Distributive Property.

$2(x + 2) = \boxed{}(x) + \boxed{}(2)$ \quad\quad Distributive Property

$\quad\quad\quad = \boxed{} + \boxed{}$ \quad\quad\quad\quad Simplify.

Example 4 Distribute Over Addition

Use the Distributive Property to expand 4(−3x + 6).

$4(-3x + 6) = \boxed{}(-3x) + \boxed{}(6)$ \quad\quad Distributive Property

$\quad\quad\quad\quad = -12x + 24$ \quad\quad\quad\quad Simplify.

So, $4(-3x + 6) =$ _____.

Check

Use the Distributive Property to expand $9(-5a + 3b)$.

Show your work here

 Go Online You can complete an Extra Example online.

💭 **Think About It!**

What terms will be multiplied by 4 when you expand the expression?

🗨 **Talk About It!**

In the second step, why is it helpful to use parentheses when expanding the expression?

Copyright © McGraw-Hill Education

Example 5 Distribute Over Subtraction

Use the Distributive Property to expand $(2x - 5y)3$.

$(2x - 5y)3 = [(2x + (-5y)]3$ Rewrite subtraction as addition.

$\qquad\quad = 3(2x) + 3(-5y)$ Distributive Property

$\qquad\quad = 6x + (-15y)$ Multiply.

$\qquad\quad = 6x - 15y$ Because parentheses are in the expression, it is not simplified. Rewrite using subtraction.

So, $(2x - 5y)3 =$ _____.

Check

Use the Distributive Property to expand $(-4w - 7)5$.

Talk About It!

What is another way you can expand the expression $(2x - 5y)3$ without first rewriting the subtraction to addition?

Example 6 Distribute Negative Numbers

Use the Distributive Property to expand $-5(2x - 9)$.

$-5(2x - 9) = -5[2x + (-9)]$ Rewrite subtraction as addition.

$\qquad\quad = -5(2x) + -5(-9)$ Distributive Property

$\qquad\quad = -10x + 45$ Simplify.

So, $-5(2x - 9) =$ _____.

Check

Use the Distributive Property to expand $-6(-8y + 10)$.

Talk About It!

What mistake might be made if you do not rewrite subtraction as addition?

Go Online You can complete an Extra Example online.

Apply Geometry

The side lengths of two triangles are shown. Represent the perimeter of each triangle with an expression in simplest form. Which triangle has a greater perimeter if $x = 3$? Will this be true if $x = 2$? Justify your response.

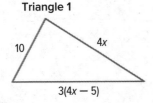

Triangle 1

10   4x

3(4x − 5)

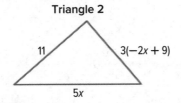

Triangle 2

11   3(−2x + 9)

5x

1 What is the task?

Make sure you understand exactly what question to answer or problem to solve. You may want to read the problem three times. Discuss these questions with a partner.

First Time Describe the context of the problem, in your own words.
Second Time What mathematics do you see in the problem?
Third Time What are you wondering about?

2 How can you approach the task? What strategies can you use?

Record your observations here

3 What is your solution?

Use your strategy to solve the problem.

Show your work here

4 How can you show your solution is reasonable?

Write About It! Write an argument that can be used to defend your solution.

Talk About It!
What properties did you use when solving this problem?

Check

The side lengths of two triangles are shown. Represent the perimeter of each triangle with an expression in simplest form. Which triangle has a greater perimeter if $x = 5$?

Triangle 1

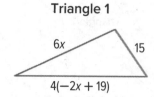

6x   15

4(−2x + 19)

Triangle 2

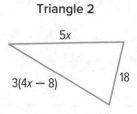

5x

3(4x − 8)   18

Show your work here

Go Online You can complete an Extra Example online.

Pause and Reflect

Where in the lesson did you feel most confident? Why?

Record your observations here

Practice

Go Online You can complete your homework online.

1. The cost of a set of DVDs after a 25% markup can be represented by the expression $c + 0.25c$. Simplify the expression. (Example 1)

2. The cost of a new robotic toy after an 8% markup can be represented by the expression $r + 0.08r$. Simplify the expression. (Example 1)

Simplify each expression. (Examples 2 and 3)

3. $-y + 9z - 16y - 25z + 4$

4. $8z + x - 5 - 9z + 2$

5. $5c - 3d - 12c + d - 6$

6. $-\frac{3}{4}x - \frac{1}{3} + \frac{7}{8}x - \frac{1}{2}$

7. $\frac{1}{4} + \frac{9}{10}y - \frac{3}{5}y + \frac{7}{8}$

8. $-\frac{1}{2}a + \frac{2}{5} + \frac{5}{6}a - \frac{1}{10}$

Use the Distributive Property to expand each expression. (Examples 4–6)

9. $2(-3x + 5)$

10. $6(-4x + 3y)$

11. $(3y - 2z)5$

12. $(-2x - 7)4$

13. $-7(x - 2)$

14. $-3(8x - 4)$

Test Practice

15. **Table Item** The table shows the side lengths of a triangle. The perimeter of the triangle is $6a + 3$. Write an expression in simplest form for the length of Side 3.

Triangle Side	Length (units)
1	$2(a + 3)$
2	$3a - 1$
3	

Apply

16. The side lengths of two triangles are shown. Represent the perimeter of each triangle with an expression in simplest form. Which triangle has a greater perimeter if $x = 4$? Will this be true if $x = 5$? Justify your response.

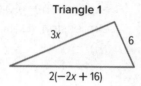

Triangle 1

$3x$ 6

$2(-2x + 16)$

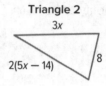

Triangle 2

$3x$

$2(5x - 14)$ 8

17. The side lengths of two quadrilaterals are shown. Represent the perimeter of each quadrilateral with an expression in simplest form. Which quadrilateral has a greater perimeter if $x = 3$? Will this be true if $x = 4$? Justify your response.

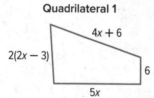

Quadrilateral 1

$4x + 6$

$2(2x - 3)$

6

$5x$

Quadrilateral 2

$3(3x - 7)$

$3x$ 12

$2x + 4$

18. Create Write an expression with at least three unlike terms and then simplify the expression.

19. 🔵 **Find the Error** A student simplified the expression $5x - 3(x + 4)$ to $2x + 12$. Find the student's error and correct it.

20. 🔵 **Justify Conclusions** Is the following statement *true* or *false*? If false, explain.

When using the Distributive Property, if the term outside the parentheses is negative, then the sign of each term inside the parentheses will not change.

Add Linear Expressions

I Can... use different methods to add linear expressions.

Explore Add Expressions

Online Activity You will use Web Sketchpad to explore how to add linear expressions.

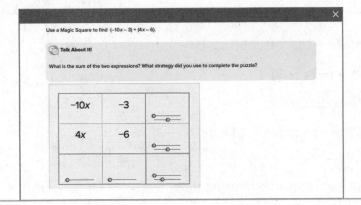

Use a Magic Square to find $(-10x - 3) + (4x - 6)$.

Talk About It!

What is the sum of the two expressions? What strategy did you use to complete the puzzle?

−10x	−3
4x	−6

Learn Add Linear Expressions

A **linear expression** is an algebraic expression in which each term is a constant or the product of a constant and the variable raised to the first power. When simplified, a linear expression cannot contain a variable in the denominator of a fraction.

Sort the expressions by writing each one in the appropriate bin. Examples of each type are given.

$$-5x^2 \qquad -4x + 3 \qquad \frac{1}{2}x - 5 \qquad \frac{6}{x} \qquad x^3 + 2$$

Linear Expressions	Nonlinear Expressions
5x	5mn
3x + 2	$x^4 - 7$

What Vocabulary Will You Learn?
linear expression

Talk About It!
Why are $-5x^2$ and $\frac{6}{x}$ *not* linear expressions?

(continued on next page)

When you add the expressions $4x + 2$ and $5x - 7$, the answer is $9x + (-5)$ or $9x - 5$. Why can we rewrite $9x + (-5)$ as $9x - 5$?

Think About It!

How would you begin finding the sum?

Go Online Watch the animation to learn how to add linear expressions.

The animation shows how to add the expressions $(4x + 2) + (5x - 7)$.

$(4x + 2) + (5x - 7)$

$= (4x + 2) + [5x + (-7)]$ Rewrite each subtraction as addition.

$=$ $4x + 2$ Arrange like terms in columns.
$(+)$ $\underline{5x + (-7)}$

     $9x + (-5)$ or $9x - 5$ Add the like terms in each column.

Example 1   Add Linear Expressions

Find $(4x - 2) + (-7x - 3)$.

Method 1 Use algebra tiles.

Use algebra tiles to add $(4x - 2) + (-7x - 3)$.

Step 1 Model each expression using tiles.

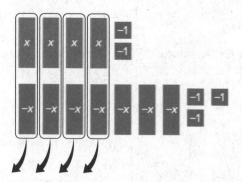

Step 2 Remove zero pairs.

There are three $-x$-tiles and five -1-tiles remaining.

So, $(4x - 2) + (-7x - 3) = $ _____.

(continued on next page)

Method 2 Arrange terms in columns.

$(4x - 2) + (-7x - 3)$　　　　　Write the original expression.

　　$= [4x + (-2)] + [-7x + (-3)]$　　Rewrite subtraction as addition.

　　$= 4x + (-2)$　　　　　　　　Arrange like terms in columns.
$(+)\ \underline{-7x + (-3)}$
　　　$-3x + (-5)$　　　　　　　Add.

So, $(4x - 2) + (-7x - 3) =$ _____.

Check

Find $(5x - 4) + (-2x + 2)$.

(Show your work here)

🡒 **Go Online** You can complete an Extra Example online.

Pause and Reflect

Are you ready to move on to the next Example? If yes, what have you learned that you think will help you? If no, what questions do you still have? How can you get those questions answered?

(Record your observations here)

Talk About It!

How are the processes for Method 1 and Method 2 similar?

Could you simplify this expression another way? Explain.

Example 2 Add Linear Expressions

Find $\left(\frac{1}{3}x + 9\right) + \left(\frac{5}{12}x - 4\right)$.

Step 1 Rewrite subtraction as addition.

$\left(\frac{1}{3}x + 9\right) + \left(\frac{5}{12}x - 4\right)$

$= \left(\frac{1}{3}x + 9\right) + \left[\frac{5}{12}x + (-4)\right]$

Write the original expression.

Rewrite subtraction as addition.

Step 2 Arrange like terms in columns.

$\frac{1}{3}x + 9 \qquad \rightarrow \qquad \boxed{}\,x + 9$

Rewrite using a common denominator.

$+\ \frac{5}{12}x + (-4) \qquad \rightarrow \qquad +\ \boxed{}\,x + (-4)$

$\frac{9}{12}x + 5$ Add.

$\boxed{}\,x + 5$ Simplify.

So, $\left(\frac{1}{3}x + 9\right) + \left(\frac{5}{12}x - 4\right) = \frac{3}{4}x + 5$.

Check

Find $\left(\frac{1}{4}x + 4\right) + \left(\frac{2}{3}x - 8\right)$.

Show your work here

 Go Online You can complete an Extra Example online.

Pause and Reflect

Compare what you learned today with something similar you learned in an earlier module or grade. How are they similar? How are they different?

Record your observations here

Apply Theater

The drama club is selling tickets for their latest production. They are also accepting additional cash donations. They plan to save 20% of the money from all ticket sales and donations for their spring trip. Ticket sales and donations from two performances are represented in the table, where t represents the cost of a ticket. If tickets cost $9, how much money will the drama club have available for the spring trip?

Performance	Ticket Sales and Donations
Friday Night	$92t + 109$
Saturday Night	$34t + 13$

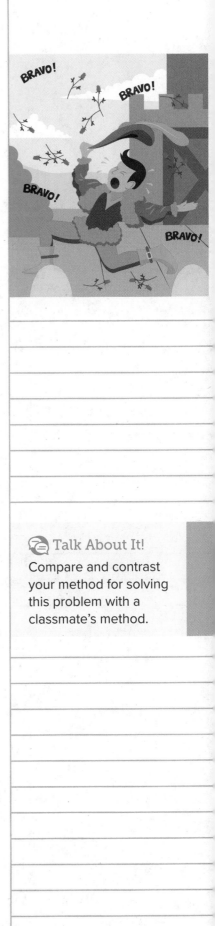

1 What is the task?

Make sure you understand exactly what question to answer or problem to solve. You may want to read the problem three times. Discuss these questions with a partner.

First Time Describe the context of the problem, in your own words.
Second Time What mathematics do you see in the problem?
Third Time What are you wondering about?

2 How can you approach the task? What strategies can you use?

Record your observations here

3 What is your solution?

Use your strategy to solve the problem.

Show your work here

4 How can you show your solution is reasonable?

 Write About It! Write an argument that can be used to defend your solution.

Talk About It!
Compare and contrast your method for solving this problem with a classmate's method.

Check

The football team is selling tickets for their next two home games. They also sell food at the concession stand during the games. They plan to save 30% of the money from all ticket sales and concession stand sales for their summer camp. Ticket and concession sales for two home games are represented in the table, where t represents the cost of a ticket. If tickets cost $5, how much money will the football team have for their summer camp?

Home Game	Ticket and Concession Sales
Week 1	$(75t + 130)$
Week 2	$(82t + 115)$

Show your work here

Go Online You can complete an Extra Example online.

Foldables It's time to update your Foldable, located in the Module Review, based on what you learned in this lesson. If you haven't already assembled your Foldable, you can find the instructions on page FL1.

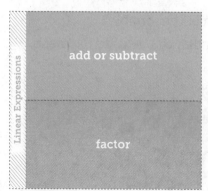

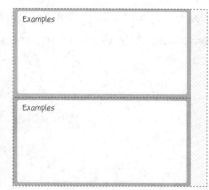

Practice

▶ **Go Online** You can complete your homework online.

Add. (Examples 1 and 2)

1. $(8x + 9) + (-6x - 2)$

2. $(5x + 4) + (-8x - 2)$

3. $(-7x + 1) + (4x - 5)$

4. $(-3x - 9) + (4x + 8)$

5. $(-5x + 4) + (-9x - 3)$

6. $(-2x + 10) + (-8x - 1)$

7. $\left(\frac{1}{4}x - 3\right) + \left(\frac{3}{16}x + 5\right)$

8. $\left(\frac{1}{2}x - 3\right) + \left(\frac{1}{6}x + 1\right)$

9. $\left(4x + \frac{3}{4}\right) + \left(-3x - \frac{5}{12}\right)$

10. $\left(-9x - \frac{4}{5}\right) + \left(2x + \frac{2}{3}\right)$

11. $\left(\frac{1}{3}x - 3\right) + \left(-\frac{3}{4}x - 5\right)$

12. $\left(-5x - \frac{2}{3}\right) + \left(-4x - \frac{1}{9}\right)$

Test Practice

13. Open Response The table shows the length and width of a rectangle. Write a simplified expression for the perimeter of the rectangle.

Dimension	Measurement (units)
Length	$3x + 6$
Width	$2x - 4$

Apply

14. Jade and Chet get a weekly allowance plus x dollars for each time the pair walks the dog. They plan to save 40% of their combined earnings in one week to purchase a new app for their smart tablet. Their earnings in a certain week are represented in the table. If their parents pay $2.50 each time they walk the dog, how much money will they have to purchase the app?

	Earnings ($)
Jade	$8 + 2x$
Chet	$4x + 6$

15. Elsa is selling bracelets at craft shows to raise money for an animal shelter. She is also accepting additional cash donations. She plans to give 75% of the money from all bracelet sales and donations to the shelter. She will use the remaining money to buy more supplies. Bracelet sales and donations from the first craft show are represented by the expression $24n + 32$, where n represents the amount Elsa charges for each bracelet. The second craft show sales and donations is represented by the expression $40n + 56$. If Elsa charges $4 for each bracelet, how much money will she donate to the animal shelter?

16. 🔲 **Identify Structure** Write two linear expressions that have a sum of $6x + 9$.

17. 🔲 **Identify Structure** What linear expression would you need to add to $(-6x + 3)$ to have a sum of $-x$?

18. **Which One Doesn't Belong?** Identify the linear expression that is not equivalent to the other three. Explain your reasoning.

a. $(2x - 1) + (-3x + 7)$

b. $(-5x + 3) + (4x + 3)$

c. $(5x - 6) + (-6x + 12)$

d. $(-5x - 1) + (6x + 7)$

19. 🔲 **Reason Inductively** When will the sum of two linear expressions with only x-terms be zero?

Subtract Linear Expressions

I Can... use different methods to subtract linear expressions.

Learn Additive Inverses of Expressions

When two expressions have a sum of zero, the expressions are additive inverses.

To find the additive inverse of an expression, use the Distributive Property to multiply the expression by -1.

Find the additive inverse of $4x + 2$.

$$-1(4x + 2) = \boxed{}(4x) + \left(\boxed{}\right)(2) \qquad \text{Multiply the expression by } -1.$$

$$= \boxed{} + \left(\boxed{}\right) \qquad \text{Simplify.}$$

$$= -4x - 2 \qquad \text{Definition of subtraction}$$

You can check to see if $4x + 2$ and $-4x - 2$ are additive inverses by adding them together to see if their sum is zero.

$$\begin{array}{ll} \quad\ 4x + 2 & \text{Arrange like terms in columns.} \\ (+)\ \underline{-4x - 2} & \\ \quad\ 0x + 0 \text{ or } 0 & \text{Add.} \end{array}$$

Because $4x + 2$ and $-4x - 2$ have a sum of zero, they are additive inverses.

Talk About It!

How could you find the additive inverse of an expression mentally?

Pause and Reflect

Give at least one example of an expression and its additive inverse. Then show how you would check your work.

> Record your observations here

Example 1 Find the Additive Inverse of Expressions

Find the additive inverse of $5x - 7$.

To find the additive inverse, multiply the expression by -1.

$-1(5x - 7) = -1[5x + (-7)]$ Rewrite subtraction as addition.

$\quad\quad\quad = -1(5x) + (-1)(-7)$ Distributive Property

$\quad\quad\quad = -5x + 7$ Simplify.

So, the additive inverse of $5x - 7$ is $-5x + 7$.

Check

Find the additive inverse of $-7x + 3$.

Show
your work
here

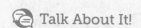**Go Online** You can complete an Extra Example online.

Talk About It!

How is the process for subtracting linear expressions the same as adding? How is it different?

Learn Subtract Linear Expressions

When subtracting integers, you add the opposite, or the additive inverse. The same process is used when subtracting linear expressions.

Go Online Watch the animation to learn how to subtract linear expressions.

The animation shows how to subtract $(8y - 5) - (3y + 4)$.

$\quad\quad (8y - 5)$ Arrange like terms in columns.
$- \underline{(3y + 4)}$

$\quad\quad\quad 8y - 5$
$(+) \underline{-3y - 4}$ Add the additive inverse.
$\quad\quad\quad 5y - 9$ Add.

Example 2 Subtract Linear Expressions

Find $(6x + 5) - (3x - 2)$.

Method 1 Use algebra tiles.

Step 1 Model $6x + 5$ using algebra tiles.

Step 2 To subtract $3x - 2$, add the additive inverse, or $-3x + 2$. Then, remove any zero pairs.

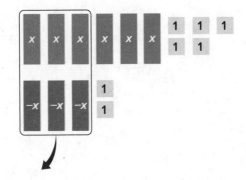

So, $(6x + 5) - (3x - 2) = 3x + 7$.

Method 2 Arrange terms in columns.

$(6x + 5) - (3x - 2)$ Write the expression.

$= (6x + 5) + (-3x + 2)$ The additive inverse of $(3x - 2)$ is $(-3x + 2)$.

$$\begin{array}{r} = 6x + 5 \\ (+) \ \underline{-3x + 2} \\ 3x + 7 \end{array}$$ Arrange like terms in columns.

Add.

So, $(6x + 5) - (3x - 2) = $ _____.

Check

Find $(-3x + 9) - (8x - 7)$.

Go Online You can complete an Extra Example online.

Think About It!

How would you begin finding the difference?

Talk About It!

Compare the methods for subtracting linear expressions.

- Method 1: Use algebra tiles to subtract linear expressions.
- Method 2: Arrange terms in columns.

Example 3 Subtract Linear Expressions

Find $\left(\frac{2}{3}x + \frac{1}{2}\right) - \left(\frac{1}{6}x - \frac{3}{8}\right).$

Rewrite using the additive inverse.

$\left(\frac{2}{3}x + \frac{1}{2}\right) - \left(\frac{1}{6}x - \frac{3}{8}\right)$ Write the expression.

$= \left(\frac{2}{3}x + \frac{1}{2}\right) \boxed{} \left(\boxed{} \right)$ The additive inverse of $\left(\frac{1}{6}x - \frac{3}{8}\right)$ is $\left(-\frac{1}{6}x + \frac{3}{8}\right)$.

Arrange like terms in columns.

$\frac{2}{3}x + \frac{1}{2}$ $\quad\rightarrow\quad$ $\frac{4}{6}x + \frac{4}{8}$ Rewrite using the

$(+) -\frac{1}{6}x + \frac{3}{8}$ $\quad\rightarrow\quad$ $(+) -\frac{1}{6}x + \frac{3}{8}$ common denominator.

$\boxed{}\,x + \boxed{}$ Add.

$\boxed{}\,x + \boxed{}$ Simplify.

So, $\left(\frac{2}{3}x + \frac{1}{2}\right) - \left(\frac{1}{6}x - \frac{3}{8}\right) = \frac{1}{2}x + \frac{7}{8}.$

Check

Find $\left(-\frac{1}{6}x + \frac{3}{4}\right) - \left(\frac{1}{2}x - \frac{1}{8}\right).$

Show your work here

🔾 **Go Online** You can complete an Extra Example online.

Pause and Reflect

Show how rewriting an expression using the additive inverse is beneficial when subtracting linear expressions.

Record your observations here

🌐 Apply Sales

Bonnie owns a T-shirt shop where she sells plain T-shirts and printed T-shirts. The table shows the cost per shirt and the number of each type of shirt sold over w weeks. After 25 weeks, how much more did she earn in sales of printed T-shirts than in sales of plain T-shirts?

Style	Cost ($)	Number Sold
Plain	15	$5w - 4$
Printed	15	$8w + 3$

1 What is the task?

Make sure you understand exactly what question to answer or problem to solve. You may want to read the problem three times. Discuss these questions with a partner.

First Time Describe the context of the problem, in your own words.
Second Time What mathematics do you see in the problem?
Third Time What are you wondering about?

2 How can you approach the task? What strategies can you use?

Record your observations here

3 What is your solution?

Use your strategy to solve the problem.

Show your work here

4 How can you show your solution is reasonable?

✏️ **Write About It!** Write an argument that can be used to defend your solution.

💬 Talk About It!

How do you know just by looking at the expressions that Bonnie sold more printed T-shirts than plain T-shirts?

Check

The table shows a bakery's sales of sugar cookies and chocolate chip cookies sold in h hours.

Cookie Sales		
Flavor	Cost ($)	Number Sold
Sugar	1.15	$6h - 5$
Chocolate Chip	1.15	$10h + 6$

After 15 hours, how much more did the bakery earn in sales of chocolate chip cookies than in sales of sugar cookies?

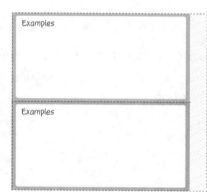
Go Online You can complete an Extra Example online.

Foldables It's time to update your Foldable, located in the Module Review, based on what you learned in this lesson. If you haven't already assembled your Foldable, you can find the instructions on page FL1.

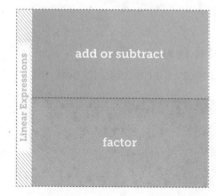

Copyright © McGraw-Hill Education

Practice

Go Online You can complete your homework online.

Find the additive inverse of each linear expression. (Example 1)

1. $3x - 6$

2. $-9x + 3$

3. $-4x - 8$

Subtract. (Examples 2 and 3)

4. $(8x + 9) - (6x - 2)$

5. $(3x - 4) - (x - 5)$

6. $(-5x - 9) - (-6x - 1)$

7. $(-7x - 14) - (x - 5)$

8. $(-8x + 2) - (-5x + 7)$

9. $\left(\frac{3}{5}x + \frac{3}{4}\right) - \left(\frac{1}{3}x - \frac{1}{8}\right)$

10. $\left(\frac{1}{10}x - \frac{2}{3}\right) - \left(\frac{4}{5}x - \frac{1}{6}\right)$

11. $\left(-\frac{5}{6}x - \frac{7}{12}\right) - \left(-\frac{1}{3}x - \frac{1}{4}\right)$

12. $\left(-\frac{1}{2}x + \frac{7}{10}\right) - \left(-\frac{3}{4}x - \frac{1}{5}\right)$

Test Practice

13. Open Response The table shows the scores of two teams in a trivia challenge at the end of the first half. How many more points did the Huskies score than the Bobcats?

Team	Points Scored
Bobcats	$2x - 7$
Huskies	$5x - 3$

Apply

14. The table shows the sales of plain and Asiago cheese bagels at a bakery for h hours. After 6 hours, how much more will the bakery have made in sales of Asiago cheese bagels than the sales of plain bagels?

Bagel Sales		
Bagel	Cost ($)	Number Sold After h hours
Asiago Cheese	1.50	$12h + 7$
Plain	1.50	$7h - 4$

15. Derek owns a snack shop where he sells tins of buttered and caramel popcorn. The table shows the number of each type of popcorn sold over w weeks. After 12 weeks, how much more will he have made in sales of buttered popcorn than the sales of caramel popcorn?

Popcorn Sales		
Popcorn	Cost ($)	Number Sold Over w Weeks
Buttered	11	$8w + 9$
Caramel	11	$6w - 1$

16. **MP Identify Structure** Write two linear expressions that have a difference of $x + 1$.

17. **MP Identify Structure** What linear expression would you need to subtract from $(5x + 3)$ to have a difference of $-x$?

18. **Which One Doesn't Belong?** Identify the linear expression that does not belong with the other three. Explain your reasoning.

a. $(-5x + 3) - (-7x - 1)$

b. $(-3x + 3) - (-5x - 2)$

c. $(x - 6) - (-x - 10)$

d. $(-7x + 2) - (-9x - 2)$

19. **MP Find the Error** A student simplified the expression $(6x - 2) - (-x + 5)$ to $7x + 3$. Find the student's error and correct it.

Factor Linear Expressions

I Can... use GCF to factor linear expressions.

Learn Greatest Common Factor of Monomials

To **factor** a number means to write it as a product of its factors. A monomial can be factored using the same method you would use to factor a number.change

The **greatest common factor** (GCF) of two monomials is the greatest monomial that is a factor of both. The greatest common factor also includes any variables that the monomials have in common.

Number	Monomial
The GCF of 25 and 30 is 5.	The GCF of $25x$ and $30xy$ is $5x$.

Example 1 Find the GCF of Monomials

Find the GCF of $12y$ and $30y$.

Step 1 Find the GCF of the coefficients.

Circle the factors of 12 and underline the factors of 30.

1	2	3	4	5	6	7	8	9	10
11	12	13	14	15	16	17	18	19	20
21	22	23	24	25	26	27	28	29	30

The common factors of 12 and 30 are numbers that you both circled and underlined. Find the greatest of these numbers.

The greatest common factor of the coefficients is 6.

Step 2 Find the GCF of the variables.

The common variable of $12y$ and $30y$ is y.

Step 3 Find the GCF of the monomials.

To find the GCF of the monomials, multiply the GCF of the coefficients, _____, by the common variable, _____.

So, the GCF of $12y$ and $30y$ is $6y$.

Copyright © McGraw-Hill Education

What Vocabulary Will You Learn?
factor
factored form
greatest common factor

Talk About It!
Based on what you know about finding the GCF of numbers, how do you think you can find the GCF of monomials?

Check

Find the GCF of $24a$ and $32a$.

Example 2 Find the GCF of Monomials

Use prime factorization to find the GCF of $18a$ and $20ab$.

Step 1 Identify common factors.

Complete the prime factorization for each term.

$18a = 2 \cdot 3 \cdot 3 \cdot a$

$20ab = 2 \cdot 2 \cdot 5 \cdot a \cdot b$

The common factors in each term are 2 and a.

Step 2 Multiply the common factors.

The GCF of $18a$ and $20ab$ is $2 \cdot a$ or _____.

Check

Use prime factorization to find the GCF of $42xy$ and $14y$.

Think About It!

What variable(s) are common in each term?

Talk About It!

The factor 2 is common in each term. Why can only one 2 of the term $20ab$ be selected as a common factor?

🔵 **Go Online** You can complete Extra Examples online.

Explore Factor Linear Expressions

Online Activity You will use algebra tiles to explore how to factor linear expressions.

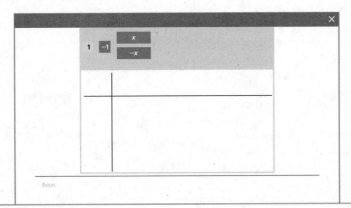

Learn Factor Linear Expressions

You can use the Distributive Property and work backward to express a linear expression as a product of its factors. A linear expression is in **factored form** when it is expressed as the product of its factors.

Go Online Watch the animation to learn how to factor $12a + 6b$.

Step 1 Find the GCF of the terms.

$12a = 2 \cdot 2 \cdot 3 \cdot a$	Write the prime factorization of each term.
$6b = 2 \cdot 3 \cdot b$	Circle the common factors.
$2 \cdot 3$ or 6	Multiply the common factors to find the GCF.

Step 2 Write each term as a product with the GCF as a factor.

$$12a + 6b = 6(2a) + 6(b)$$

Step 3 Apply the Distributive Property.

$$6(2a) + 6(b) = 6(2a + b)$$

So, the factored form of $12a + 6b$ is $6(2a + b)$.

Think About It!

What is the GCF of $3x$ and 9?

 Talk About It!

How can you check to see if the factored form is correct?

 Think About It!

How do you find the GCF of two monomials?

 Talk About It!

Write another expression that cannot be factored. Explain why it cannot be factored.

Example 3 Factor Linear Expressions

Factor $3x + 9$.

$3x = 3 \cdot x$ Write the prime factorization of each term.

$9 = 3 \cdot 3$ Circle the common factors.

3 Multiply, if necessary, the common factors to find the GCF.

$3x + 9 = 3\left(\boxed{}\right) + 3\left(\boxed{}\right)$ Write each term as a product of its factors.

$= 3(x + 3)$ Distributive Property

So, $3x + 9$ is _____.

Check

Factor $6x + 14$.

Show your work here

Example 4 Expressions With No Common Factors

Factor $12x + 7y$.

Write the prime factorization of each term.

$12x = 2 \cdot 2 \cdot 3 \cdot x$

$7y = 7 \cdot y$

What, if any, are the common factors?

_____

Because there are no common factors, $12x + 7y$ cannot be factored.

Check

Factor $12x + 11$.

Show your work here

 Go Online You can complete Extra Examples online.

Example 5 Factor Linear Expressions

Factor $\frac{1}{4}x + \frac{1}{4}$.

To factor $\frac{1}{4}x + \frac{1}{4}$, write each term as a product of the GCF and its remaining factors. The GCF of the terms is $\frac{1}{4}$.

$$\frac{1}{4}x + \frac{1}{4} = \frac{1}{4}\left(\boxed{}\right) + \frac{1}{4}\left(\boxed{}\right)$$

Write each term as a product of its factors.

$$= \frac{1}{4}\left(\boxed{}\right)$$

Distributive Property

So, $\frac{1}{4}x + \frac{1}{4} = \frac{1}{4}(x + 1)$.

Check

Factor $\frac{2}{3}x - \frac{2}{3}$.

Show your work here

 Go Online You can complete an Extra Example online.

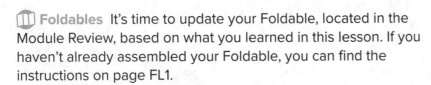 **Foldables** It's time to update your Foldable, located in the Module Review, based on what you learned in this lesson. If you haven't already assembled your Foldable, you can find the instructions on page FL1.

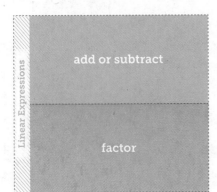

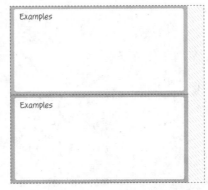

Talk About It!
When you factor out the common factor of $\frac{1}{4}$, where does the 1 come from in the expression $\frac{1}{4}(x + 1)$?

Pause and Reflect

Create a graphic organizer to record the steps for factoring linear expressions. Be sure to include examples of expressions that can be factored and expressions that cannot be factored.

Record your observations here

Practice

Go Online You can complete your homework online.

Find the GCF of each pair of monomials. (Example 1)

1. $4y, 12y$

2. $48x, 32x$

3. $16mn, 24m$

Use prime factorization to find the GCF of each pair of monomials. (Example 2)

4. $8xy, 12x$

5. $14ab, 28ab$

6. $27cd, 72cde$

Factor each expression. If the expression cannot be factored, write cannot be factored. (Examples 3–5)

7. $5x + 35$

8. $8x - 14$

9. $3x + 11y$

10. $32x - 15$

11. $72x - 18xy$

12. $45xy - 81y$

13. $25x + 14y$

14. $\frac{1}{3}x - \frac{1}{3}$

15. $\frac{1}{2}x + \frac{1}{2}$

Test Practice

16. Multiselect Select all of the expressions that cannot be factored.

☐ $7x - 14y$

☐ $27x - 18y$

☐ $9x + 31$

☐ $15x - 28y$

☐ $4x - 5y + 2z$

☐ $24x + 12x$

Apply

17. The total cost for Baydan and three of her friends to go ice skating can be represented by the expression $4x + 36$. The four friends pay an amount x to rent the ice skates and an admission fee. How much is the admission fee for one person?

18. The amount, in dollars, of Marisa's savings account can be represented by the expression $5x + 40$. Marisa saved the same amount x each month for a period of 5 months. Her mother contributed an additional amount each month to Marisa's savings account. How much did her mother contribute each month?

19. MP **Identify Structure** Write two monomials whose greatest common factor is $3x$.

20. MP **Identify Structure** What expression, in factored form, is $3x(3 + 7y)$?

21. **Which One Doesn't Belong?** Identify the expression that does not belong with the other three. Explain your reasoning.

   a. $7x + 35$

   b. $3x - 27$

   c. $7x + 3$

   d. $7x + 21$

22. MP **Find the Error** A student is factoring $18x + 6x$. Find the student's mistake and correct it.

   $18x + 6x = 6x(3)$

   $= 18x$

Combine Operations with Linear Expressions

I Can... combine operations to simplify linear expressions.

Example 1 Combine Operations to Simplify Expressions

Simplify $-2(x + 3) + 8x$. Write your answer in factored form.

$$-2(x + 3) + 8x = -2x - 6 + 8x \qquad \text{Distributive Property}$$
$$= 6x - 6 \qquad \text{Combine like terms.}$$
$$= 6(x - 1) \qquad \text{Write in factored form.}$$

So, $-2(x + 3) + 8x = $ _____.

Think About It!
How would you begin simplifying the expression?

Check

Simplify $-3(4x - 9) + 30x$. Write your answer in factored form.

Talk About It!
Why was the Distributive Property performed first?

 **Go Online** You can complete an Extra Example online.

Example 2 Combine Operations to Simplify Expressions

Simplify $\frac{3}{8}x + \frac{1}{2}\left(\frac{1}{4}x - 2\right)$.

$$\frac{3}{8}x + \frac{1}{2}\left(\frac{1}{4}x - 2\right) \qquad \text{Write the expression.}$$
$$= \frac{3}{8}x + \left(\frac{1}{2}\right)\left(\frac{1}{4}x\right) - \left(\frac{1}{2}\right)(2) \qquad \text{Distributive Property}$$
$$= \frac{3}{8}x + \frac{1}{8}x - 1 \qquad \text{Multiply.}$$
$$= \frac{4}{8}x - 1 \qquad \text{Combine like terms.}$$
$$= \frac{1}{2}x - 1 \qquad \text{Simplify.}$$

So, $\frac{3}{8}x + \frac{1}{2}\left(\frac{1}{4}x - 2\right) = \frac{1}{2}x - 1$.

Talk About It!
Why is the answer $\frac{4}{8}x - 1$ not completely simplified?

Check

Simplify $\frac{5}{12}a - \frac{1}{4}\left(\frac{2}{3}a - 8\right)$.

Show your work here

Example 3 Combine Operations to Simplify Expressions

Simplify $\frac{2}{3}(18x - 12) - (6x + 7)$. Write your answer in factored form.

$\frac{2}{3}(18x - 12) - (6x + 7)$

$= (12x - 8) - (6x + 7)$	Distributive Property
$= (12x - 8) + (-6x - 7)$	Add the additive inverse.
$= 12x - 8$	Arrange like terms in
$\underline{(+) -6x - 7}$	columns.
$6x - 15$	Add.
$= 3(2x - 5)$	Write in factored form.

So, $\frac{2}{3}(18x - 12) - (6x + 7) = $ _____ .

Check

Simplify $\frac{2}{3}(27x - 45) - (4x - 9)$. Write your answer in factored form.

Show your work here

 Go Online You can complete an Extra Example online.

Pause and Reflect

What part(s) of the lesson made you want to learn more? Why?

Record your observations here

🌐 Apply Gardening

A garden consists of a rectangular sitting region surrounded by a flower border. The sitting region has a length of 3x feet and a width of 5 feet. The flower border is 3 feet wide. Write an expression, in factored form, that represents the area of the flower border.

Go Online watch the animation.

1 What is the task?

Make sure you understand exactly what question to answer or problem to solve. You may want to read the problem three times. Discuss these questions with a partner.

First Time Describe the context of the problem, in your own words.
Second Time What mathematics do you see in the problem?
Third Time What are you wondering about?

2 How can you approach the task? What strategies can you use?

Record your observations here

3 What is your solution?

Use your strategy to solve the problem.

Show your work here

Talk About It!

Why can't you find a numerical value for the area of the flower border?

4 How can you show your solution is reasonable?

✏️ **Write About It!** Write an argument that can be used to defend your solution.

Check

The diagram shows a walkway that is 5 feet wide surrounding a rectangular swimming pool. Write an expression, in factored form, that represents the area of the walkway.

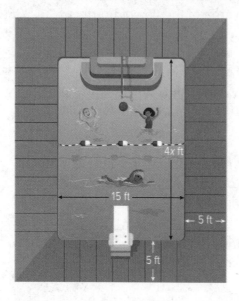

4x ft

15 ft

5 ft

5 ft

Show your work here

🢒 **Go Online** You can complete an Extra Example online.

Pause and Reflect

Write a real-world problem that uses the concepts from today's lesson. Explain how you came up with that problem. Exchange problems with a classmate and solve each other's problem.

Record your observations here

Practice

◐ **Go Online** You can complete your homework online.

Simplify each expression. For Exercises 1—4 and 9—12, write your answer in factored form. (Examples 1–3)

1. $3(x + 4) + 5x$

2. $-4(x + 1) + 6x$

3. $-5(2x - 6) + 25x$

4. $2(-8x - 3) + 18x$

5. $\frac{1}{6}x + \frac{3}{4}\left(\frac{1}{2}x - 4\right)$

6. $\frac{2}{3}\left(6x - \frac{1}{6}\right) + 3x$

7. $\frac{5}{8}x + \frac{1}{2}\left(\frac{1}{4}x + 10\right)$

8. $\frac{2}{5}\left(10x + \frac{3}{4}\right) - 2x$

9. $\frac{3}{4}(24x + 28) - (4x - 1)$

10. $-\frac{1}{2}(32x - 40) + (20x - 4)$

11. $\frac{2}{3}(9x - 15) - (-6x + 2)$

12. $-\frac{4}{5}(30x - 40) + (42x + 4)$

Test Practice

13. Equation Editor The table shows the area of two rugs a preschool teacher has in her room. She places the two rugs together to make one big rug. What is the area in square units of the new rug in factored form?

Rug	Area (square units)
Blue	$8\left(x + \frac{1}{2}\right)$
Red	$10\left(x + \frac{2}{5}\right)$

Apply

14. The diagram shows a border that is 4 inches wide surrounding Annie's painting. Write an expression, in factored form, that represents the area of the border.

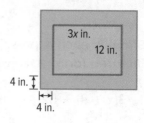

3x in.

12 in.

4 in.

4 in.

15. Raul's hamster's house sits in the corner of its cage as shown in the diagram. The area around the house is 5 inches wide. Write an expression, in factored form, that represents the area around the house.

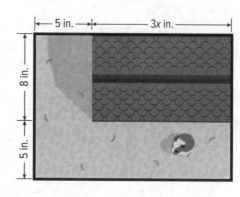

5 in.   3x in.

8 in.

5 in.

16. Create Write and simplify a linear expression with more than one operation.

17. **Find the Error** A student is simplifying the expression below. Find the student's error and correct it.

$$-3(x + 2) + 6x = -3x + 2 + 6x$$
$$= 3x + 2$$

18. A student said that $\frac{4}{6}x + 1$ is written in simplest form. Is the student correct? Explain why or why not.

19. **Persevere with Problems** Write each expression in factored form.

a. $\frac{1}{2}x + 6$

b. $\frac{3}{4}x - 18$

📙 **Foldables** Use your Foldable to help review the module.

Linear Expressions

Explanation

Explanation

Rate Yourself!

Complete the chart at the beginning of the module by placing a checkmark in each row that corresponds with how much you know about each topic after completing this module.

Write about one thing you learned.

_____
_____
_____
_____
_____
_____
_____

Write about a question you still have.

_____
_____
_____
_____
_____
_____
_____

Reflect on the Module

Use what you learned about expressions to complete the graphic organizer.

e **Essential Question**

Why is it beneficial to rewrite expressions in different forms?

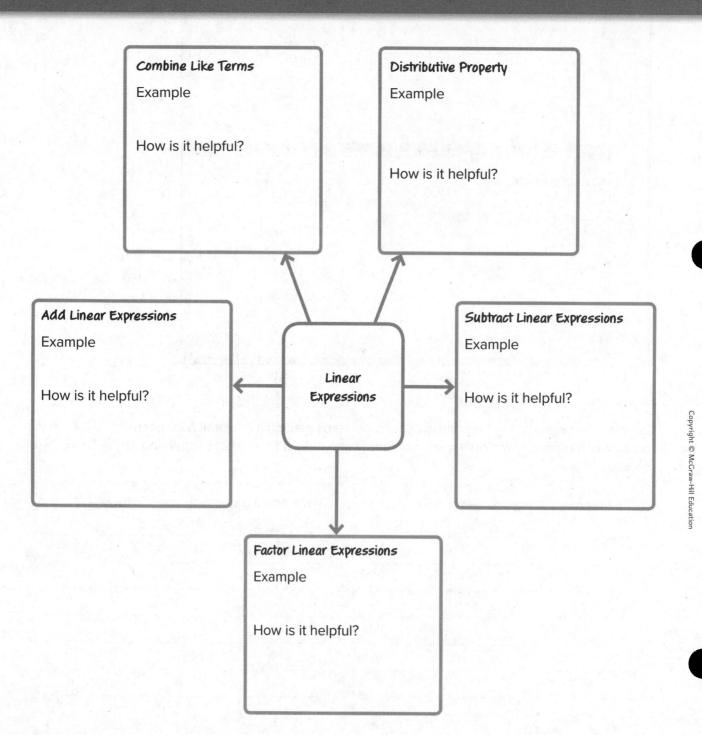

Combine Like Terms

Example

How is it helpful?

Distributive Property

Example

How is it helpful?

Add Linear Expressions

Example

How is it helpful?

Linear Expressions

Subtract Linear Expressions

Example

How is it helpful?

Factor Linear Expressions

Example

How is it helpful?

Test Practice

1. Equation Editor The cost of Noah's lunch, c, after a 15% tip can be represented by the expression $c + 0.15c$. What is this expression written in simplest form? **(Lesson 1)**

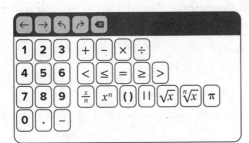

2. Multiple Choice Which of the following correctly shows the result when $-6(5a - 3b)$ is expanded using the Distributive Property? **(Lesson 1)**

Ⓐ $-30a - 18b$

Ⓑ $-30a + 18b$

Ⓒ $-6a - 30b$

Ⓓ $-a - 9b$

3. Open Response Two triangles are shown. **(Lesson 1)**

Triangle 1

17 / $4(-3x + 22)$ / $6x$

Triangle 2

$5x$ / 24 / $3(5x - 18)$

A. Represent the perimeter of each triangle as an algebraic expression written in simplest form.

B. If $x = 5$, which triangle has a greater perimeter?

4. Multiselect Which of the following expressions are linear? Select all that apply. **(Lesson 2)**

☐ $8x - 1$ ☐ $9x^2$

☐ $-3x$ ☐ $4x + 3$

☐ $7xy$ ☐ $\frac{1}{x}$

5. Multiple Choice What is the simplest form of $(-7x + 3) + (11x - 4)$? **(Lesson 2)**

Ⓐ $4x - 7$

Ⓑ $4x + 7$

Ⓒ $4x - 1$

Ⓓ $18x + 1$

6. Equation Editor What is $(-3x + 5) - (-7x + 2)$? **(Lesson 3)**

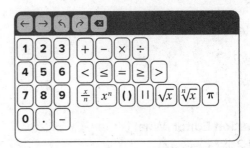

7. Open Response What is $\left(\frac{1}{4}x - 7\right) - \left(\frac{3}{4}x - 5\right)$? Explain how you found your answer. **(Lesson 3)**

8. Open Response The table shows the sales of retractable leashes and standard leashes at a pet store for *w* weeks. (Lesson 3)

Style	Price ($)	Number Sold
Retractable	17	$8w + 5$
Standard	9	$3w + 2$

A. Write an expression that represents the difference between the number of retractable leashes sold and the number of standard leashes sold.

B. After 15 weeks, how many more retractable leashes has the store sold?

9. Multiple Choice Which of the following is the GCF of $16m$ and $40mn$? (Lesson 4)

Ⓐ 8

Ⓑ $4m$

Ⓒ $8m$

Ⓓ $4mn$

10. Equation Editor What is the GCF of $20x$ and $25x$? (Lesson 4)

11. Multiple Choice What is $10xy - 15y$ written in factored form? (Lesson 4)

Ⓐ $5y(2x - 3)$

Ⓑ $5(2x - 3y)$

Ⓒ $10y(x - 3)$

Ⓓ cannot be factored

12. Open Response Factor $12m + 5n$. Explain how you found your answer. (Lesson 4)

13. Equation Editor Simplify $3(2x - 1) - 4x$. Write your answer in factored form. (Lesson 5)

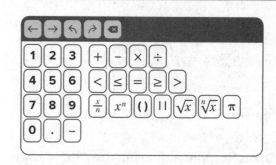

14. Open Response A swimming pool has a 3-foot wide cement walkway around its perimeter as shown. (Lesson 5)

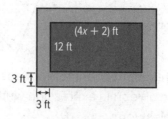

Write an expression, in factored form, that represents the area (in square feet) of the cement walkway.

Module 7
Equations and Inequalities

e Essential Question
How can equations be used to solve everyday problems?

What Will You Learn?

Place a checkmark (✓) in each row that corresponds with how much you already know about each topic **before** starting this module.

KEY	Before			After		
■ — I don't know. ◆ — I've heard of it. ★ — I know it!	■	◆	★	■	◆	★
writing and solving two-step equations in the form of $px + q = r$						
writing and solving two-step equations in the form of $p(x + q) = r$						
writing and solving equations with variables on each side						
writing and solving multi-step equations						
creating and solving equations with infinitely many solutions						
creating and solving equations with no solution						
graphing inequalities						
writing and solving one-step addition and subtraction inequalities						
writing and solving one-step multiplication and division inequalities						
writing and solving two-step inequalities						

📖 Foldables Cut out the Foldable and tape it to the Module Review at the end of the module. You can use the Foldable throughout the module as you learn about writing and solving equations and inequalities.

What Vocabulary Will You Learn?

Check the box next to each vocabulary term that you may already know.

☐ Addition Property of Inequality   ☐ order of operations

☐ Division Property of Inequality   ☐ Subtraction Property of Inequality

☐ inequality   ☐ two-step equation

☐ Multiplication Property of Inequality   ☐ two-step inequality

Are You Ready?

Study the Quick Review to see if you are ready to start this module.
Then complete the Quick Check.

Quick Review	
Example 1 **Solve one-step addition and subtraction equations.** Solve $34 = x - 12$. $\begin{aligned} 34 &= x - 12 \quad \text{Write the equation.} \\ +12 &= +12 \quad \text{Addition Property of Equality} \\ 46 &= x \qquad\quad \text{Simplify.} \end{aligned}$	**Example 2** **Solve one-step multiplication and division equations.** Solve $-5n = 35$. $\begin{aligned} -5n &= 35 \quad \text{Write the equation.} \\ \frac{-5n}{-5} &= \frac{35}{-5} \quad \text{Division Property of Equality} \\ n &= -7 \quad \text{Simplify.} \end{aligned}$

Quick Check	
1. Ana has 8 stamps. Together, Ricky and Ana have 23 stamps. The equation $23 = r + 8$ represents this situation, where r is the number of stamps Ricky has. Solve the equation to find the number of stamps Ricky has.	**2.** A certain number of plates will be placed on 8 tables. Each table will have 6 plates. The equation $\frac{p}{8} = 6$ represents this situation, where p is the total number of plates. Solve the equation to find the total number plates.

How Did You Do?

Which exercises did you answer correctly in the Quick Check?
Shade those exercise numbers at the right.

Write and Solve Two-Step Equations: $px + q = r$

I Can... Write two-step equations of the form $px + q = r$ and use inverse operations to solve the equations.

What Vocabulary Will You Learn?
order of operations

two-step equation

Explore Solve Two-Step Equations Using Algebra Tiles

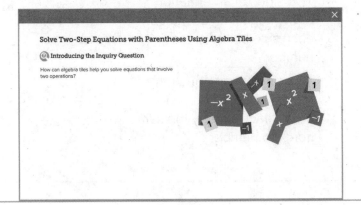 **Online Activity** You will use algebra tiles to explore how to represent and solve two-step equations.

Learn Two-Step Equations

A **two-step equation**, such as $5x + 3 = 13$, has two operations paired with the variable. In this case, the operations are multiplication and addition. To solve a two-step equation, undo the operations in reverse order of the **order of operations**.

 Go Online Watch the animation to see how to solve the two-step equation $5x + 3 = 13$.

Talk About It!
Compare and contrast the equations.
$5x + 3 = 13$ and $5x = 10$.

Equation	Steps
$5x + 3 = 13$ $ -3 = -3$ $\overline{ 5x = 10}$	Undo the addition.
$\dfrac{5x}{5} = \dfrac{10}{5}$	Undo the multiplication.
$x = 2$	Simplify.
$5(2) + 3 \overset{?}{=} 13$ $13 = 13 \checkmark$	Check the solution.

Learn Properties of Equality

You can use the properties of equality to solve equations algebraically.

Property	Words	Symbols
Addition Property of Equality	Two sides of an equation remain equal when you add the same number to each side.	If $a = b$, then $a + c = b + c$.
Subtraction Property of Equality	Two sides of an equation remain equal when you subtract the same number from each side.	If $a = b$, then $a - c = b - c$.
Multiplication Property of Equality	Two sides of an equation remain equal if you multiply each side by the same number.	If $a = b$, then $ac = bc$.
Division Property of Equality	Two sides of an equation remain equal when you divide each side by the same nonzero number.	If $a = b$, and $c \neq 0$, then $\frac{a}{c} = \frac{b}{c}$.

Example 1 Solve Two-Step Equations

Solve $-2y - 7 = 3$. Check your solution.

$$-2y - 7 = 3 \qquad \text{Write the equation.}$$

$$\underline{+7 +7} \qquad \text{Addition Property of Equality}$$

$$-2y = \boxed{} \qquad \text{Simplify.}$$

$$\frac{-2y}{-2} = \frac{10}{-2} \qquad \text{Division Property of Equality}$$

$$y = \boxed{} \qquad \text{Simplify.}$$

So, the solution of the equation is $y = -5$.

Check your solution by substituting -5 for y in the equation.

$$-2(-5) - 7 \stackrel{?}{=} 3$$

$$10 - 7 = 3 \checkmark$$

Because $10 - 7 = 3$ is a true statement, the solution is correct.

Copyright © McGraw-Hill Education

🎂 **Think About It!**

What two operations are paired with the variable?

♻ **Talk About It!**

In the fourth line of the solution, why was each side of the equation divided by -2 instead of 2?

Check

Solve $5w - 8 = -3$.

Show
your work
here

Example 2 Solve Two-Step Equations

Solve $4 + \frac{1}{5}r = -1$. Check your solution.

$4 + \frac{1}{5}r = -1$	Write the equation.
$\underline{-4 \qquad\qquad -4}$	Subtraction Property of Equality
$\frac{1}{5}r = -5$	Simplify.
$5 \cdot \frac{1}{5}r = 5 \cdot (-5)$	Multiplication Property of Equality
$r = -25$	Simplify.

So, the solution of the equation is $r = -25$.

Check your solution by substituting -25 for r in the equation.

$4 + \frac{1}{5}(-25) \stackrel{?}{=} -1$

$4 + (-5) \quad = -1 \checkmark$

Because $4 + (-5) = -1$ is a true statement, the solution is correct.

Check

Solve $-2 + \frac{2}{3}w = 10$.

Show
your work
here

 Go Online You can complete an Extra Example online.

Pause and Reflect

Create a two-step equation involving a fractional coefficient. Trade your equation with a partner. Solve each other's equations and explain to each other how you handled the fractional coefficient.

Record your
observations
here

Think About It!

What do you notice about this equation?

Talk About It!

In the fourth line of the solution, why was each side of the equation multiplied by 5? Describe another strategy you can use to solve the equation.

Learn Two-Step Equations: Arithmetic Method and Algebraic Method

Using only numbers and operations to solve a problem is an arithmetic method.

Using variables to solve the problem is an algebraic method. Consider the following problem.

Natalie is participating in a jog-a-thon. Her goal is to raise $125. She has collected $80 and will receive $3 for each lap around the track she jogs. How many laps does she need to jog to reach her goal?

Arithmetic Method

The amount Natalie still needs to reach her goal can be represented by the following equation.

$125 - $80 = $45

She will earn $3 for every lap she jogs. So, she needs to jog $45 ÷ 3, or 15 laps to reach her goal.

Algebraic Method

Let x represent the number of laps Natalie needs to jog around the track to reach her goal. Write and solve an equation.

$$125 = 80 + 3x \qquad \text{Write the equation.}$$

$$\underline{-80 \quad -80} \qquad \text{Subtract 80 from each side.}$$

$$\frac{45}{3} = \frac{3x}{3} \qquad \text{Divide each side by 3.}$$

$$15 = x \qquad \text{Simplify.}$$

So, using either method, Natalie needs to jog 15 laps to reach her goal.

Pause and Reflect

How are the arithmetic and algebraic methods similar and different?

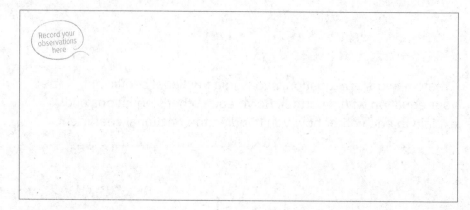

Record your observations here

(continued on next page)

Consider the following problem.

Rashan is saving money to buy a skateboard that costs $85. He has already saved $40. He plans to save the same amount each week for three weeks. How much should Rashan save each week?

Arithmetic Method

Solve the problem using the arithmetic method. Describe the steps you used.

Algebraic Method

Solve the problem using the algebraic method.

Let x represent the amount saved each week. The equation $40 + 3x = 85$ represents this situation. Solve the equation to find how much Rashan should save each week.

$$40 + 3x = 85$$

Compare and contrast the arithmetic method and algebraic method used to solve the problem.

So, using either method, Rashan should save $15 each week.

Pause and Reflect

Where did you encounter struggle in this lesson, and how did you deal with it? Write down any questions you still have.

Lesson 7-1 • Write and Solve Two-Step Equations: $px + q = r$ **387**

Online Activity You will use bar diagrams to explore how to write two-step equations to model and solve real-world problems.

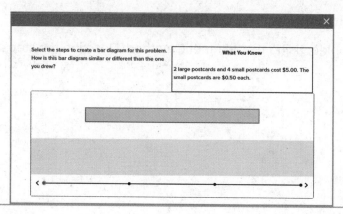

Learn Write Two-Step Equations

Some real-world situations can be represented by two-step equations. Consider the following problem.

A caterer is preparing a dinner for a party. She charges an initial fee of $16 and $8.25 per person. How many people can attend a dinner that costs $131.50?

The table shows how to model the problem with a two-step equation.

Words
Describe the mathematics of the problem.
The initial fee of $16 plus $8.25 per person equals $131.50.
Variable
Define the variable to represent the unknown quantity.
Let p represent the number of people.
Equation
Translate the words into an algebraic equation.
$16 + 8.25p = 131.50$

Example 3 Write and Solve Two-Step Equations

Toya had her birthday party at the movies. It cost $27 for pizza and $8.50 per friend for the movie tickets.

Write and solve an equation to determine how many friends Toya had at her party if she spent $78.

Part A Write an equation.

Words
The cost of pizza plus the cost per friend times the number of friends equals $78.
Variable
Let n represent the number of friends.
Bar Diagram

```
|------------------- $78 -----------------|
 ┌─────────────┬──────────────────────────┐
 │   pizza     │         tickets          │
 └─────────────┴──────────────────────────┘
|---- $27 ----|------- $8.50 • n --------|
```

Equation
$27 + 8.50n = 78$

Talk About It!

How can you use the bar diagram to check your answer?

Part B Solve the equation.

$27 + 8.50n = 78$	Write the equation.
$\underline{-27 \qquad\qquad -27}$	Subtraction Property of Equality
$8.50n = 51$	Simplify.
$\dfrac{8.50n}{8.50} = \dfrac{51}{8.50}$	Division Property of Equality
$n = 6$	Simplify.

So, Toya had _____ friends at her party.

Talk About It!

Why was it important to subtract $27 before dividing by $8.50?

Check the solution.

If Toya had _____ friends at her party, then the cost of the movie tickets was 6($8.50), or $51. Adding the cost of the pizza means the total cost was $51 + $27 or $78, which is what Toya spent. The solution is correct.

Check

Cassidy went to a football game with some of her friends. The tickets cost $6.50 each, and they spent $17.50 on snacks. The total amount paid was $63.00. Write and solve an equation to determine the number of people p that went to the game.

Show your work here

 Example 4 Write and Solve Two-Step Equations

Diego's aquarium contains $30\frac{1}{2}$ gallons of water. He drains the water at a rate of 5 gallons per minute for cleaning.

Write and solve an equation to determine in how many minutes the amount of water will reach $10\frac{1}{2}$ gallons.

Part A Write an equation.

Words
$30\frac{1}{2}$ gallons minus 5 gallons per minute equals $10\frac{1}{2}$ gallons.
Variable
Let m represent the number of minutes.
Equation
$30\frac{1}{2} - 5m = 10\frac{1}{2}$

Part B Solve the equation.

$$30\frac{1}{2} - 5m = 10\frac{1}{2}$$ Write the equation.

$$\underline{-30\frac{1}{2} \qquad\qquad -30\frac{1}{2}}$$ Subtraction Property of Equality

$$-5m = -20$$ Simplify.

$$\frac{-5m}{-5} = \frac{-20}{-5}$$ Division Property of Equality

$$m = 4$$ Simplify.

So, it will take _____ minutes to drain the tank to $10\frac{1}{2}$ gallons.

Check

Amelia started with $54, and spent $6 each day at camp. She has $18 left. Write and solve an equation to find how many days d Amelia was at camp.

Show your work here

 Think About It!

What is the unknown in this problem?

Talk About It!

How can you check your answer for reasonableness?

Go Online You can complete an Extra Example online.

Copyright © McGraw-Hill Education

🌐 Apply Budgets

The students at Worthingway Middle School would like to rent a moon bounce for an end-of-year party. The graph shows the cost to rent the moon bounce for locations within the regular delivery area. To deliver the moon bounce to locations beyond the regular delivery area, the company charges a $50 delivery fee. Worthingway Middle School is located beyond the regular delivery area. The students have $200 to spend. For how many full hours can they rent the moon bounce?

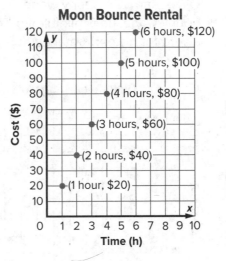

Moon Bounce Rental

- (6 hours, $120)
- (5 hours, $100)
- (4 hours, $80)
- (3 hours, $60)
- (2 hours, $40)
- (1 hour, $20)

Cost ($) vs Time (h)

🡕 **Go Online** Watch the animation.

1 What is the task?

Make sure you understand exactly what question to answer or problem to solve. You may want to read the problem three times. Discuss these questions with a partner.

First Time Describe the context of the problem, in your own words.
Second Time What mathematics do you see in the problem?
Third Time What are you wondering about?

2 How can you approach the task? What strategies can you use?

Record your observations here

3 What is your solution?

Use your strategy to solve the problem.

Show your work here

4 How can you show your solution is reasonable?

🖊 **Write About It!** Write an argument that can be used to defend your solution.

💬 **Talk About It!**
How can you use the graph to determine the answer?

Check

Oliver earns an hourly wage, as shown in the graph. In addition to his hourly wage, he is eligible for bonuses. If he received a $100 bonus award for his performance and worked 20 hours, how much did he earn?

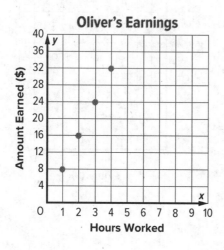

Oliver's Earnings

Show your work here

Go Online You can complete an Extra Example online.

Foldables It's time to update your Foldable, located in the Module Review, based on what you learned in this lesson. If you haven't already assembled your Foldable, you can find the instructions on page FL1.

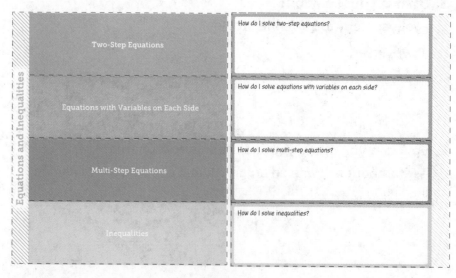

Practice

Go Online You can complete your homework online.

Solve each equation. Check your solution. (Examples 1–3)

1. $5x + 2 = 17$

2. $-6x - 7 = 17$

3. $-5 = 3x - 14$

4. $3.8 = 2x - 11.2$

5. $2 + \frac{1}{6}x = -4$

6. $-9 = \frac{2}{7}x + 5$

Write and solve an equation for each exercise. Check your solution. (Examples 4 and 5)

7. Easton went to a concert with some of his friends. The tickets cost $29.50 each, and they spent a total of $15 on parking. The total amount spent was $133. Determine how many people went to the concert.

8. Ishi bought a $6.95 canvas and 8 tubes of paint. She spent a total of $24.95 on the canvas and paints. Determine the cost of each tube of paint.

9. A hot air balloon is at an altitude of $100\frac{1}{5}$ yards. The balloon's altitude decreases by $10\frac{4}{5}$ yards every minute. Determine the number of minutes it will take the balloon to reach an altitude of 57 yards.

10. The current temperature is 48°F. It is expected to drop 1.5°F each hour. Determine in how many hours the temperature will be 36°F.

Test Practice

11. Open Response The table shows the costs of a membership and fruit baskets at a discount warehouse club. Mrs. Williams paid a total of $105 for her annual membership fee and several fruit baskets as gifts for her coworkers. Solve the equation $15x + 30 = 105$ to find the number of fruit baskets Mrs. Williams purchased.

Item	Cost ($)
Membership Fee	30
Fruit Basket	15

Apply

12. A face painting artist's fee for parties is shown in the table. Customers are also charged a $25 reservation fee. A school has $175 to spend on a face painting artist for their carnival. For how many full hours can they hire the artist?

Hours	Fee ($)
2	49.00
3	73.50
4	98.00
5	122.50

13. The table shows the cost of boarding a dog at a dog kennel. Owners are also charged a $10 registration fee. The Kittle family boarded their dog at the kennel for 7 days. What was the total, with registration fee, of boarding their dog?

Days	Cost ($)
3	97.50
4	130.00
5	162.50
6	195.00

14. Write a real-world problem that could be represented by the equation $2x + 5 = 35$. Then solve the equation.

15. (MP) **Persevere with Problems** Mia discovered that if she takes half her age and adds 3, it produces the same results as taking one-fourth of her age and adding 7. How old is Mia?

16. (MP) **Multiple Representations** At the start of a school trip, a minibus contains 28 gallons of gas. Each hour, the minibus uses 6 gallons of gas. The bus will stop when there are 4 gallons of gas left.
a. Make a table to show how many gallons of gas are remaining after 1, 2, and 3 hours.

b. Write and solve an equation to find how many hours will pass before the minibus will have to stop for gas.

17. (MP) **Find the Error** A student is solving $-5 + 2x = 15$. Find the student's mistake and correct it.

$$-5 + 2x = 15$$
$$-5 + (-5) + 2x = 15 + (-5)$$
$$2x = 10$$
$$x = 5$$

Write and Solve Two-Step Equations: $p(x + q) = r$

I Can... write two-step equations of the form $p(x + q) = r$ and use inverse operations to solve the equations.

Explore Solve Two-Step Equations Using Algebra Tiles

Online Activity You will use algebra tiles to explore how to model and solve two-step equations with parentheses.

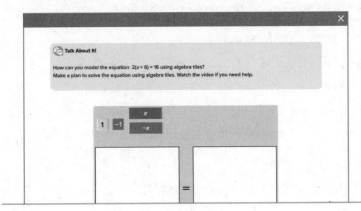

Learn Two-Step Equations

An equation like $2(x + 6) = 14$ is in the form $p(x + q) = r$. It contains two factors, p and $(x + q)$, and is considered a two-step equation because two steps are needed to solve the equation.

Go Online Watch the animation to learn how to solve two-step equations with parentheses.

Follow the steps to solve the equation $3(x + 2) = -18$ using the Distributive Property.

Equation	Steps
$3(x + 2) = -18$	
$3x + 6 = -18$	Distributive Property
$\underline{-6 \quad -6}$	Subtraction Property of Equality
$\dfrac{3x}{3} = \dfrac{-24}{3}$	Division Property of Equality
$x = -8$	

(continued on next page)

Copyright © McGraw-Hill Education

Follow the steps to solve the equation using the properties of equality.

Equation	Steps
$\dfrac{3(x+2)}{3} = \dfrac{-18}{3}$	Division Property of Equality
$x + 2 = -6$	Simplify.
$\underline{-2-2}$	Subtraction Property of Equality
$x = -8$	

Think About It!

How is the equation $3(x + 5) = 45$ different from the equation $3x + 5 = 45$?

Example 1 Solve Two-Step Equations

Solve $3(x + 5) = 45$. Check your solution.

Method 1 Use the Division Property of Equality first.

$3(x + 5) = 45$	Write the equation.
$\dfrac{3(x+5)}{3} = \dfrac{45}{3}$	Division Property of Equality
$x + 5 = 15$	Simplify.
$\underline{-5-5}$	Subtraction Property of Equality
$x = 10$	Simplify.

Talk About It!

Compare and contrast the two methods used to solve the equation.

Method 2 Use the Distributive Property first.

$3(x + 5) = 45$	Write the equation.
$3x + 15 = 45$	Distributive Property
$\underline{-15-15}$	Subtraction Property of Equality
$3x = 30$	Simplify.
$\dfrac{3x}{3} = \dfrac{30}{3}$	Division Property of Equality
$x = 10$	Simplify.

So, using either method, the solution of the equation is $x = 10$.

Check your solution by substituting the result back into the original equation.

$3(x + 5) = 45$	Write the original equation.
$3(10 + 5) \overset{?}{=} 45$	Replace x with 10.
$45 = 45$ ✓	The sentence is true.

Check

Solve $2(x + 4) = -20$.

Show your work here

Think About It!

How would you begin solving the equation?

Example 2 Solve Two-Step Equations

Solve $5(n - 2) = -30$. Check your solution.

Method 1 Use the Division Property of Equality first.

$5(n - 2) = -30$	Write the equation.
$\dfrac{5(n - 2)}{5} = \dfrac{-30}{5}$	Division Property of Equality
$n - 2 = -6$	Simplify.
$\underline{+2 \quad +2}$	Addition Property of Equality
$n = -4$	Simplify.

Method 2 Use the Distributive Property.

$5(n - 2) = -30$	Write the equation.
$5n - 10 = -30$	Distributive Property
$\underline{+10 \quad +10}$	Addition Property of Equality
$5n = -20$	Simplify.
$\dfrac{5n}{5} = \dfrac{-20}{5}$	Division Property of Equality
$n = -4$	Simplify.

Talk About It!

When might it be advantageous to use one method over the other?

So, using either method, the solution of the equation is $n = -4$.

Check your solution by substituting the result back into the original equation.

$5(n - 2) = -30$	Write the original equation.
$5(-4 - 2) \stackrel{?}{=} -30$	Replace n with -4.
$-30 = -30$ ✓	The sentence is true.

Check

Solve $3(x - 6) = -12$.

Show your work here

 Go Online You can complete an Extra Example online.

How can you use the
reciprocal of $\frac{2}{3}$ to
solve this equation?

🧠 **Think About It!**

How can you use the
reciprocal of $\frac{2}{3}$ to
solve this equation?

💬 **Talk About It!**

Why is the use of
reciprocals important
in solving the
equation?

Example 3 Solve Two-Step Equations

Solve $\frac{2}{3}(n + 6) = 10$. Check your solution.

$$\frac{2}{3}(n + 6) = 10 \qquad \text{Write the equation.}$$

$$\frac{3}{2} \cdot \frac{2}{3}(n + 6) = \frac{3}{2} \cdot 10 \qquad \text{Multiplication Property of Equality}$$

$$(n + 6) = \frac{3}{2} \cdot \frac{10}{1} \qquad \frac{3}{2} \cdot \frac{2}{3} = 1; \text{ write 10 as } \frac{10}{1}.$$

$$n + 6 = 15 \qquad \text{Simplify.}$$

$$\underline{\quad -6 \quad -6 \quad} \qquad \text{Subtraction Property of Equality}$$

$$n = 9 \qquad \text{Simplify.}$$

So, the solution to the equation is $n = 9$.

Check your solution by substituting the result back into the equation.

$$\frac{2}{3}(n + 6) = 10 \qquad \text{Write the original equation.}$$

$$\frac{2}{3}(9 + 6) \stackrel{?}{=} 10 \qquad \text{Replace } n \text{ with 9.}$$

$$10 = 10 \checkmark \qquad \text{The sentence is true.}$$

Check

Solve $\frac{1}{4}(d - 3) = -15$.

🔖 **Go Online** You can complete an Extra Example online.

Learn Two-Step Equations: Arithmetic Method and Algebraic Method

Using only numbers and operations to solve a problem is an arithmetic method. Using variables to solve the problem is an algebraic method. Consider the following problem.

Three friends went to a band party at a local farm. Each student spent the same amount of money and a total of $21 altogether. Each student bought a hot dog for $5. If they each also bought a hay ride ticket, how much did each hay ride ticket cost?

(continued on next page)

The table shows how to use these methods to solve this problem.

Arithmetic Method	Algebraic Method
Divide the total amount spent by 3 to find the amount each friend spent. $$\$21 \div \$3 = \$7$$ Then subtract the cost of a hot dog from $7 to find the cost of a hay ride ticket: $$\$7 - \$5 = \$2$$ So, each hay ride ticket cost $2.	Let x represent the cost of a hay ride ticket. The equation $3(x + 5) = 21$ represents this situation. Solve the equation to find the cost of a hay ride ticket. $$3(x + 5) = 21$$ $$\frac{3(x + 5)}{3} = \frac{21}{3}$$ $$x + 5 = 7$$ $$\underline{-5 \quad -5}$$ $$x = 2$$

Talk About It!

Compare and contrast the arithmetic method and algebraic method used to solve the problem.

Explore Write Two-Step Equations

Online Activity You will use bar diagrams to write two-step equations with parentheses to represent real-world problems.

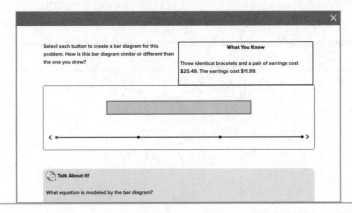

Learn Write Two-Step Equations

Some real-world situations can be modeled by two-step equations of the form $p(x + q) = r$. Consider the following problem.

Mr. Vargas takes his class of 24 students ice skating. Each student pays an entrance fee to enter the rink and a $4.75 fee to rent skates. The total cost for the students to enter the rink and rent skates is $234. What is the ice-skating rink's entrance fee?

Words
24 times the total cost for each student is $234.

Variable
Let f represent the entrance fee. So, $f + 4.75$ is the total cost for each student.

Equation
$24(f + 4.75) = 234$

Example 4 Write and Solve Two-Step Equations

Think About It!

What will the variable represent in this problem?

Mackenzie drives the same distance to and from school each day. She also drives $1\frac{3}{4}$ miles round trip each day to go to the library. During a 5-day school week, Mackenzie drives a total of 50 miles.

Write and solve an equation to determine the total distance to and from school.

Part A Write an equation.

Words
5 times the total distance each day equals 50 miles.

Variable
Let d represent the total distance to and from school. So, $d + 1\frac{3}{4}$ is the total distance she drives each day.

Bar Diagram
Complete the bar diagram to assist you in writing the equation.

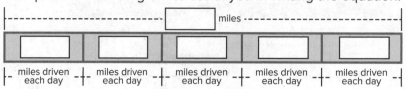

| miles | miles driven each day | miles driven each day | miles driven each day | miles driven each day | miles driven each day |

Equation
The equation is $5\left(d + 1\frac{3}{4}\right) = 50$.

Talk About It!

How does the bar diagram help you write the equation

$5\left(d + 1\frac{3}{4}\right) = 50$?

Part B Solve the equation.

$5\left(d + 1\frac{3}{4}\right) = 50$ Write the equation.

$\dfrac{5\left(d + 1\frac{3}{4}\right)}{5} = \dfrac{50}{5}$ Division Property of Equality

$d + 1\frac{3}{4} = 10$ Simplify.

$\underline{-1\frac{3}{4} \quad -1\frac{3}{4}}$ Subtraction Property of Equality

$d = 8\frac{1}{4}$ Simplify.

So, the distance to and from school is _____ miles.

Check

Mrs. Byers is making 5 costumes that each require $1\frac{3}{8}$ yards of blue fabric and a certain amount of red fabric. She will use $8\frac{3}{4}$ yards in all. Write and solve an equation to determine the number of yards of red fabric r she will need for each costume.

Show your work here

🖱 **Go Online** You can complete an Extra Example online.

Pause and Reflect

Explain how you wrote and solved the equation in the Check problem. Did you use a bar diagram? Why or why not? Can you solve the problem in a different way from what was shown in the Example? If so, explain.

Record your observations here

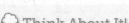 Example 5 Write and Solve Two-Step Equations

Jamal and two cousins received the same amount of money to go to a movie. Each boy spent $15. Afterward, the boys had $30 altogether.

Write and solve an equation to find the amount of money each boy received.

Part A Write an equation.

Words
3 times the amount of money each boy has left to spend equals $30.
Variable
Let m represent the amount of money each boy received. So, $m - 15$ is the amount of money each boy has left to spend.
Equation
The equation is $3\left(\boxed{}\right) = 30$.

Part B Solve the equation.

 Talk About It!

Does the solution make sense in the context of the problem?

$3(m - 15) = 30$ Write the equation.

$\dfrac{3(m - 15)}{3} = \dfrac{30}{3}$ Division Property of Equality

$m - 15 = 10$ Simplify.

$\underline{+\ 15 \quad +\ 15}$ Addition Property of Equality

$m = 25$ Simplify.

So, each boy received $ _____ .

Check

Mr. Singh had three sheets of stickers. He gave 20 stickers from each sheet to his students and has 12 total stickers left. Write and solve an equation to find the total number of stickers s there were originally on each sheet.

 Go Online You can complete an Extra Example online.

🌐 Apply Perimeter

Pierre uses two rectangular pieces of paper as bookmarks. The width of Bookmark B is equal to the length of Bookmark A. The length of Bookmark B is equal to half the perimeter of Bookmark A. The perimeter of Bookmark A is 14 centimeters. What is the perimeter of Bookmark B?

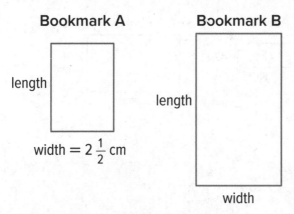

Bookmark A

length

width = $2\frac{1}{2}$ cm

Bookmark B

length

width

1 What is the task?

Make sure you understand exactly what question to answer or problem to solve. You may want to read the problem three times. Discuss these questions with a partner.

First Time Describe the context of the problem, in your own words.
Second Time What mathematics do you see in the problem?
Third Time What are you wondering about?

2 How can you approach the task? What strategies can you use?

Record your observations here

3 What is your solution?

Use your strategy to solve the problem.

Show your work here

4 How can you show your solution is reasonable?

✏️ **Write About It!** Write an argument that can be used to defend your solution.

💬 **Talk About It!**
How can you find the length of Bookmark A?

Check

Two rectangular swimming pools are shown. The length of Pool A is equal to the width of Pool B. The width of Pool A is $\frac{1}{4}$ the length of Pool B. The perimeter of Pool B is 120 feet. What is the perimeter of Pool A?

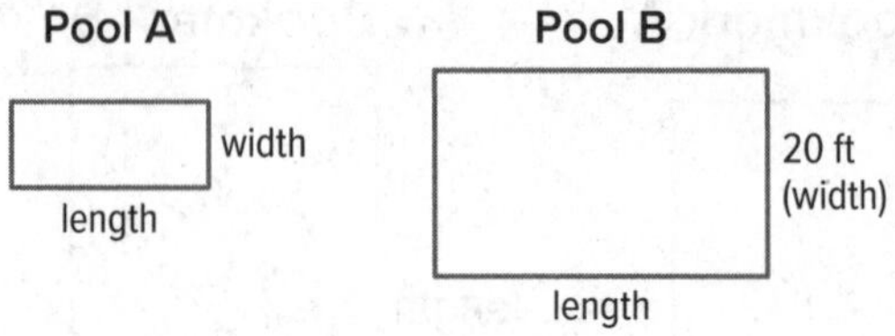

Pool A
width
length
Pool B
20 ft
(width)
length

Show your work here

Go Online You can complete an Extra Example online.

Foldables It's time to update your Foldable, located in the Module Review, based on what you learned in this lesson. If you haven't already assembled your Foldable, you can find the instructions on page FL1.

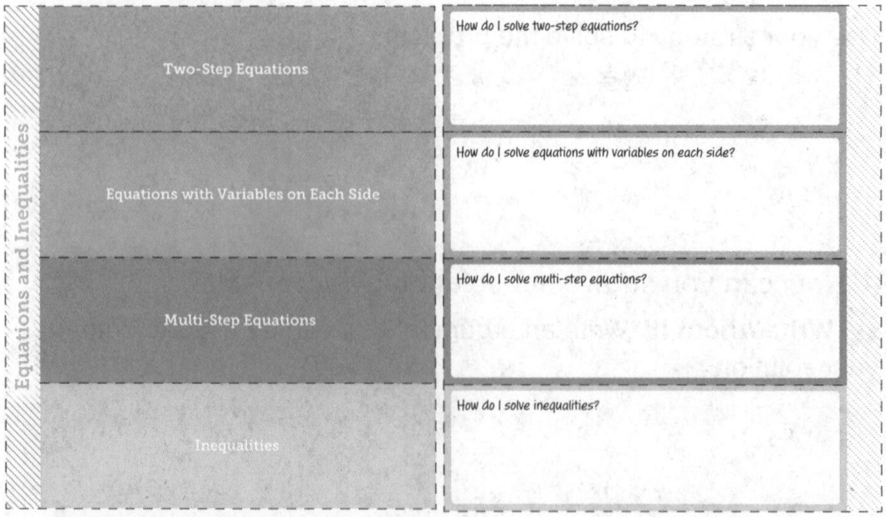

Equations and Inequalities
Two-Step Equations
How do I solve two-step equations?
Equations with Variables on Each Side
How do I solve equations with variables on each side?
Multi-Step Equations
How do I solve multi-step equations?
Inequalities
How do I solve inequalities?

Practice

⊙ **Go Online** You can complete your homework online.

Solve each equation. Check your solution. (Examples 1–3)

1. $-2(x + 4) = 18$

2. $10(x - 5) = -80$

3. $-0.25(8 + x) = 14$

4. $-0.8(10 - x) = 36$

5. $\frac{1}{2}(x - 4) = 5$

6. $-\frac{7}{9}(x + 3) = 14$

Write and solve an equation for each exercise. Check your solution. (Examples 4 and 5)

7. Ayana is making 6 scarves that each require $1\frac{1}{4}$ yards of purple fabric and a certain amount of blue fabric. She will use 10 yards in all. Determine how many yards of blue fabric are needed for each scarf.

8. Sara is making 3 batches of chocolate chip cookies and 3 batches of oatmeal cookies. Each batch of chocolate chip cookies uses $2\frac{1}{4}$ cups of flour. She will use $12\frac{3}{4}$ cups of flour for all six batches. Determine how many cups of flour are needed for each batch of oatmeal cookies.

9. Javier bought 3 bags of balloons for a party. He used 8 balloons from each bag. Determine how many balloons were originally in each bag if there were 21 balloons left over.

10. Vera and her three sisters received the same amount of money to go to the school festival. Each girl spent $12. Afterward, the girls had $24 altogether. Determine the amount of money each girl received.

Test Practice

11. Open Response Mrs. James buys 5 hat and glove sets for charity. She has coupons for $1.50 off the regular price of each set. After using the coupons, the total cost is $48.75. Determine the regular price of a hat and glove set.

Item	Cost ($)
Hat and glove set	p
Scarf	9.99

12. Olive and Ryan have picture frames with the same perimeter. Olive's picture frame has a width 1.25 times the width of Ryan's picture frame. What is the length ℓ of Olive's picture frame?

Ryan's Picture Frame

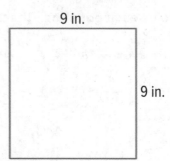

9 in.

9 in.

13. A landscape architect designed a flower garden in the shape of a trapezoid. The area of the garden is 13.92 square meters. A fence is planned around the perimeter of the garden. How many meters of fencing are needed?

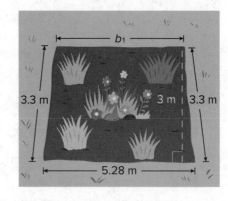

b_1

3.3 m   3 m   3.3 m

5.28 m

14. **Justify Conclusions** Suppose for some value of x the solution to the equation $2.5(y - x) = 0$ is $y = 6$. What must be true about x? Justify your conclusion.

15. **Persevere with Problems** Keith is 5 years older than Trina. Two times the sum of their ages is 62. Write and solve an equation to find Keith's age.

16. **Find the Error** A student is solving $-2(x - 5) = 12$. Find the student's mistake and correct it.

$$-2(x - 5) = 12$$
$$-2x - 5 = 12$$
$$-2x - 5 + 5 = 12 + 5$$
$$-2x = 17$$
$$x = -8.5$$

17. **Create** Write a real-world problem that could be represented by the equation $12(x + 2.50) = 78$. Then solve the equation.

Write and Solve Equations with Variables on Each Side

I Can... use the properties of equality to write and solve equations with variables on each side that have rational coefficients.

Explore Equations with Variables on Each Side

Online Activity You will use Web Sketchpad to explore how using a balance can help you solve equations with variables on each side.

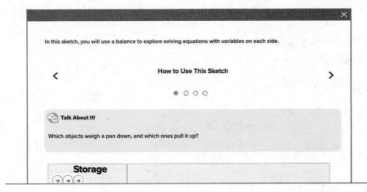

In this sketch, you will use a balance to explore solving equations with variables on each side.

How to Use This Sketch

‹   ● ○ ○ ○   ›

Talk About It!

Which objects weigh a pan down, and which ones pull it up?

Storage

Learn Equations with Variables on Each Side

Some equations, like $8 + 3x = 5x + 2$, have variables on each side of the equals sign. To solve, use the properties of equality to write an equivalent equation with variables on one side of the equals sign. Then solve the equation.

$8 + 3x = 5x + 2$	Write the equation.
$-3x = -3x$	Subtraction Property of Equality
$8 = 2x + 2$	Simplify.
$-2 = -2$	Subtraction Property of Equality
$\frac{6}{2} = \frac{2x}{2}$	Division Property of Equality
$3 = x$	Simplify.

Copyright © McGraw-Hill Education

(continued on next page)

Go Online Watch the animation to complete the steps to solve the equation $8 + 3x = 5x + 2$ using algebra tiles.

Step 1 Model the equation.

Draw eight 1-tiles and three x-tiles to model the $8 + 3x$ on the left side of the mat. Draw five x-tiles and two 1-tiles to model $5x + 2$ on the right side of the mat.

Step 2 Draw the remaining tiles after removing three x-tiles from each side of the mat.

The resulting equation is $8 = 2x + 2$.

Step 3 Draw the remaining tiles after removing two 1-tiles from each side of the mat.

The resulting equation is $6 = 2x$.

Step 4 Draw the tiles so that the tiles are separated into two equal groups.

There are three 1-tiles in each group, so $x = 3$.

Example 1 Solve Equations with Variables on Each Side

Solve $6n - 1 = 4n - 5$. Check your solution.

$$6n - 1 = 4n - 5$$ Write the equation.

$$\underline{-4n \quad\quad = \boxed{}}$$ Subtraction Property of Equality

$$\boxed{} - 1 = -5$$ Simplify.

$$\underline{\quad +1 = +1\quad}$$ Addition Property of Equality

$$2n = \boxed{}$$ Simplify.

$$\frac{2n}{2} = \frac{-4}{2}$$ Division Property of Equality

$$n = \boxed{}$$ Simplify.

So, the solution to the equation is $n = -2$.

Check the solution.

$$6n - 1 = 4n - 5$$ Write the equation.

$$6(\boxed{}) - 1 = 4(\boxed{}) - 5$$ Replace n with -2.

$$-12 - 1 = -8 - 5$$ Multiply.

$$\boxed{} = \boxed{}$$ Simplify. The solution, -2, is correct.

Check

Solve $10 - 3x = -5 + 2x$.

Copyright © McGraw-Hill Education

> **Talk About It!**
> Describe how you can use algebra tiles to solve the equation. Does it matter that the variable is n?

🌐 **Go Online** You can complete an Extra Example online.

Example 2 Solve Equations with Rational Coefficients

Solve $\frac{2}{3}x - 1 = 9 - \frac{1}{6}x$. **Check your solution.**

Method 1 Solve the equation using fractions.

$\frac{2}{3}x - 1 = 9 - \frac{1}{6}x$ — Write the equation.

$\frac{\boxed{}}{6}x - 1 = 9 - \frac{\boxed{}}{6}x$ — The common denominator of the coefficients is 6.

$\underline{+\frac{1}{6}x \quad\quad = \quad\quad +\frac{1}{6}x}$ — Addition Property of Equality

$\dfrac{\boxed{}}{\boxed{}}x - 1 = \boxed{}$ — Simplify.

$\underline{+1 = +1}$ — Addition Property of Equality

$\frac{5}{6}x \quad\quad = \boxed{}$ — Simplify.

$\left(\frac{6}{5}\right)\frac{5}{6}x = 10\left(\frac{6}{5}\right)$ — Multiplication Property of Equality

$x = \boxed{}$ — Simplify.

Method 2 Solve the equation by using the LCD to eliminate the fractions.

$\frac{2}{3}x - 1 = 9 - \frac{1}{6}x$ — Write the equation.

$\frac{4}{6}x - 1 = 9 - \frac{1}{6}x$ — Rewrite with common denominators.

$6\left(\frac{4}{6}x - 1\right) = \left(9 - \frac{1}{6}x\right)6$ — Multiply by 6 to eliminate fractions.

$\boxed{} - \boxed{} = \boxed{} - \boxed{}$ — Distributive Property

$\underline{+x \quad\quad = \quad\quad +x}$ — Addition Property of Equality

$\boxed{} - 6 = \ 54$ — Simplify.

$\underline{+6 = +6}$ — Addition Property of Equality

$5x \quad\quad = \boxed{}$ — Simplify.

$\frac{5x}{5} = \frac{60}{5}$ — Division Property of Equality

$x = \boxed{}$ — Simplify.

So, using either method, the solution to the equation is $x = 12$.

(continued on next page)

Think About It!

What do you notice about the coefficients in the equation?

Talk About It!

Compare the two methods used to solve the equation.

Check the solution.

$$\frac{2}{3}x - 1 = 9 - \frac{1}{6}x$$ Write the equation.

$$\frac{2}{3} \cdot \boxed{} - 1 = 9 - \frac{1}{6} \cdot \boxed{}$$ Replace x with 12.

$$\boxed{} - 1 = 9 - \boxed{}$$ Multiply.

$$7 = 7$$ Simplify. The solution, 12, is correct.

Check

Solve $\frac{2}{3}x + 5 = \frac{2}{5}x - 3$.

Show
your work
here

Go Online You can complete an Extra Example online.

Explore Write and Solve Equations with Variables on Each Side

Online Activity You will explore how to write an equation with variables on each side to solve a real-world problem.

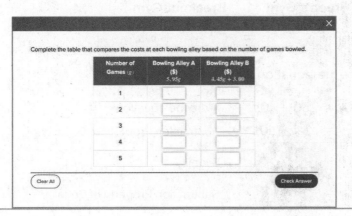

Complete the table that compares the costs at each bowling alley based on the number of games bowled.

Number of Games (g)	Bowling Alley A ($) 5.95g	Bowling Alley B ($) 4.45g + 3.00
1		
2		
3		
4		
5		

Clear All   Check Answer

Learn Write and Solve Equations with Variables on Each Side

You can represent many real-world problems using equations.

A music streaming website offers two plans. The first plan costs $0.99 per song plus an initial fee of $25. The second plan costs $1.50 per song plus an initial fee of $10. For how many songs will the two plans cost the same?

Words
a fee of $25 plus $0.99 per song is the same as a fee of $10 plus $1.50 per song
Variables
Let *s* represent the number of songs.
Equation
$25 + 0.99s = 10 + 1.50s$

 ## Example 3 Write and Solve Equations with Variables on Each Side

Green's Gym charges a one-time application fee of $50 plus $30 per session for a personal trainer. Breakout Gym charges an annual fee of $250 plus $10 for each session with a trainer.

For how many sessions is the cost of the two plans the same? Write and solve an equation to represent this problem. Check your solution.

Part A Write an equation to represent the problem.

Let *s* represent the number of _____. Write an equation that models when the cost of the two plans are equal to each other.

Green's Gym		Breakout Gym
50 + ☐	=	☐ + 10s

Part B Solve the equation.

$50 + 30s$	$= 250 + 10s$	Write the equation.
$-10s$	$= \quad -10s$	Subtraction Property of Equality
$50 + \boxed{}$	$= \boxed{}$	Simplify.
-50	$= -50$	Subtraction Property of Equality
$\boxed{}$	$= \boxed{}$	Simplify.
$\dfrac{20s}{20}$	$= \dfrac{200}{20}$	Division Property of Equality
s	$= \boxed{}$	Simplify.

So, the cost is the same for 10 personal trainer sessions.

(continued on next page)

 Talk About It!

What will a solution to the equation represent within the context of the problem?

Think About It!

What is the unknown in this problem?

Talk About It!

Suppose a customer plans to have a personal training session once a month, for one year. Which gym is more cost effective for them to choose? Explain.

Check the solution.

Green's Gym

$$50 + 30s = 50 + 30(10)$$

Replace *s* with 10.

$$= 50 + 300$$

Multiply.

$$= \boxed{}$$

Simplify.

Breakout Gym

$$250 + 10s = 250 + 10(10)$$

$$= 250 + 100$$

$$= \boxed{}$$

Check

A container has 130 gallons of water and is being filled at a rate of $\frac{1}{4}$ gallon each second. Another container has 200 gallons of water and is draining at a rate of $\frac{1}{3}$ gallon each second. Write and solve an equation that could be used to determine *s*, the number of seconds, when the two containers have the same amount of water.

 Go Online You can complete an Extra Example online.

🌐 **Example 4** Write and Solve Equations with Variables on Each Side

Ryan's Rentals charges $40 per day plus $0.25 per mile. Road Trips charges $25 per day plus $0.45 per mile.

For what number of miles is the daily cost of renting a car the same? Write and solve an equation to represent this problem. Check your solution.

Part A Write an equation.

Let *m* represent the number of _____. Write an equation that models when the cost of the two rentals are equal.

Ryan's Rentals   Road Trips

$$40 + \boxed{} = \boxed{} + 0.45m$$

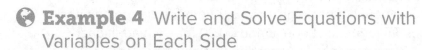

(continued on next page)

Part B Solve the equation.

$40 + 0.25m = 25 + 0.45m$ Write the equation.

$ - 0.25m = - 0.25m$ Subtraction Property of Equality

$\boxed{} = 25 + \boxed{}$ Simplify.

$- 25 = - 25$ Subtraction Property of Equality

$\boxed{} = \boxed{}$ Simplify.

$\dfrac{15}{0.20} = \dfrac{0.20m}{0.20}$ Division Property of Equality

$\boxed{} = m$ Simplify.

So, the cost is the same for 75 miles in one day.

Check the solution.

Ryan's Rentals

$40 + 0.25m$

$= 40 + (0.25)75$ Replace m with 75.

$= 40 + \boxed{}$ Multiply.

$= \boxed{}$ Simplify.

Road Trips

$25 + 0.45m$

$= 25 + (0.45)75$

$= 25 + \boxed{}$

$= \boxed{}$

Check

Annie is comparing the cost to ship a package. One shipping company charges $7 for the first pound and $0.20 for each additional pound a package weighs. Another shipping company charges $5 for the first pound and $0.30 for each additional pound. Write and solve an equation that could be used to determine p, the number of pounds, when the costs for the two shipping companies are the same.

Show your work here

Talk About It!

Which company is more cost effective to use if you plan to drive 50 miles in one day? 150 miles? Explain.

 Go Online You can complete an Extra Example online.

⊕ Apply Home Improvement

Suppose you are replacing the carpet in a living room where the length of the living room is five feet shorter than twice its width, *w*. Tack strip is placed around the perimeter of the room, which is equal to five times the width. If carpet costs $2.99 a square foot, what is the total cost to carpet the living room?

1 What is the task?

Make sure you understand exactly what question to answer or problem to solve. You may want to read the problem three times. Discuss these questions with a partner.

First Time Describe the context of the problem, in your own words.
Second Time What mathematics do you see in the problem?
Third Time What are you wondering about?

2 How can you approach the task? What strategies can you use?

3 What is your solution?

Use your strategy to solve the problem.

4 How can you show your solution is reasonable?

⊘ **Write About It!** Write an argument that can be used to defend your solution.

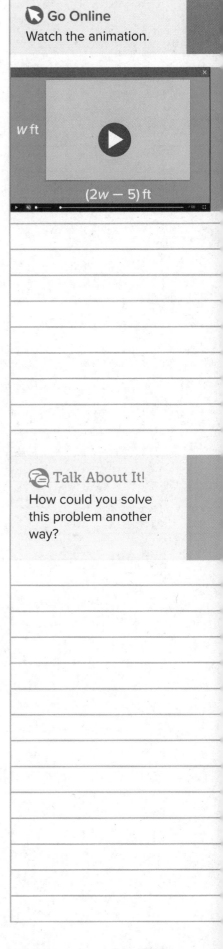

w ft

(2*w* − 5) ft

💬 **Talk About It!**
How could you solve this problem another way?

Check

A rectangular bathroom with the side lengths shown is being covered with tiles, where x is the length, in feet, of a square tile. The perimeter is equal to $48x - 6$. If each square foot of tile costs $8.49, what is the total cost to tile the bathroom?

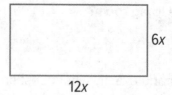

6x

12x

Show your work here

Go Online You can complete an Extra Example online.

Foldables It's time to update your Foldable, located in the Module Review, based on what you learned in this lesson. If you haven't already assembled your Foldable, you can find the instructions on page FL1.

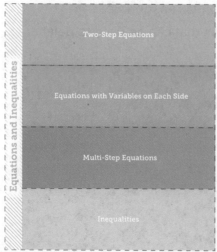

Equations and Inequalities

Two-Step Equations

Equations with Variables on Each Side

Multi-Step Equations

Inequalities

How do I solve two-step equations?

How do I solve equations with variables on each side?

How do I solve multi-step equations?

How do I solve inequalities?

Practice

Solve each equation. Check your solution. (Examples 1 and 2)

1. $-2a - 9 = 6a + 15$

2. $\frac{1}{2}x - 5 = 10 - \frac{3}{4}x$

3. $\frac{2}{3}y + 1 = \frac{1}{6}y + 8$

4. $5.4p + 13.1 = -2.6p + 3.5$

Write and solve an equation for each exercise. Check your solution.
(Examples 3 and 4)

5. Marko has 45 comic books in his collection, and Tamara has 61 comic books. Marko buys 4 new comic books each month and Tamara buys 2 comic books each month. After how many months will Marko and Tamara have the same number of comic books?

6. A fish tank has 150 gallons of water and is being drained at a rate of $\frac{1}{2}$ gallon each second. A second fish tank has 120 gallons of water and is being filled at a rate of $\frac{1}{4}$ gallon each second. After how many seconds will the two fish tanks have the same amount of water?

Test Practice

7. **Open Response** Deanna and Lulu are playing games at the arcade. Deanna starts with $15, and the machine she is playing costs $0.75 per game. Lulu starts with $13, and her machine costs $0.50 per game. After how many games will the two friends have the same amount of money remaining? Let g represent the number of games.

Apply

8. Aiden is replacing the tile in a rectangular kitchen. The length of the kitchen is nine feet shorter than three times its width, *w*. The perimeter of the kitchen is six times the width. If tiles cost $1.69 a square foot, what is the total cost to tile the kitchen?

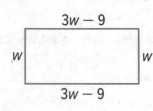

9. Hailey is putting up a fence in the shape of an isosceles triangle in her backyard. The fence has side lengths as shown, where *x* represents the number of feet in each fence section. The perimeter of the fence can be covered using 8 total fence sections represented by the expression 8*x*. If fencing costs $6.50 a foot, what would be the total cost of the fence?

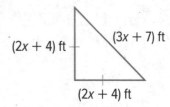

10. MP **Identify Structure** Explain how the Distributive Property can be used to eliminate the fractions in the equation $\frac{1}{5}x + 8 = \frac{1}{10}x + 9$.

11. Ling worked three more hours on Tuesday than she did on Monday. On Wednesday, she worked one hour more than twice the number of hours that she worked on Monday. The total number of hours is two more than five times the number of hours worked on Monday. Write and solve an equation to find the number of hours she worked on Monday.

12. MP **Find the Error** A student wrote the equation $22 + 4 = 6s + 12s$ to represent the problem shown at the right. Find his mistake and correct it.

Darnell and Emma are college students. Darnell currently has 22 credits and he plans on taking 6 credits per semester. Emma has 4 credits and plans to take 12 credits per semester. After how many semesters, *s*, will Darnell and Emma have the same number of credits?

Write and Solve Multi-Step Equations

I Can... write and solve multi-step linear equations with rational coefficients by using the Distributive Property and combining like terms.

Learn Solve Multi-Step Equations

Some equations contain expressions with grouping symbols on one or both sides of the equals sign.

To solve equations like this, first expand the expressions that contain grouping symbols. Then solve the equation, combining any like terms and using the Properties of Equality.

 Go Online Watch the animation to see how to solve the multi-step equation $-5(2x + 3) - x = 4(x + 11) + 1$.

The animation shows that you can use the Distributive Property and the Properties of Equality to solve a multi-step equation that contains expressions with grouping symbols.

$-5(2x + 3) - x = 4(x + 11) + 1$	Write the equation.
$-10x - 15 - x = 4x + 44 + 1$	Expand the expressions using the Distributive Property.
$-11x - 15 = \quad 4x + 45$	Combine the like terms $-10x$ and $-x$. Combine the like terms 44 and 1.
$\underline{+ 11x \qquad = + 11x}$	Addition Property of Equality
$-15 = 15x + 45$	Simplify.
$\underline{-45 = \qquad -45}$	Subtraction Property of Equality.
$-60 = 15x$	Simplify.
$\dfrac{-60}{15} = \dfrac{15x}{15}$	Division Property of Equality
$-4 = x$	Simplify.

The solution of the equation is $x = $ _____.

Talk About It!

How can you make sure that you solved the equation correctly?

Example 1 Solve Multi-Step Equations

Solve $3(8x + 12) - 15x = 2(3 - 3x)$. Check your solution.

$$3(8x + 12) - 15x = 2(3 - 3x)$$ Write the equation.

$$\boxed{} + \boxed{} - 15x = \boxed{} - \boxed{}$$ Distributive Property

$$\boxed{} + 36 = 6 - 6x$$ Combine like terms.

$$+ 6x \qquad = \quad + 6x$$ Addition Property of Equality

$$\boxed{} + 36 = 6$$ Simplify.

$$- 36 = -36$$ Subtraction Property of Equality

$$15x = \boxed{}$$ Simplify.

$$\frac{15x}{\boxed{}} = \frac{-30}{\boxed{}}$$ Division Property of Equality

$$x = \boxed{}$$ Simplify.

So, the solution to the equation is $x = -2$.

Check the solution.

$$3(8x + 12) - 15x = 2(3 - 3x)$$ Write the equation.

$$3\left[8\left(\boxed{}\right) + 12\right] - 15\left(\boxed{}\right) = 2\left[3 - 3\left(\boxed{}\right)\right]$$ Replace x with -2.

$$-12 - (-30) = 2(9)$$ Multiply.

$$\boxed{} = \boxed{}$$ Simplify.

Check

Solve $8(-2 + x) - 3x = 6x + 13$.

Show your work here

🔴 **Go Online** You can complete an Extra Example online.

Example 2 Solve Multi-Step Equations

Solve 0.3(10 − 5x) = 31.5 − (8x + 9). Check your solution.

$0.3(10 - 5x) = 31.5 - (8x + 9)$	Write the equation.
$\boxed{} - \boxed{} = 31.5 - (8x + 9)$	Distributive Property
$3 - 1.5x = 31.5 + (-8x - 9)$	Rewrite using the additive inverse.
$3 - 1.5x = \boxed{} - 8x$	Combine like terms.
$\underline{-3 \qquad\quad = -3}$	Subtraction Property of Equality
$\boxed{} = \boxed{} - 8x$	Simplify.
$\underline{+8x = \qquad\quad +8x}$	Addition Property of Equality
$\boxed{} = 19.5$	Simplify.
$\dfrac{6.5x}{\boxed{}} = \dfrac{19.5}{}$	Division Property of Equality
$x = \boxed{}$	Simplify.

So, the solution to the equation is $x = 3$.

Check the solution.

To check your solution, replace x with 3 in the original equation.

$0.3(10 - 5x) = 31.5 - (8x + 9)$	Write the equation.
$0.3\left(10 - 5 \cdot \boxed{}\right) = 31.5 - \left(8 \cdot \boxed{} + 9\right)$	Substitute.
$0.3\left(10 - \boxed{}\right) = 31.5 - \left(\boxed{} + 9\right)$	Multiply.
$0.3\left(\boxed{}\right) = 31.5 - \boxed{}$	Simplify.
$\boxed{} = \boxed{}$	Simplify.

Check

Solve $15 - (7x + 20) = 0.5(6x - 14)$.

Show your work here

Go Online You can complete an Extra Example online.

Think About It!

How would you begin solving the equation?

Talk About It!

How does the Additive Inverse Property allow the parentheses to be removed in order to simplify the right side of the equation?

Example 3 Solve Multi-Step Equations

Solve $\frac{1}{2}(6x - 4) + 6x = -12\left(\frac{1}{6}x + 2\right)$. Check your solution.

$\frac{1}{2}(6x - 4) + 6x = -12\left(\frac{1}{6}x + 2\right)$ Write the equation.

$\boxed{} - \boxed{} + 6x = \boxed{} - \boxed{}$ Distributive Property

$\boxed{} - 2 = -2x - 24$ Combine like terms.

$\underline{ + 2x = + 2x }$ Addition Property of Equality

$\boxed{} - 2 = -24$ Simplify.

$\underline{ + 2 = + 2 }$ Addition Property of Equality

$11x = \boxed{}$ Simplify.

$\dfrac{11x}{\boxed{}} = \dfrac{-22}{\boxed{}}$ Division Property of Equality

$x = \boxed{}$ Simplify.

So, the solution to the equation is $x = -2$.

Check

Solve $\frac{3}{4}(-5x - 8) = -6(x + 4) + \frac{1}{4}x$.

Show your work here

👆 **Go Online** You can complete an Extra Example online.

Explore Translate Problems into Equations

👆 **Online Activity** You will explore how to model a real-world problem with a multi-step equation with variables on both sides.

Complete the table to show the cost, in dollars, of shirts for the number of players shown.

Number of Players	Cost of Shirt with Name, $
1	$20 + n$
2	$2(20 + n)$
3	$3(20 + n)$
4	
5	
6	

Learn Write and Solve Multi-Step Equations

You can represent many real-world problems using multi-step equations.

Four friends bought zoo tickets and spent $9.50 each on wristbands for the zoo rides. The same day, a group of 5 different friends bought zoo tickets and spent $3 each on the sting ray exhibit. The group of 5 friends also rented a locker for $8. If both groups spent the same amount, how much did each person pay for a zoo ticket?

Words
4 times the cost of a ticket and a $9.50 wrist band is the same as an $8 locker plus five times the cost of a ticket and a $3 exhibit
Variables
Let t represent the cost of each ticket.
Equation
$4(t + 9.50) = 8 + 5(t + 3)$

 Talk About It!

Describe another real-world situation in which you might need to solve a multi-step equation in order to solve a problem.

🌐 Example 4 Write and Solve Multi-Step Equations

Mrs. Hill is designing a rectangular vegetable garden for her backyard. The width of the garden is $11\frac{1}{2}$ feet shorter than twice its length.

If the perimeter of the garden is 37 feet, what is the length of the garden? Write and solve an equation. Check your solution.

Part A Write an equation.

Let ℓ represent the _____, in feet, of the garden.

The width of the garden is twice the length _____ $11\frac{1}{2}$ feet.

So, the expression $2\ell - 11\frac{1}{2}$ represents the _____ of the garden.

Write an equation that models the perimeter.

$P = 2\ell + 2w$

$\boxed{} = 2\ell + 2\left(\boxed{}\right)$

 Think About It!

What is the unknown in the problem?

 Talk About It!

What is another equation you can write to represent the perimeter of the garden, other than $P = 2\ell + 2w$?

(continued on next page)

Part B Solve the equation.

$$37 = 2\ell + 2\left(2\ell - 11\tfrac{1}{2}\right)$$ Write the equation.

$$37 = 2\ell + \boxed{} - \boxed{}$$ Distributive Property

$$37 = \boxed{} - 23$$ Combine like terms.

$$+\,23 = +\,23$$ Addition Property of Equality

$$\boxed{} = 6\ell$$ Simplify.

$$\frac{60}{6} = \frac{6\ell}{6}$$ Division Property of Equality

$$\boxed{} = \ell$$ Simplify.

So, the length of the garden is 10 feet.

Check the solution.

Length: $\boxed{}$ feet

Width: $2\ell - 11\tfrac{1}{2} = 2(10) - 11\tfrac{1}{2}$ or $\boxed{}$ feet

Perimeter: $2\ell - 2w = 2(10) + 2(8.5)$ or $\boxed{}$ feet

Check

The hallway in Oscar's house is in the shape of a rectangle. The length of the hallway is $\tfrac{1}{2}$ foot longer than twice its width. The perimeter of the hallway is 31 feet.

Part A

Which equation(s) can be used to find the width w of the hallway? Select all that apply.

☐ $31 = \tfrac{1}{2} + 4w$ ☐ $31 = 1 + 6w$ ☐ $31 = 2 + 2w$

☐ $31 = 2\left(\tfrac{1}{2} + 2w\right) + 2w$ ☐ $31 = \tfrac{1}{2} + 2w$ ☐ $31 = 2\left(\tfrac{1}{2} + 3w\right)$

Part B

What is the width of the hallway?

Show your work here

 Go Online You can complete an Extra Example online.

⊕ Example 5 Write and Solve Multi-Step Equations

Mr. Murphy's class of 20 students is going on a field trip to the science center. They will also watch the 3-D movie. Mrs. Todd's class of 15 students is going on a field trip to the art museum and will take the audio tour. Admission to the art museum is 2.5 times that of the science center's entry fee, as shown in the table.

If the total cost is the same at both the science center and the art museum, what is the entry fee per student to the science center? Check your solution.

Science Center	Art Museum
Entry fee: $x per student	Admission: $2.5x per student
3-D movie: $2.50 per student	Audio Tour: $1 per student

Think About It!
What quantity will the variable represent?

Part A Write an equation.

Let x represent the cost of the entry fee per student to the science center. The cost of the trip to the science center can be represented by the expression $(x + 2.50)$ for each student to pay the entry fee and 3-D movie. The cost of the trip to the Art Museum can be represented by the expression $(2.5x + 1)$ for admission and the audio tour for each student. Complete the equation that models the total cost.

$$\boxed{}(x + 2.50) = \boxed{}(2.5x + 1)$$

Part B Solve the equation.

$20(x + 2.50) = 15(2.5x + 1)$	Write the equation.
$\boxed{} + \boxed{} = \boxed{} + \boxed{}$	Distributive Property
$\underline{-20x \qquad\quad = -20x}$	Subtraction Property of Equality
$50 = \boxed{} + 15$	Simplify.
$\underline{-15 = \qquad\quad -15}$	Subtraction Property of Equality
$\boxed{} = 17.5x$	Simplify.
$\dfrac{35}{\boxed{}} = \dfrac{17.5x}{\boxed{}}$	Division Property of Equality
$\boxed{} = x$	Simplify.

So, the entry fee per student to the science center is $2.

Check the solution.

Science Center: $20(x + 2.5) = 20(2 + 2.50)$ or $\boxed{}$

Art Museum: $15(2.5x + 1) = 15(2.5(2) + 1)$ or $\boxed{}$

Talk About It!
Can you solve the equation without using the Distributive Property? Why is the Distributive Property helpful?

Copyright © McGraw-Hill Education

Check

A group of 6 friends went to a basketball tournament game. They each bought a ticket for x dollars and spent $8.50 each on snacks. Another group of 4 friends went to the championship game and paid twice as much for each of their tickets as the first group. The group of 4 friends also spent $6.50 each on snacks. The two groups spent the same amount of money. Write and solve an equation that can be used to find the cost of each ticket for the group of 6 friends.

Show your work here

 Go Online You can complete an Extra Example online.

Pause and Reflect

Write one sentence about today's lesson for each of the categories: Who, What, Where, How, and Why.

Record your observations here

Business Finance

The table shows the hours worked by employees at a coffee shop last month. If each employee earns $15 per hour and the total number of hours worked is represented by $7m + 19$ where m represents the number of hours Mai worked.

Employee	Hours Worked
Shantel	48
Lorenzo	$2m + 7$
Jamie	$3.5(m - 6)$
Mai	m

What was the total payroll, or amount paid to the employees?

1 What is the task?

Make sure you understand exactly what question to answer or problem to solve. You may want to read the problem three times. Discuss these questions with a partner.

First Time Describe the context of the problem, in your own words.
Second Time What mathematics do you see in the problem?
Third Time What are you wondering about?

2 How can you approach the task? What strategies can you use?

Record your observations here

3 What is your solution?

Use your strategy to solve the problem.

Show your work here

4 How can you show your solution is reasonable?

Write About It! Write an argument that can be used to defend your solution.

Talk About It!
What method did you use to solve the problem? Explain why you chose that method.

Check

The table shows the number of minutes Declan participated in various activities last week. The total number of minutes he participated in all of the activities was $5x + 15.75$. What is the ratio of the number of minutes Declan rode his bike to the number of minutes he had soccer practice?

Activity	Time (minutes)
Ride Bike	x
Soccer	$2.5(x + 3) + 2.5$
Swim	15.5
Dog Walk	$3(x - 13.25)$

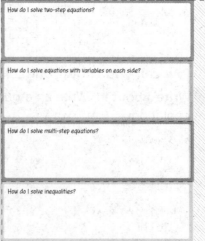

Go Online You can complete an Extra Example online.

Foldables It's time to update your Foldable, located in the Module Review, based on what you learned in this lesson. If you haven't already assembled your Foldable, you can find the instructions on page FL1.

Practice

Go Online You can complete your homework online.

Solve each equation. Check your solution. (Examples 1–3)

1. $-g + 2(3 + g) = -4(g + 1)$

2. $0.6(4 - 2x) = 20.5 - (3x + 10)$

3. $\frac{1}{2}(-4 + 6n) = \frac{1}{3}n + \frac{2}{3}(n + 9)$

4. $\frac{1}{5}(5x - 5) + 3x = -9\left(\frac{1}{3}x + 4\right)$

Write and solve an equation for each exercise. Check your solution.
(Examples 4 and 5)

5. Mr. Reed is drawing a blueprint of a rectangular patio. The width of the patio is $40\frac{3}{4}$ feet shorter than twice its length. The perimeter of the patio is $86\frac{1}{2}$ feet. What is the length of the patio?

6. The Yearbook Club is going to an amusement park, and each of their 12 members will pay for admission and will also help pay for parking. The Robotics Club is going to a waterpark, and each of their 14 members will pay for admission and will also purchase a meal ticket. Admission to the amusement park is 1.5 times that of the waterpark's admission, as shown in the table. If the total cost is the same at both the amusement park and the waterpark, what is the admission per student to the waterpark?

Amusement Park	Waterpark
Admission: $1.5x$ per student	Admission: x per student
Parking: $2 per student	Meal Ticket: $10.50 per student

Test Practice

7. Equation Editor Solve the equation shown for q.

$$2\left(\frac{1}{2}q + 1\right) = -3(2q - 1) + 8q + 4$$

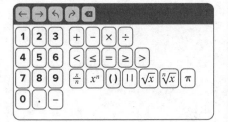

Apply

8. Four siblings have a dog walking business. The table shows the hours worked by each sibling. Each sibling earns $25.50 per hour and the total number of hours worked is represented by $10h + 15$, where h represents the number of hours Michael worked. What was the total amount the siblings earned?

Sibling	Hours Worked
Martin	$2.5h + 3$
Emilio	$4(h - 2)$
Michael	h
Mario	31

9. The triangle and the square shown have the same perimeter. Write and solve an equation to find the value of x. Then find the length of one side of the square.

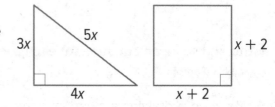

10. Ⓜ️ **Find the Error** A student solved the equation $3(-4 + x) - 5x = 7x + 15$. Find her mistake and correct it.

$$3(-4 + x) - 5x = 7x + 15$$
$$-12 + 3x - 15x = 7x + 15$$
$$-12 - 12x = 7x + 15$$
$$-12 - 12x + 12x = 7x + 12x + 15$$
$$-12 = 19x + 15$$
$$-12 - 15 = 19x + 15 - 15$$
$$-27 = 19x$$
$$-1\frac{8}{19} = x$$

11. Ⓜ️ **Persevere with Problems** Elijah put $2x + 3$ dollars in the bank the first week. The following week he doubled the first week's savings and put that amount in the bank. The next week, he doubled what was in the bank and put that amount in the bank. He now has $477 in the bank. Write and solve an equation to find how much money he put in the bank the first week.

12. Ⓜ️ **Identify Structure** Describe the role of the Distributive Property when solving multi-step equations that contain expressions with grouping symbols.

Determine the Number of Solutions

I Can... identify the number of solutions of a linear equation in one variable by simplifying each side and comparing coefficients and constants.

Explore Number of Solutions

Online Activity You will use Web Sketchpad to explore equations with one solution, no solution, and infinitely many solutions.

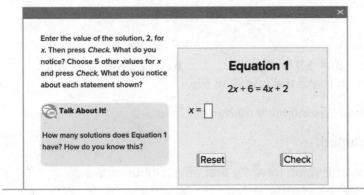

Enter the value of the solution, 2, for x. Then press *Check*. What do you notice? Choose 5 other values for x and press *Check*. What do you notice about each statement shown?

Talk About It!

How many solutions does Equation 1 have? How do you know this?

Equation 1

$$2x + 6 = 4x + 2$$

$x = \boxed{}$

Reset   Check

Learn Number of Solutions

The solution to an equation is the value of the variable that makes the equation true. Some equations have one solution, while some equations have no solution. When this occurs, no value will make the equation true. Other equations may have infinitely many solutions. When this occurs, the equation is true for every value of the variable.

	No Solution	**One Solution**	**Infinite Solutions**
Symbols	$a = b$	$x = a$	$a = a$
Examples	$3x + 4 = 3x$ $4 = 0$	$2x = 20$ $x = 10$	$4x + 2 = 4x + 2$ $2 = 2$
	Since $4 \neq 0$, there is no solution.	Since $x = 10$, there is one solution.	Since $2 = 2$, all of the values of x are solutions.

Talk About It!

Study the structure of the equations with no solution compared to those with infinitely many solutions. What do you notice?

Your Notes

🗨 Think About It!

How will you know if the equation has one solution, no solution, or infinitely many solutions?

🗨 Talk About It!

After simplifying each side of the equation, the equation becomes $6x - 8 = 6x - 8$. Why is it not necessary to finish solving to find the number of solutions?

Example 1 Equations with Infinitely Many Solutions

Solve $6(x - 3) + 10 = 2(3x - 4)$. Determine whether the equation has one solution, no solution, or infinitely many solutions. Check your solution.

$$6(x - 3) + 10 = 2(3x - 4)$$ Write the equation.

$$\boxed{} - \boxed{} + 10 = \boxed{} - \boxed{}$$ Distributive Property

$$6x - \boxed{} = 6x - 8$$ Combine like terms.

$$-6x \qquad = -6x$$ Subtract $6x$ from each side.

$$-8 = -8$$ Simplify.

The equation $-8 = -8$ is _____ true because any value can be substituted to make the equation true.

So, the equation has infinitely many solutions.

Check the solution.

Replace x with any value to verify that any solution will work.

$$6(x - 3) + 10 = 2(3x - 4)$$ Write the equation.

$$6\left(\boxed{} - 3\right) + 10 = 2\left(3 \cdot \boxed{} - 4\right)$$ Replace x with 4.

$$16 = 16$$ Simplify.

Check

Which equation has an infinite number of solutions?

Ⓐ $2(3c - 6) - 2c = 4(c + 4)$

Ⓑ $3(2p - 1) = 2(p + 10) + 1$

Ⓒ $8(x - 9) = 6(2x - 12) - 4x$

Ⓓ $5(x - 2) - 20 = -5(x - 6)$

Show your work here

🅑 **Go Online** You can complete an Extra Example online.

Example 2 Equations with No Solution

Solve $8(4 - 2x) = 4(3 - 5x) + 4x$. Determine whether the equation has one solution, no solution, or infinitely many solutions.

$8(4 - 2x) = 4(3 - 5x) + 4x$ Write the equation.

$\boxed{} - \boxed{} = \boxed{} - \boxed{} + 4x$ Distributive Property

$32 - 16x = 12 - \boxed{}$ Combine like terms.

$\underline{ + 16x = + 16x }$ Addition Property of Equality

$\boxed{} = 12$ Simplify.

The equation $32 = 12$ is _____ true because no values can be substituted to make the equation true.

So, the equation has no solution.

Check

Which equation has no solution?

Ⓐ $5(b + 3) - 2b = 2b + 3$

Ⓑ $4(d - 3) + 5 = 3(d + 2) - 7$

Ⓒ $3(x + 5) = 5(x + 3) - 2x$

Ⓓ $-10y + 18 = -3(5y - 7) + 5y$

Show your work here

🅡 **Go Online** You can complete an Extra Example online.

Think About It!
What are some possible first steps to solving this equation?

Talk About It!
After simplifying each side of the equation, the equation becomes $32 - 16x = 12 - 16x$. Without continuing to solve, how can you determine that the equation has no solution just by studying it?

Talk About It!

What do you notice about the relationship between the number of solutions and the coefficients of each side?

Learn Analyze Equations to Determine the Number of Solutions

It is possible to determine the number of solutions to an equation without actually solving it. The number of solutions to an equation can be found after each side has been simplified.

Complete the table to indicate whether the coefficients and constants are *the same* or *different*.

	No Solution	One Solution	Infinite Solutions
Equation	$6x + 3 = 6x + 1$	$6x + 3 = 4x + 1$	$6x + 3 = 6x + 3$
Coefficients	the same		the same
Constants		different or the same	

Example 3 Create Equations with Infinitely Many Solutions

What numbers would complete the equation so that it has infinitely many solutions?

$$6x - x + 4 + 2x = \boxed{?}\, x + \boxed{?}$$

$$6x - x + 4 + 2x = \boxed{?}\, x + \boxed{?} \qquad \text{Write the equation.}$$

$$\boxed{} + 4 = \boxed{?}\, x + \boxed{?} \qquad \text{Combine like terms.}$$

$$7x + 4 = \boxed{}\, x + \boxed{} \qquad \begin{array}{l}\text{The coefficients and constants}\\ \text{must be the same on each side.}\end{array}$$

So, $6x - x + 4 + 2x = 7x + 4$ is the equation with infinitely many solutions.

Talk About It!

How can you verify that the equation $6x - x + 4 + 2x = 7x + 4$ has infinitely many solutions?

Check

Complete the equation with values that will result in an equation with infinitely many solutions.

$$4x - 2(x + 5) = \boxed{}\, x - \boxed{}$$

Show your work here

🔵 **Go Online** You can complete an Extra Example online.

Example 4 Create Equations with No Solution

What numbers would complete the equation so that it has no solution?

$3(2x + 4) - x = \boxed{?}\, x + \boxed{?}$

$$3(2x + 4) - x = \boxed{?}\, x + \boxed{?} \qquad \text{Write the equation.}$$

$$\boxed{} + \boxed{} - x = \boxed{?}\, x + \boxed{?} \qquad \text{Distributive Property}$$

$$\boxed{} + 12 = \boxed{?}\, x + \boxed{?} \qquad \text{Combine like terms.}$$

So, $5x + 12 = 5x + 8$ has no solution since they have the same coefficient and different constants.

Check

Complete the equation with values that will result in an equation with no solution.

$-3x + 8x - 6 - x = \boxed{}\, x - \boxed{}$

Show your work here

 Go Online You can complete an Extra Example online.

Pause and Reflect

How does knowing the structure of equations with infinitely many solutions and equations with no solution help you create these types of equations?

> Record your observations here

💬 **Talk About It!**

How can you verify that the equation
$3(2x + 4) - x = 5x + 8$
has no solution?

💬 **Talk About It!**

What are some other expressions, other than $5x + 8$, that would result in the equation having no solution?

Pause and Reflect

Create a graphic organizer that will help you study the concepts you learned today in class.

> Record your observations here

🌐 Apply School

Caden, Lilly, and Amelia are analyzing the expressions $\frac{1}{2}(11x + 24)$ and $-\left(\frac{1}{2}x + 5\right) + 6(x + 8)$, for all values of x. Each student claims the expressions are related according to the results shown in the table. Which student is correct?

Caden	$\frac{1}{2}(11x + 24)$	$=$	$-\left(\frac{1}{2}x + 5\right) + 6(x + 8)$
Lilly	$\frac{1}{2}(11x + 24)$	$>$	$-\left(\frac{1}{2}x + 5\right) + 6(x + 8)$
Amelia	$\frac{1}{2}(11x + 24)$	$<$	$-\left(\frac{1}{2}x + 5\right) + 6(x + 8)$

1 What is the task?

Make sure you understand exactly what question to answer or problem to solve. You may want to read the problem three times. Discuss these questions with a partner.

First Time Describe the context of the problem, in your own words.
Second Time What mathematics do you see in the problem?
Third Time What are you wondering about?

2 How can you approach the task? What strategies can you use?

Record your observations here

3 What is your solution?

Use your strategy to solve the problem.

Show your work here

4 How can you show your solution is reasonable?

✍️ **Write About It!** Write an argument that can be used to defend your solution.

💬 **Talk About It!**

For all values of x,

$\frac{1}{2}(11x + 24) < -\left(\frac{1}{2}x + 5\right) + 6(x + 8)$.

Generate two different expressions, A and B, for the right side of the inequality, according to the guidelines below.

$\frac{1}{2}(11x + 24) = A$

$\frac{1}{2}(11x + 24) > B$

Check

Olivia and Harry are analyzing the expressions $\frac{1}{4}(8 + 3x) + 5$ and $-2\left(\frac{3}{4}x + 7\right) + 2\frac{1}{4}x + 24$, for all values of x. Olivia claims that the value of these expressions is always equal. Harry claims that the value of the first expression is always less than the value of the second expression. Which student is correct?

 Go Online You can complete an Extra Example online.

Pause and Reflect

What questions do you still have about equations with infinitely many solutions and equations with no solution?

Record your observations here

Practice

🔁 **Go Online** You can complete your homework online.

Solve each equation. Determine whether the equation has one solution, no solution, or infinitely many solutions. (Examples 1 and 2)

1. $4(x - 8) + 12 = 2(2x - 9)$

2. $3(2k - 5) = 6(k - 4) + 9$

3. $-4y - 3 = \frac{1}{3}(12y - 9) - 8y$

4. $6(3 - 5w) = 5(4 - 2w) - 20w$

Complete each equation so that it has infinitely many solutions. (Example 3)

5. $2x - 7(x + 10) = \boxed{}\, x - \boxed{}$

6. $12x - x + 8 + 3x = \boxed{}\, x + \boxed{}$

Complete each equation so that it has no solution. (Example 4)

7. $-15x + 4x + 2 - x = \boxed{}\, x + \boxed{}$

8. $9(x - 4) - 5x = \boxed{}\, x - \boxed{}$

Test Practice

9. Multiple Choice Which of the following explains why $\frac{2}{3}(x + 3) = \frac{2}{3}(x - 6)$ has no solution?

Ⓐ The coefficients are different, and the constants are different.

Ⓑ The coefficients are the same, and the constants are the same.

Ⓒ The coefficients are different, and the constants are the same.

Ⓓ The coefficients are the same, and the constants are different.

Apply

10. Three students in the Math Club are analyzing the expressions $\frac{1}{4}(10x + 8)$ and $-(\frac{1}{2}x + 13) + 3(x + 5)$, for all values of x. Each student claims the expressions are related according to the results shown in the table. Which student is correct?

Student 1	$\frac{1}{4}(10x + 8)$	$=$	$-(\frac{1}{2}x + 13) + 3(x + 5)$
Student 2	$\frac{1}{4}(10x + 8)$	$>$	$-(\frac{1}{2}x + 13) + 3(x + 5)$
Student 3	$\frac{1}{4}(10x + 8)$	$<$	$-(\frac{1}{2}x + 13) + 3(x + 5)$

11. Daniel and Fatima are analyzing the expressions below for all values of x. Daniel claims that the value of these expressions is always equal. Fatima claims that the value of the expression on the left is always greater than the value of the expression on the right. Which student is correct?

$0.6(-7x + 9)$ and $4(x - 3) - (-8 + 8.2x)$

12. (MP) **Find the Error** A student solved the equation and determined that the solution was -2. Find her error and correct it.

$$1.5x - 2 = -2 + 1.5x$$
$$-2 = -2$$

13. (MP) **Make an Argument** Suppose the solution to an equation is $x = 0$. Explain why it is incorrect to conclude that the equation has no solution.

14. Determine if the statement is *true* or *false*. Justify your response.

An equation will always have at least one solution.

15. (MP) **Identify Structure** What values of a, b, c, and d will make the equation have one solution? Then alter your equation so that it has no solution. Finally, alter the equation again so that it has infinitely many solutions.

$$ax + b = cx + d$$

Write and Solve One-Step Addition and Subtraction Inequalities

I Can... write one-step addition and subtraction inequalities from real-world situations and use inverse operations to solve the inequalities.

What Vocabulary Will You Learn?
Addition Property of Inequality
inequality
Subtraction Property of Inequality

Explore Addition and Subtraction Properties of Inequality

Online Activity You will use Web Sketchpad to explore the effects of adding or subtracting a number from each side of an inequality.

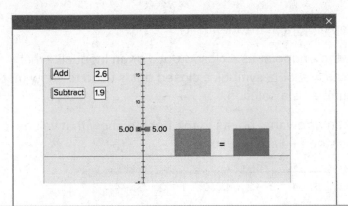

Learn Inequalities

An **inequality** is a mathematical sentence that compares quantities. Inequalities contain the symbols <, >, ≤, or ≥.

The table shows the meaning of each inequality symbol.

Symbol	Meaning
<	is less than 4 < 8
>	is greater than 3 > −2
≤	is less than or equal to −6 ≤ 1 or 5 ≤ 5
≥	is greater than or equal to 9 ≥ 6 or −7 ≥ −7

Talk About It!

When graphing an inequality on the number line, how do you know whether to use an open dot or a closed dot?

Learn Graph Inequalities

Go Online Watch the animation to learn how to graph inequalities on the number line.

Follow the steps to graph the inequality $x < 3.5$.

Step 1 Place the endpoint as an open dot. For an inequality that contains a $<$ or $>$ symbol, an open dot is used to show that the number is not a solution.

Step 2 Draw an arrow that points to the left to show that numbers less than 3.5 are part of the solution.

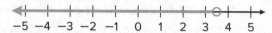

Follow the steps to graph the inequality $x \geq -1$.

Step 1 Place the endpoint as a closed dot. For an inequality that contains a $\leq$ or $\geq$ symbol, a closed dot is used to show that the number is a solution.

Step 2 Draw an arrow that points to the right to show that values greater than -1 are solutions.

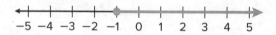

Talk About It!

If you were asked to graph the solution $x > 35$, what range of values can you use to create the number line? Explain.

Learn Subtraction and Addition Properties of Inequality

Solving an inequality means finding values for the variable that make the inequality true. You can solve addition inequalities by using the **Subtraction Property of Inequality**. You can solve subtraction inequalities by using the **Addition Property of Inequality**.

Talk About It!

Are the Addition and Subtraction Properties of Inequality true for the symbols $\geq$ and $\leq$? Explain.

Words	When you add or subtract the same number from each side of an inequality, the inequality remains true.
Symbols	For all numbers a, b, and c, if $a > b$, then $a - c > b - c$ and $a + c > b + c$, if $a < b$, then $a - c < b - c$ and $a + c < b + c$.
Examples	$\begin{array}{r} 2 < 5 \\ \underline{-4 \quad -4} \\ -2 < 1 \end{array}$ $\qquad$ $\begin{array}{r} 4 > -1 \\ \underline{+3 \quad +3} \\ 7 > 2 \end{array}$

These properties are also true for $a \geq b$ and $a \leq b$.

Example 1 Solve and Graph Addition Inequalities

Solve 0.4 + y ≤ −9.6. Check your solution. Then graph the solution set on a number line.

Part A Solve the inequality.

$0.4 + y \leq -9.6$	Write the inequality.
$\underline{-0.4 \qquad\qquad -0.4}$	Subtraction Property of Inequality
$y \leq -10$	Simplify.

The solution of the inequality $0.4 + y \leq -9.6$ is $y \leq$ _____.

You can check the solution $y \leq -10$ by substituting a number less than or equal to −10 into the original inequality.

$0.4 + y \leq -9.6$	Write the original inequality.
$0.4 + (-10) \overset{?}{\leq} -9.6$	Substitute −10 for y.
$-9.6 \leq -9.6$	Simplify.

Part B Graph the solution set on a number line.

To graph $y \leq -10$, draw a closed dot at −10 and an arrow pointing to the left. This shows that −10 is part of the solution, and values less than −10 are part of the solution.

```
<--+----+----+----+----+----+----+----+-->
  -13  -12  -11  -10  -9   -8   -7   -6   -5
```

Check

Solve $x + 1.3 < 5.4$ and graph the solution set.

Part A Solve $x + 1.3 < 5.4$.

Show your work here

Part B Graph the solution set.

```
<--+----+----+----+----+----+----+----+-->
  3.7  3.8  3.9  4.0  4.1  4.2  4.3  4.4  4.5
```

Go Online You can complete an Extra Example online.

Math History Minute

The Inca empire of 15th and 16th century South America used knotted and colored strings called *quipus* to keep complex records of everything from the empire's population to the amount of food a village had in store for lean seasons.

💬 Talk About It!

Why does the range of the number line extend from −13 to −5? Can you use a different range?

Example 2 Solve and Graph Subtraction Inequalities

Solve $-6 \geq n - 5$. Check your solution. Then graph the solution set on a number line.

Part A Solve the inequality.

$-6 \geq n - 5$		Write the inequality.
$\underline{+5 \qquad\quad +5}$		Addition Property of Inequality
$\boxed{} \geq \boxed{}$		Simplify.

The solution of the inequality $-6 \geq n - 5$ is $-1 \geq n$ or $n \leq -1$.

You can check this solution by substituting a number less than or equal to -1 into the original inequality.

$-6 \geq n - 5$	Write the original inequality.
$-6 \overset{?}{\geq} -3 - 5$	Replace n with -3.
$-6 \geq -8$	Simplify.

Part B Graph the solution set on a number line.

To graph $n \leq -1$, draw a closed dot at -1 and an arrow pointing to the left. This shows that -1 is part of the solution, and values less than -1 are part of the solution.

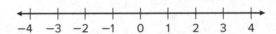

Check

Solve $x - 2.3 \geq -8.5$ and graph the solution set.

Part A Solve $x - 2.3 \geq -8.5$.

Show your work here

Part B Graph the solution set.

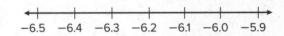

Go Online You can complete an Extra Example online.

Talk About It!
You can write $-1 \geq n$ as $n \leq -1$. Which way will help you visualize the solution on a number line? Explain your reasoning.

Talk About It!
Why is $-1 \geq n$ equivalent to $n \leq -1$? Substitute values for n that verify that the two inequalities are equivalent.

Learn Write Inequalities

Inequalities can be used to represent real-world situations. The table shows common phrases that describe each inequality.

Symbols	Phrases
<	is less than is fewer than
>	is greater than is more than
≤	is less than or equal to is no more than is at most
≥	is greater than or equal to is no less than is at least

The table outlines the steps used to write an inequality from a real-world situation.

Words	Describe the mathematics of the problem.
Variable	Define a variable to represent the unknown quantity.
Inequality	Translate the words into an inequality.

For example, a certain semi-truck can hold no more than 25 tons of material. This can be represented by the inequality $x \leq 25$, where x represents the weight of the material the truck is carrying.

Talk About It!
How do you know which inequality symbol to use when representing a real-world situation?

🌐 Example 3 Write and Solve Addition Inequalities

Dylan can spend at most $18 to ride go-karts and play games at the state fair. Suppose the go-karts cost $5.50.

Write and solve an inequality to determine the amount Dylan can spend on games, if he rides go-karts once. Interpret the solution.

Part A Write the inequality.

Words	Cost of go-kart ride plus cost of games is less than or equal to the total amount he can spend.
Variable	Let x represent the cost of the games.
Inequality	$5.5 + x \leq 18$

(continued on next page)

Part B Solve the inequality.

$$5.5 + x \le 18 \qquad \text{Write the inequality.}$$
$$\underline{-5.5 \qquad\quad -5.5} \qquad \text{Subtraction Property of Inequality}$$
$$x \le 12.5 \qquad \text{Simplify.}$$

The solution of the inequality $5.5 + x \le 18$ is _____.

Part C Interpret the solution.

Graph the solution set of $x \le 12.50$ on the number line.

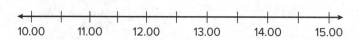

10.00   11.00   12.00   13.00   14.00   15.00

The greatest value that is part of the solution set is 12.50.

So, the most Dylan can spend on games is $12.50.

Check

Hannah's exercise goal is to walk at least $6\frac{1}{2}$ miles this week. She has already walked $2\frac{3}{4}$ miles this week. Write and solve an inequality to determine the distance Hannah needs to walk to meet or exceed her goal. Then interpret the solution.

Show your work here

Go Online You can complete an Extra Example online.

🌐 Example 4 Write and Solve One-Step Subtraction Inequalities

Caleb owns two types of dogs. The difference in height of his Yorkshire Terrier and his Great Dane is at least 25 inches. Caleb's Yorkshire Terrier has a height of $6\frac{1}{4}$ inches.

Write and solve an inequality to determine the possible height of the Great Dane. Then interpret the solution.

Part A Write an inequality.

Words	The Great Dane's height **minus** the Yorkshire Terrier's height **is at least** 25 inches.
Variable	Let h represent the height of the Great Dane.
Inequality	$h - 6\frac{1}{4} \geq 25$

Think About It!

What inequality symbol will you use to write the inequality?

Part B Solve the inequality.

$h - 6\frac{1}{4} \geq 25$ Write the inequality.

$+\boxed{} \quad +\boxed{}$ Addition Property of Inequality

$h \quad\quad \geq 31\frac{1}{4}$ Simplify.

The solution of the inequality $h - 6\frac{1}{4} \geq 25$ is $h \geq 31\frac{1}{4}$.

Part C Interpret the solution.

Graph the solution set.

30	$30\frac{1}{4}$	$30\frac{1}{2}$	$30\frac{3}{4}$	31	$31\frac{1}{4}$	$31\frac{1}{2}$	$31\frac{3}{4}$	32

So, the height of the Great Dane is at least $31\frac{1}{4}$ inches.

Check

Gabriella paid $17.25 for a sweatshirt that was on sale. The difference between the original price and the sale price was at most $8.50. Write and solve an inequality to determine the possible price of the sweatshirt. Then interpret the solution.

Part A Write an inequality that can be used to find the highest original price of the sweatshirt.

Part B What is the solution of the inequality in Part A?

(Show your work here)

Part C The original price of the sweatshirt was at _____.

 Go Online You can complete an Extra Example online.

Pause and Reflect

Compare and contrast solving one-step addition and one-step subtraction inequalities to solving one-step addition and one-step subtraction equations.

(Record your observations here)

🌐 **Apply** Elevators

The maximum weight capacity of the elevator in Maya's apartment building is 900 pounds. One morning she and five other people are on the elevator. Then two more passengers get on the elevator. If Maya weighs 108 pounds, what could be the maximum sum of the weights of the two additional passengers without exceeding the maximum weight capacity?

Passenger	Weight (lb)
1	126
2	182
3	78
4	135
5	63

🖱 **Go Online** Watch the animation.

1 What is the task?

Make sure you understand exactly what question to answer or problem to solve. You may want to read the problem three times. Discuss these questions with a partner.

First Time Describe the context of the problem, in your own words.
Second Time What mathematics do you see in the problem?
Third Time What are you wondering about?

2 How can you approach the task? What strategies can you use?

3 What is your solution?

Use your strategy to solve the problem.

4 How can you show your solution is reasonable?

⚡ **Write About It!** Write an argument that can be used to defend your solution.

💬 Talk About It!

Are the following sets of values possible solutions to the problem? Explain why or why not.

- 115 and 93
- 120 and 90
- 35 and 165

Check

Hayley's reading log is shown. She is required to read at least 2 hours each week. Which could be the amount of time she needs to read on Saturday and Sunday to meet her reading goal?

Day	Number of Minutes
Monday	20
Tuesday	25
Wednesday	15
Thursday	10
Friday	20
Saturday	x
Sunday	y

(A) Saturday: 5 minutes; Sunday: 10 minutes

(B) Saturday: 10 minutes; Sunday: 10 minutes

(C) Saturday: 10 minutes; Sunday: 15 minutes

(D) Saturday: 15 minutes; Sunday: 20 minutes

Show your work here

Go Online You can complete an Extra Example online.

Foldables It's time to update your Foldable, located in the Module Review, based on what you learned in this lesson. If you haven't already assembled your Foldable, you can find the instructions on page FL1.

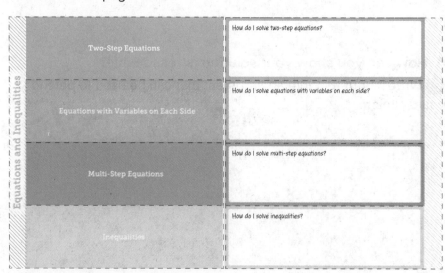

Practice

 Go Online You can complete your homework online.

Solve each inequality. Graph the solution set on a number line. (Examples 1 and 2)

1. $x + 8 \geq 14$

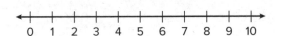

2. $x + 5.4 < -1.6$

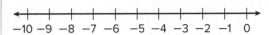

3. $3 \leq \frac{1}{3} + x$

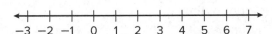

4. $4 \leq x - 7$

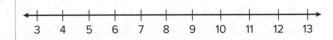

5. $x - 3 \leq -8$

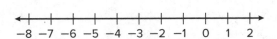

6. $6.9 < x - 2.3$

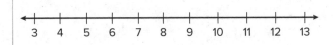

Solve each problem by first writing an inequality. (Examples 3 and 4)

7. A dolphin is swimming at a depth of −50 feet and then ascends a certain number of feet to a depth above −35 feet. Determine the number of feet the dolphin ascended. Then interpret the solution.

8. Linda has two cats. The difference in weight of her Maine Coon and Siberian is at least 6 pounds. Linda's Siberian has a weight of $8\frac{3}{4}$ pounds. Determine the possible weight of the Maine Coon. Then interpret the solution.

9. The difference between the monthly high and low temperatures was less than 27° Fahrenheit. The monthly low temperature was −2° Fahrenheit. Determine the possible monthly high temperature. Then interpret the solution.

Test Practice

10. Open Response Teddy has two piggy banks. The difference in the amount of money between the two banks is no more than $10. One piggy bank has $7.31 in it. Determine the possible amount of money in the other piggy bank. Then interpret the solution.

11. Zeg has $9.20 left on a gift card to the candy store. He has the following items in his shopping basket: 2 giant lollipops, 3 popcorn balls, 1 candy bar, and 5 candy sticks. Zeg wants to buy two more items. Name two more possible items he can buy using his gift card.

Item	Cost ($)
Candy Bars	0.99
Candy Stick	0.45
Giant Lollipops	1.75
Popcorn Ball	0.50

12. To prepare for a dance competition, a dance team needs to practice at least 12.75 hours a week. The team has already practiced 10.5 hours this week. Solve the inequality $10.5 + x \geq 12.75$ to find the amount of time x, in hours, the team has left to practice. What is the minimum number of minutes the team needs to practice?

13. Write a real-world problem that could have the solution $x \leq 10$.

14. ⓂⓅ **Reason Inductively** There is space for 120 students to go on a field trip. Currently, 74 students have signed up. Can 46 more students sign up for the field trip? Explain your reasoning.

15. Create Write and solve a real-world problem that involves a one-step addition inequality.

16. ⓂⓅ **Persevere with Problems** William is 3 feet 1 inch tall and would like to ride a roller coaster. Riders must be at least 42 inches tall to ride the coaster. Write and solve an addition inequality to determine how much taller William must be to ride the coaster.

Write and Solve One-Step Multiplication and Division Inequalities

I Can... use inverse operations to solve one-step multiplication and division inequalities with positive and negative coefficients.

What Vocabulary Will You Learn?
Division Property of Inequality
Multiplication Property of Inequality

Explore Multiplication and Division Properties of Inequality

Online Activity You will use Web Sketchpad to determine if multiplying or dividing each side of an inequality by the same positive number keeps the inequality true.

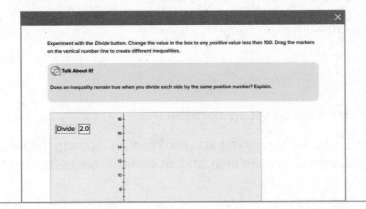

Experiment with the *Divide* button. Change the value in the box to any *positive* value less than 100. Drag the markers on the vertical number line to create different inequalities.

Talk About It!

Does an inequality remain true when you divide each side by the same positive number? Explain.

Divide 2.0

Learn Division and Multiplication Properties of Inequality

You can solve multiplication inequalities by using the **Division Property of Inequality**. You can solve division inequalities by using the **Multiplication Property of Inequality**.

The table outlines the process of using the Division and Multiplication Properties of Inequality with *positive* coefficients.

Words	An inequality remains true when you divide or multiply each side of the inequality by the same positive number.	
Symbols	For all numbers a, b, and c, where $c > 0$, if $a > b$, then $\frac{a}{c} > \frac{b}{c}$ and $ac > bc$. if $a < b$, then $\frac{a}{c} < \frac{b}{c}$ and $ac < bc$.	
Examples	$9 < 15$ $\frac{9}{3} < \frac{15}{3}$ $3 < 5$	$10 > 7$ $2(10) > 2(7)$ $20 > 14$

These properties are also true for $a \geq b$ and $a \leq b$.

Talk About It!

The Division Property of Inequality states:

For all numbers a, b, and c, where $c > 0$,

1. if $a > b$, then $\frac{a}{c} > \frac{b}{c}$.

2. if $a < b$, then $\frac{a}{c} < \frac{b}{c}$.

What does the inequality $c > 0$ mean?

Example 1 Solve and Graph Multiplication Inequalities

Solve $8x \leq 40$. Check your solution. Then graph the solution set on a number line.

Part A Solve the inequality.

$8x \leq 40$ Write the inequality.

$\dfrac{8x}{8} \leq \dfrac{40}{8}$ Division Property of Inequality

$x \leq 5$ Simplify.

The solution of the inequality $8x \leq 40$ is _____.

You can check this solution by substituting a number less than or equal to 5 into the original inequality. Try using 4.

$8x \leq 40$ Write the inequality.

$8(4) \overset{?}{\leq} 40$ Replace x with 4.

$32 \leq 40$ Simplify.

Part B Graph the solution set on a number line.

To graph $x \leq 5$, place a closed dot at 5 and draw an arrow to the left. This shows that the values less than, and including, 5 are part of the solution.

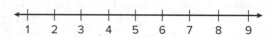

Check

Solve $12x \geq -36$ and graph the solution set.

Part A Solve $12x \geq -36$.

Show your work here

Part B Graph the solution set.

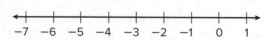

🅝 **Go Online** You can complete an Extra Example online.

Example 2 Solve and Graph Division Inequalities

Solve $\frac{d}{2} > 7$. Check your solution. Then graph the solution set on a number line.

Part A Solve the inequality.

$\frac{d}{2} > 7$ Write the inequality.

$2\left(\frac{d}{2}\right) > 2(7)$ Multiplication Property of Inequality

$d > 14$ Simplify.

The solution of the inequality $\frac{d}{2} > 7$ is _____.

You can check this solution by substituting a number greater than 14 into the original inequality. Try using 15.

$\frac{d}{2} > 7$ Write the inequality.

$\frac{15}{2} \overset{?}{>} 7$ Replace d with 15.

$7.5 > 7$ Simplify.

Part B Graph the solution set on a number line.

To graph $d > 14$, place an open dot on 14 and draw an arrow pointing to the right to show that all values greater than, but not including, 14 are part of the solution.

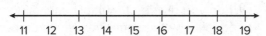

Check

Solve $-6 \geq \frac{x}{7}$ and graph the solution set.

Part A Solve $-6 \geq \frac{x}{7}$.

Show your work here

Part B Graph the solution set.

 Go Online You can complete an Extra Example online.

Explore Multiply and Divide Inequalities by Negative Numbers

Online Activity You will use Web Sketchpad to explore how multiplying and dividing each side of an inequality by the same negative number affects the inequality.

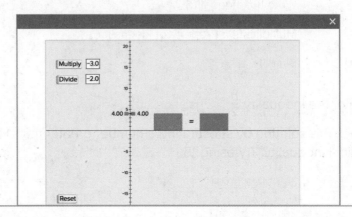

Learn Division and Multiplication Properties of Inequality

The table outlines the process of using the Division and Multiplication Properties of Inequality when dividing or multiplying each side of an inequality by a *negative* number.

Words	When you divide or multiply each side of an inequality by the same *negative* number, the inequality symbol must be reversed for the inequality to remain true.	
Symbols	For all numbers a, b, and c, where $c < 0$, if $a < b$, then $\dfrac{a}{c} > \dfrac{b}{c}$ and $ac > bc$, if $a > b$, then $\dfrac{a}{c} < \dfrac{b}{c}$ and $ac < bc$.	
Examples	$18 > -12$ $\dfrac{18}{-3} < \dfrac{-12}{-3}$ $-6 < 4$	$-4 < 5$ $-4(-3) > 5(-3)$ $12 > -15$

These properties are also true for $a \leq b$ and $a \geq b$.

(continued on next page)

Go Online Watch the animation to understand why you need to reverse the inequality symbol when you divide each side of an inequality by the same negative number.

Step 1 Start by graphing the numbers on each side of the true inequality $-6 < 4$.

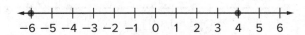

Step 2 Divide each side of the inequality by the same negative number, for example, -2.

$$\frac{-6}{-2} \overset{?}{<} \frac{4}{-2}$$

Step 3 Simplify each side of the inequality. Is the inequality still true?

$$3 \overset{?}{<} -2$$

Step 4 Graph each side of the inequality to determine that it is NOT true. The number 3 is not less than -2.

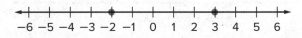

Step 5 Make the inequality true by reversing the inequality symbol. The number 3 is greater than -2.

$$3 > -2$$

(continued on next page)

(continued on next page)

Talk About It!

Will the inequality symbol need to be reversed when solving the inequality $3x > -9$? Why or why not?

Pause and Reflect

You just observed that dividing each side of the inequality by -2 results in a false inequality, unless the inequality symbol is reversed. Do you think that this will always be true for any inequality and/or for any negative divisor? Create your own inequalities with which to experiment and verify whether or not this will always be true.

Record your observations here

Copyright © McGraw-Hill Education

Go Online Watch the animation to understand why you need to reverse the inequality symbol when you multiply each side of an inequality by the same negative number.

Step 1 Start by graphing the numbers on each side of the true inequality $-2 < 4$.

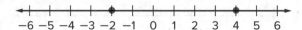

Step 2 Multiply each side of the inequality by the same negative number, for example, -1.

$$-1(-2) \overset{?}{<} -1(4)$$

Step 3 Simplify each side of the inequality. Is the inequality still true?

$$2 \overset{?}{<} -4$$

Step 4 Graph each side of the inequality to determine that it is NOT true. The number 2 is not less than -4.

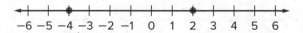

Step 5 Make the inequality true by reversing the inequality symbol. The number 2 is greater than -4.

$$2 > -4$$

Pause and Reflect

You just observed that multiplying each side of the inequality by -1 results in a false inequality, unless the inequality symbol is reversed. Do you think that this will always be true for any inequality and/or for any negative number? Create your own inequalities with which to experiment and verify whether or not this will always be true.

Record your observations here

Talk About It!

Use your understanding of opposites to explain why the inequality symbol is reversed when multiplying each side of the inequality by -1.

Example 3 Multiplication Inequalities with Negative Coefficients

Solve $-2x < 10$. Check your solution. Then graph the solution set on a number line.

Part A Solve the inequality.

$-2x < 10$	Write the inequality.
$\dfrac{-2x}{-2} > \dfrac{10}{-2}$	Divide each side by -2 and reverse the inequality symbol.
$x > -5$	Simplify.

The solution of the inequality $-2x < 10$ is $x > -5$.

You can check the solution by substituting a number greater than -5 into the original inequality.

$-2x < 10$	Write the inequality.
$-2(-3) \overset{?}{<} 10$	Replace x with -3.
$6 < 10$	Simplify.

Part B Graph the solution set on a number line.

$-10\ -9\ -8\ -7\ -6\ -5\ -4\ -3\ -2$

Check

Solve $-4.1x > 12.3$ and graph the solution set.

Part A Solve $-4.1x > 12.3$.

Show your work here

Part B Graph the solution set.

$-7\ -6\ -5\ -4\ -3\ -2\ -1\ \ 0\ \ 1\ \ 2\ \ 3$

Go Online You can complete an Extra Example online.

Think About It!

What is important to remember when multiplying or dividing each side of an inequality by a negative number?

Talk About It!

How can you use the graph of the solution to check your work?

Example 4 Division Inequalities with Negative Coefficients

Solve $\frac{x}{-4} \geq 3$. Check your solution. Then graph the solution set on a number line.

Part A Solve the inequality.

$$\frac{x}{-4} \geq 3 \qquad \text{Write the inequality.}$$

$$\boxed{}\left(\frac{x}{-4}\right) \leq \boxed{}(3) \qquad \text{Multiplication Property of Inequality}$$

$$x \leq -12 \qquad \text{Simplify.}$$

The solution of the inequality $\frac{x}{-4} \geq 3$ is _____.

You can check the solution by substituting a number less than or equal to -12 into the original inequality.

$$\frac{x}{-4} \geq 3 \qquad \text{Write the inequality.}$$

$$\frac{\boxed{}}{-4} \overset{?}{\geq} 3 \qquad \text{Replace } x \text{ with } -20.$$

$$5 \geq 3 \qquad \text{Simplify.}$$

Part B Graph the solution set on a number line.

To graph the solution $x \leq -12$, place a closed dot at -12 and draw an arrow to the left to show that all values less than, and including, -12 are part of the solution.

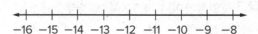

Check

Solve $\frac{x}{-2.25} \leq -4$ and graph the solution set.

Part A Solve $\frac{x}{-2.25} \leq -4$.

Show your work here

Part B Graph the solution set.

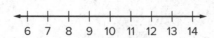

Go Online You can complete an Extra Example online.

Copyright © McGraw-Hill Education

Think About It!

What step(s) do you need to take in order to solve the inequality?

🌐 **Example 5** Write and Solve One-Step Multiplication Inequalities

Ling earns $15 per hour working at the zoo.

Write and solve an inequality to determine the number of hours Ling must work in a week to earn at least $225. Then interpret the solution.

Part A Write an inequality.

Words	The amount earned per hour **times** the number of hours **is at least** the amount earned each week.
Variable	Let x represent the number of hours.
Inequality	$15x \leq 225$

Part B Solve the inequality.

$15x \geq 225$ Write the inequality.

$\dfrac{15x}{15} \geq \dfrac{225}{15}$ Division Property of Inequality

$x \geq \boxed{}$ Simplify.

The solution of the inequality $15x \geq 225$ is $x \geq 15$.

Part C Interpret the solution.

Graph the solution set of $x \geq 15$ on the number line.

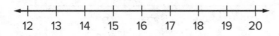

Use the graph to interpret the solution.

So, Ling must work _____ .

Check

Dominic has invited 12 friends to his birthday party. How much can he spend per person on party favors if his budget is no more than $75?

Write and solve an inequality to determine the budget for each person. Then interpret the solution.

Show your work here

🌐 Go Online You can complete an Extra Example online.

💭 **Think About It!**
What key word(s) tell you which inequality symbol to use?

🗨 **Talk About It!**
When solving the inequality, did you need to reverse the inequality symbol? Explain.

Copyright © McGraw-Hill Education

🌐 **Example 6** Write and Solve One-Step Division Inequalities

Mrs. Miller is buying wax to make candles. She wants to make at least 35 candles and needs 6.5 ounces of wax for each candle.

Write and solve an inequality to determine the amount of wax she needs to buy. Then interpret the solution.

Part A Write an inequality.

Words	The total amount of wax **divided by** the amount for each candle **is at least** the number of candles.
Variable	**Let** x **represent** the total amount of wax.
Inequality	$\dfrac{x}{6.5} \geq 35$

Part B Solve the inequality.

$$\dfrac{x}{6.5} \geq 35 \qquad \text{Write the inequality.}$$

$$\left(\boxed{}\right)\dfrac{x}{6.5} \geq \left(\boxed{}\right)35 \qquad \text{Multiplication Property of Inequality}$$

$$x \geq 227.5 \qquad \text{Simplify.}$$

The solution of the inequality $\dfrac{x}{6.5} \geq 35$ is $x \geq 227.5$.

Part C Interpret the solution.

Graph the solution set of $x \geq 227.5$ on the number line.

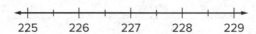

225   226   227   228   229

Use the graph to interpret the solution.

So, Mrs. Miller needs _____ ounces of candle wax.

Check

Thomas wants to make popsicles from a juice mixture. He needs 7.5 ounces of juice for each popsicle and wants to make no more than 12 popsicles. How much juice should he make?

Write and solve an inequality to determine the amount of juice he needs to make. Then interpret the solution.

Show your work here

🌐 **Go Online** You can complete an Extra Example online.

🌐 **Apply** Fundraising

The students at Westlake Middle School are raising money for buses to go on a science field trip. Each bus holds 44 students and costs $150 for the day. If at least 275 students go on the trip, how much money will they need to raise for buses?

🔗 Go Online
Watch the animation.

1 What is the task?

Make sure you understand exactly what question to answer or problem to solve. You may want to read the problem three times. Discuss these questions with a partner.

First Time Describe the context of the problem, in your own words.
Second Time What mathematics do you see in the problem?
Third Time What are you wondering about?

2 How can you approach the task? What strategies can you use?

💬 Talk About It!

If the number of students participating in the trip increases by 35 students, will more money need to be raised? Explain your reasoning.

3 What is your solution?

Use your strategy to solve the problem.

4 How can you show your solution is reasonable?

✏️ **Write About It!** Write an argument that can be used to defend your solution.

Check

Hudson needs to rent tables for a family reunion. Each table seats 8 people and costs $8.75 to rent. If at least 85 people attend the reunion, how much will it cost to rent tables?

Show your work here

Go Online You can complete an Extra Example online.

Foldables It's time to update your Foldable, located in the Module Review, based on what you learned in this lesson. If you haven't already assembled your Foldable, you can find the instructions on page FL1.

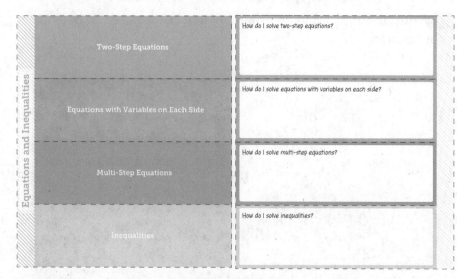

Practice

🔄 **Go Online** You can complete your homework online.

Solve each inequality. Graph the solution set on a number line. (Examples 1-4)

1. $-14 \geq 7x$

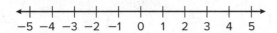

2. $2 \leq 0.25x$

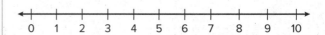

3. $\frac{x}{2} > -4$

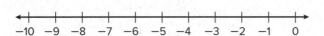

4. $-6x > 66$

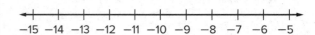

5. $-2.2x \leq -6.6$

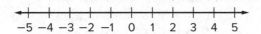

6. $\frac{x}{-5} \geq -3$

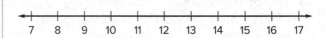

Solve each problem by first writing an inequality. (Examples 5 and 6)

7. Hermes earns $6 an hour for babysitting. He wants to earn at least $168 for a new video game system. Determine the number of hours he must babysit to earn enough money for the video game system. Then interpret the solution.

8. Sadie wants to make several batches of rolls. She has 13 tablespoons of yeast left in the jar and each batch of rolls takes $3\frac{1}{4}$ tablespoons. Determine the number of batches of rolls Sadie can make. Then interpret the solution.

9. Chase is making bookmarks. He wants to make no more than 12 bookmarks and needs 4.25 inches of fabric for each bookmark. Determine the amount of fabric he needs to buy. Then interpret the solution.

Test Practice

10. Open Response Mae wants to make more than 6 gift baskets for the school raffle. Each gift basket costs $15.50. Determine the amount of money she will spend to make the gift baskets. Then interpret the solution.

Apply

11. Greg's grandmother is knitting scarves for charity. The table shows the number of yards needed for different types of scarves. She plans to make bulky scarves and has no more than 2,250 yards of yarn. If the charity plans to sell the scarves for $24.50, how much money will the charity make?

Yarn Type	Number of Yards
Bulky	150
Light	250
Medium	200

12. At a school outing, a group decides to go rafting. Each raft holds 8 people and costs $25 for the day. If at least 70 people go rafting, how much money will they need for the rafts?

13. Write a real-world problem involving multiplication properties of an inequality that would have a solution $x \geq 25$.

14. **MP Reason Abstractly** You are asked to draw a rectangle with a width of 5 inches and an area less than 55 square inches. Can the length of the rectangle be 11 inches? Explain your reasoning.

15. **Which One Doesn't Belong?** Identify the inequality that does not belong with the other three. Explain your reasoning.

   A. $-6x \leq -36$

   B. $\dfrac{x}{2} \geq 3$

   C. $\dfrac{x}{3} \geq 2$

   D. $\dfrac{x}{-3} \geq -2$

16. **MP Find the Error** A student solved the inequality shown below. Find the student's mistake and correct it.

$$-25 \leq -5x$$

$$\frac{-25}{-5} \leq \frac{-5x}{-5}$$

$$5 \leq x$$

Write and Solve Two-Step Inequalities

I Can... write two-step inequalities from real-world situations and use inverse operations to solve the inequalities.

Learn Solve Two-Step Inequalities

A **two-step inequality** is an inequality that contains two operations.

Go Online Watch the animation to learn how to solve a two-step inequality.

The animation shows the steps used to solve the two-step inequality $-2x + 6 \geq 12$.

Steps	Example
1. Undo the addition or subtraction.	$-2x + 6 \geq 12$ $\underline{\quad -6 \quad -6}$ $-2x \qquad \geq 6$
2. Undo the multiplication or division. Reverse the inequality symbol when multiplying or dividing by a negative number.	$\dfrac{-2x}{-2} \leq \dfrac{6}{-2}$ $x \leq -3$
3. Check the solution.	$-2x + 6 \geq 12$ $-2(-4) + 6 \overset{?}{\geq} 12$ $14 \geq 12 \checkmark$

Pause and Reflect

Compare and contrast solving a two-step inequality and a two-step equation. How are they similar? How are they different?

> Record your observations here

What Vocabulary Will You Learn?
two-step inequality

Talk About It!

A student substituted the value -9 for x into the inequality $2x + 6 \geq 12$. Is this acceptable? Explain.

Think About It!

What steps do you need to take in order to solve the inequality?

Talk About It!

Suppose when Jesse solved the inequality, he claimed the solution is $x \leq -4$. Find his error and explain how to correct it.

Example 1 Solve Two-Step Inequalities

Solve $-5x - 12 \leq 8$. Check your solution. Then graph the solution set on a number line.

Part A Solve the inequality.

$-5x - 12 \leq 8$	Write the inequality.
$\underline{+\ 12\ +\ 12}$	Addition Property of Inequality
$-5x \leq 20$	Simplify.
$\dfrac{-5x}{-5} \geq \dfrac{20}{-5}$	Division Property of Inequality
$x \geq -4$	Simplify.

The solution of the inequality $-5x - 12 \leq 8$ is _____.

You can check the solution $x \geq -4$ by substituting a number greater than or equal to -4 into the original inequality.

$-5x - 12 \leq 8$	Write the inequality.
$-5(-1) - 12 \overset{?}{\leq} 8$	Replace x with -1.
$5 - 12 \overset{?}{\leq} 8$	Multiply $-5(-1)$.
$-7 \leq 8 \ \checkmark$	Simplify.

Part B Graph the solution set on a number line.

To graph $x \geq -4$ place a closed dot at -4 and an arrow to the right.

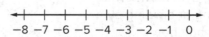

Check

Solve $-6x - 4 > 14$ and graph the solution set.

Part A Solve $-6x - 4 > 14$.

Show your work here

Part B Graph the solution set.

Go Online You can complete an Extra Example online.

Example 2 Solve Two-Step Inequalities

Solve 4.7x − 3.25 ≤ 10.85. Check your solution. Then graph the solution set on a number line.

Think About It!

What steps do you need to take in order to solve the inequality?

Part A Solve the inequality.

$4.7x - 3.25 \leq 10.85$	Write the inequality.
$\underline{+\ 3.25\ \ +3.25}$	Addition Property of Inequality
$4.7x \leq 14.1$	Simplify.
$\dfrac{4.7x}{4.7} \leq \dfrac{14.1}{4.7}$	Division Property of Inequality
$x \leq 3$	Simplify.

The solution of the inequality $4.7x - 3.25 \leq 10.85$ is _____.

You can check the solution $x \leq 3$ by substituting a number less than or equal to 3 into the original inequality.

$4.7x - 3.25 \leq 10.85$	Write the inequality.
$4.7(1) - 3.25 \overset{?}{\leq} 10.85$	Replace x with 1.
$4.7 - 3.25 \overset{?}{\leq} 10.85$	Multiply.
$1.45 \leq 10.85$ ✓	Simplify.

Part B Graph the solution set on a number line.

Check

Solve $1.3x - 3.2 \geq 4.6$ and graph the solution set.

Part A Solve $1.3x - 3.2 \geq 4.6$.

Show your work here

Part B Graph the solution set.

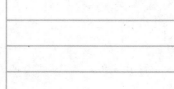

🅝 **Go Online** You can complete an Extra Example online.

Think About It!

What steps do you need to take in order to solve the inequality?

Example 3 Solve Two-Step Inequalities

Solve $\frac{3}{4}x - \frac{1}{2} > \frac{3}{8}$. **Check your solution. Then graph the solution set on a number line.**

Part A Solve the inequality.

$$\frac{3}{4}x - \frac{1}{2} > \frac{3}{8}$$ Write the inequality.

$$\frac{3}{4}x - \frac{4}{8} > \frac{3}{8}$$ Add $\frac{1}{2}$ to each side. Rewrite $\frac{1}{2}$ as $\frac{4}{8}$.

$$\underline{\quad + \frac{4}{8} \ + \frac{4}{8}\quad}$$ Addition Property of Inequality

$$\frac{3}{4}x > \frac{7}{8}$$ Simplify.

$$\frac{4}{3} \cdot \left(\frac{3}{4}x\right) > \frac{7}{8} \cdot \frac{4}{3}$$ Multiply each side by the reciprocal.

$$x > \frac{7}{6} \text{ or } 1\frac{1}{6}$$ Simplify.

The solution of the inequality $\frac{3}{4}x - \frac{1}{2} > \frac{3}{8}$ is _____.

You can check the solution $x > 1\frac{1}{6}$ by substituting a number greater than $1\frac{1}{6}$ into the original inequality.

$$\frac{3}{4}x - \frac{1}{2} > \frac{3}{8}$$ Write the inequality.

$$\frac{3}{4}(4) - \frac{1}{2} \overset{?}{>} \frac{3}{8}$$ Replace x with 4.

$$3 - \frac{1}{2} \overset{?}{>} \frac{3}{8}$$ Multiply.

$$2\frac{1}{2} > \frac{3}{8} \checkmark$$ Simplify.

Part B Graph the solution set on a number line.

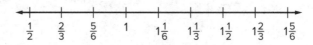

Check

Solve $\frac{2}{3}x - \frac{5}{6} < \frac{1}{2}$ and graph the solution set.

Part A Solve $\frac{2}{3}x - \frac{5}{6} < \frac{1}{2}$.

Show your work here

Part B Graph the solution set.

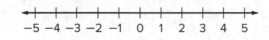

Go Online
You can complete an Extra Example online.

Example 4 Write and Solve Two-Step Inequalities

Halfway through the bowling league season, Stewart has 34 strikes. He averages 2 strikes per game. He needs at least 61 strikes to beat the league record.

Write and solve an inequality to determine the possible number of additional games Stewart should bowl to have at least 61 strikes. Then interpret the solution.

Part A Write an inequality.

Words	The number of strikes Stewart has plus two strikes per game is at least the total number of strikes needed.
Variable	Let g represent the number of games.
Inequality	$34 + 2g \geq 61$

Part B Solve the inequality.

$34 + 2g \geq 61$	Write the inequality.
$\underline{-34 \qquad -34}$	Subtraction Property of Inequality
$2g \geq 27$	Simplify.
$\dfrac{2g}{2} \geq \dfrac{27}{2}$	Division Property of Inequality
$g \geq 13.5$	Simplify.

The solution of the inequality $34 + 2g \geq 61$ is $g \geq$ _____.

Part C Interpret the solution.

Graph the solution set of $g \geq 13.5$ on the number line.

11.0 11.5 12.0 12.5 13.0 13.5 14.0 14.5 15.0 15.5 16.0

Use the graph to interpret the solution.

So, in order to beat the record, Stewart will have to bowl _____ games.

Copyright © McGraw-Hill Education

Talk About It!

When interpreting the solution of the inequality, why is the least number of games Stewart needs to bowl equal to 14, and not 13 or 13.5?

Check

Peter can spend no more than $100 on new clothes for school. He spends $35 on a new pair of shoes. Shirts cost $15.

Write and solve an inequality to determine how many shirts Peter can purchase. Then interpret the solution.

Show your work here

 Go Online You can complete an Extra Example online.

🌐 **Example 5** Write and Solve Two-Step Inequalities

Meredith is given a $50 monthly allowance to buy lunch at school. Any remaining money can be spent on entertainment. Meredith would like to have at least $12 left at the end of the month to go to the movies with her friends. It costs Meredith $2.50 per lunch that she buys at school.

Write and solve an inequality to determine the number of lunches Meredith can buy and have at least $12 left. Then interpret the solution.

Part A Write an inequality to determine the number of lunches Meredith can buy.

Words	Monthly allowance **minus** the cost per lunch **is at least** the amount remaining.
Variable	Let x represent the number of lunches.
Inequality	$50 - 2.50x \geq 12$

(continued on next page)

Part B Solve the inequality.

$$50 - 2.50x \geq 12$$ Write the inequality.

$$\underline{-50 \qquad\qquad -50}$$ Subtraction Property of Inequality

$$-2.50x \geq -38$$ Simplify.

$$\frac{-2.50x}{-2.50} \leq \frac{-38}{-2.50}$$ Division Property of Inequality

$$x \leq 15.2$$ Simplify.

The solution of the inequality $50 - 2.50x \geq 12$ is _____.

Part C Interpret the solution.

Graph the solution set of $x \leq 15.2$ on the number line.

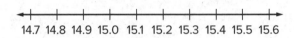

14.7 14.8 14.9 15.0 15.1 15.2 15.3 15.4 15.5 15.6

Use the graph to interpret the solution.

So, Meredith can buy _____ lunches in order to have at least $12 remaining to go to the movies.

Check

Sylvia was given $25 for her birthday and would like to use some of the money to purchase music from a music streaming website. It costs $1.20 per song she downloads. She would like to have at least $10 left.

Write and solve an inequality to determine the number of songs Sylvia can purchase and have at least $10 left. Then interpret the solution.

Show your work here

Go Online You can complete an Extra Example online.

Think About It!

What inequality symbol will you use to write the inequality?

Talk About It!

If you forgot to reverse the inequality symbol when you divided each side by -2.50, how can you know that your solution is incorrect?

Pause and Reflect

Create a graphic organizer that describes how to solve inequalities.
Be sure to include examples of solving addition, subtraction,
multiplication, division, and two-step inequalities.

Record your observations here

🌐 **Apply** School

To earn the grade she wants in her English class, Elspeth needs an average of 85% from her quiz scores. Each quiz is worth 20 points. The scores of her first four quizzes are shown in the table. If there is one more quiz, what score can she receive to earn at least an 85% average in English class?

Quiz	Score
1	18
2	16
3	19
4	14

1 What is the task?

Make sure you understand exactly what question to answer or problem to solve. You may want to read the problem three times. Discuss these questions with a partner.

First Time Describe the context of the problem, in your own words.
Second Time What mathematics do you see in the problem?
Third Time What are you wondering about?

2 How can you approach the task? What strategies can you use?

3 What is your solution?

Use your strategy to solve the problem.

4 How can you show your solution is reasonable?

✏️ **Write About It!** Write an argument that can be used to defend your solution.

💬 Talk About It!

In the inequality $\frac{67 + x}{100} \geq 0.85$ that represents Elspeth's average score, why is 100 in the denominator?

Check

In order for Ainsley to earn the grade she wants in science class, she needs an average of 85% on her quiz scores. Each quiz is worth 30 points. The scores of her first 5 quizzes are shown in the table. There will be one more quiz. What score can she receive to earn at least an 85% average in science class?

Quiz	Score
1	25
2	26
3	25
4	24
5	28

Show your work here

Go Online You can complete an Extra Example online.

Foldables It's time to update your Foldable, located in the Module Review, based on what you learned in this lesson. If you haven't already assembled your Foldable, you can find the instructions on page FL1.

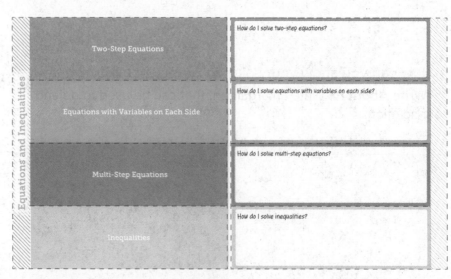

Practice

🧭 **Go Online** You can complete your homework online.

Solve each inequality. Graph the solution set on a number line. (Examples 1–3)

1. $-3x - 3 > 12$

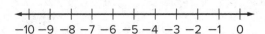
$$-10\ -9\ -8\ -7\ -6\ -5\ -4\ -3\ -2\ -1\ \ 0$$

2. $-4 \leq 4x + 8$

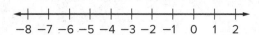
$$-8\ -7\ -6\ -5\ -4\ -3\ -2\ -1\ \ 0\ \ 1\ \ 2$$

3. $6.5x - 11.3 \leq 8.2$

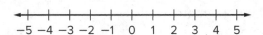
$$-5\ -4\ -3\ -2\ -1\ \ 0\ \ 1\ \ 2\ \ 3\ \ 4\ \ 5$$

4. $-2.45x + 3.2 < -6.6$

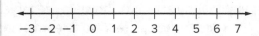
$$-3\ -2\ -1\ \ 0\ \ 1\ \ 2\ \ 3\ \ 4\ \ 5\ \ 6\ \ 7$$

5. $\frac{1}{2}x - \frac{1}{4} < \frac{5}{8}$

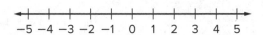
$$-5\ -4\ -3\ -2\ -1\ \ 0\ \ 1\ \ 2\ \ 3\ \ 4\ \ 5$$

6. $\frac{x}{10} + \frac{1}{4} \geq \frac{1}{5}$

$$-5\ -4\ -3\ -2\ -1\ \ 0\ \ 1\ \ 2\ \ 3\ \ 4\ \ 5$$

7. A rental company charges $15 plus $4 per hour to rent a bicycle. If Margie wants to spend no more than $27 for her rental, write and solve an inequality to determine how many hours she can rent the bicycle. Then interpret the solution. (Example 4)

8. Matilda needs at least $112 to buy a new dress. She has already saved $40. She earns $9 an hour babysitting. Write and solve an inequality to determine how many hours she will need to babysit to buy the dress. Then interpret the solution. (Example 4)

9. Douglas bought a $20 game card at a game center. The go-karts cost $3.50 each time you race. He wants to have at least $7.75 left on his card to play arcade games. Write and solve an inequality to determine how many times Douglas can race the go-karts. Then interpret the solution.

(Example 5)

Test Practice

10. **Open Response** Robin was given a $40 monthly allowance. She wants to go to the movies as many times as possible and have at least $12.50 left at the end of the month to go to a concert. A movie ticket costs $5. Write and solve an inequality to determine how many times Robin can go to the movies this month. Then interpret the solution.

Apply

11. To earn the score he wants in a trivia game, Jet needs an average of 80% after five rounds. Each round is worth 50 points. The scores of his first four rounds are shown in the table. If there is one more round, what is the minimum score he can receive to earn at least an 80% average in the trivia game?

Round	Score
1	49
2	46
3	44
4	45

12. Eden needs an average of 92% on her quiz scores to earn the grade she wants in science class. Each quiz is worth 20 points. The scores of her first four quizzes are shown in the table. There will be one more quiz. What score she can receive to earn at least a 92% average in science class?

Quiz	Score
1	20
2	18
3	19
4	17

13. Solve $-2(x + 2) < x + 8$. Then graph the solution set on a number line.

14. (MP) **Identify Structure** Write a two-step inequality that can be solved by first adding 4 to each side.

15. (MP) **Identify Structure** Explain how you can solve the inequality $-2x + 4 < 16$ without multiplying or dividing by a negative coefficient.

16. Create Write, solve, and interpret the solution to a real-world problem that involves a two-step inequality.

Foldables Use your Foldable to help review the module.

Equations and Inequalities

$2x + 7 = 9$

$14 + 3n = 5n - 6$

$-8 - x = -3(2x - 4) + 3x$

$-4x + 2 < 26$

Rate Yourself!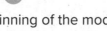

Complete the chart at the beginning of the module by placing a checkmark in each row that corresponds with how much you know about each topic after completing this module.

Reflect on the Module

Use what you learned about writing and solving equations to complete the graphic organizer.

e Essential Question

How can equations be used to solve everyday problems?

Two-Step Equations	Equations with Variables on Each Side	Multi-Step Equations
Explain how to solve $3h - 7 = 24$.	Explain how to solve $-2a - 9 = 6a + 15$.	Explain how to solve $-a + 2(3 + a) = -4(a + 1)$.
Explain how to solve $2(a + 5) = 20$.	Explain how to solve $4b + 6 = 2b + 28$.	Explain how to solve $6(x - 3) + 10 = 2(4x - 5)$.

What are the steps for writing an equation from a real-world problem?

Test Practice

1. Open Response Solve the equation $-7t - 1 = 20$. **(Lesson 1)**

2. Equation Editor Carlos wants to rent a video game console for $17.50 and some video games for $5.25 each. He has $49 to spend. How many games can he rent? **(Lesson 1)**

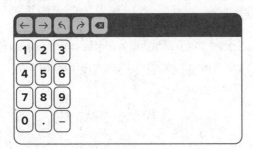

3. Equation Editor The solution to the equation $-2(x + 11) = -8$ is $x = $ _. **(Lesson 2)**

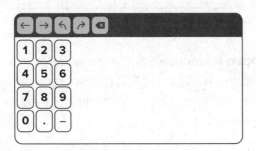

4. Open Response Solve $3n - 2 = 4n - 6$. **(Lesson 3)**

5. Open Response Monique wants to rent online movies. Movies Plus charges a one-time fee of $20 plus $5 per movie. Movies-To-Go charges $7 per movie. For how many movies is the cost of renting movies from Movies Plus and Movies-To-Go the same? **(Lesson 3)**

Write and solve an equation to represent the problem, where m is the number of movies rented.

6. Equation Editor Solve $2(6x + 4) - 3x = 5x - 4$. **(Lesson 4)**

$x = $ _____

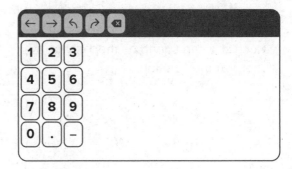

7. Multiple Choice The width of a rectangular table top is $2\frac{1}{2}$ feet shorter than three times its length. If the perimeter of the table top is 43 feet, what is the length, ℓ, of the table top? **(Lesson 4)**

Which equation represents this situation?

Ⓐ $43 = 2\ell + 2\left(3\ell - 2\frac{1}{2}\right)$

Ⓑ $43 = 2\ell + 2\left(2\ell - 2\frac{1}{2}\right)$

Ⓒ $43 = \ell + 2\left(3\ell - 2\frac{1}{2}\right)$

Ⓓ $43 = \ell + 2\left(2\ell - 2\frac{1}{2}\right)$

8. Table Item Determine whether each equation has one solution, no solution, or infinitely many solutions. (Lesson 5)

	One Solution	No Solution	Infinitely Many Solutions
$4(x + 8) = 2(4 + 2x)$			
$3(2x + 1) = 3 + 6x$			
$2(x + 5) = 5x + 1$			

9. Open Response Jessica has a department store gift card worth $75. The table shows the items that she has picked out so far. (Lesson 6)

Item	Cost ($)
bracelet	8
sunglasses	15
beach towel	12
hat	10
sandals	14

Jessica also wants to buy two T-shirts. Write and solve an inequality to represent how much she can spend on the T-shirts. Interpret the solution.

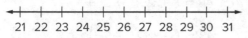

10. Grid The difference between the high and low temperatures on Sunday was at least 35°F. The low temperature on Sunday was −8°F. (Lesson 6)

A. Write and solve an inequality to find t, the high temperature on Sunday.

B. Graph the solution of the inequality.

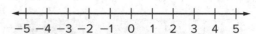
21 22 23 24 25 26 27 28 29 30 31

11. Equation Editor Write the correct symbol and value that represents the solution to the inequality $9x < 27$. (Lesson 7)

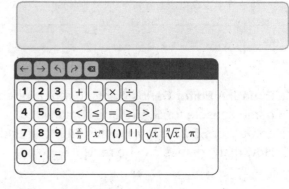

12. Grid Consider the inequality $-11x \geq 22$. (Lesson 7)

A. Solve the inequality. Explain why or why not the direction of the inequality symbol is reversed.

B. Graph the solution of the inequality.

−5 −4 −3 −2 −1 0 1 2 3 4 5

13. Open Response Solve the inequality $6.6z - 2.25 \leq -10.5$. Explain why or why not the direction of the inequality symbol is reversed. (Lesson 8)

What Are Foldables and How Do I Create Them?

Foldables are three-dimensional graphic organizers that help you create study guides for each module in your book.

Step 1 Go to the back of your book to find the Foldable for the module you are currently studying. Follow the cutting and assembly instructions at the top of the page.

Step 2 Go to the Module Review at the end of the module you are currently studying. Match up the tabs and attach your Foldable to this page. Dotted tabs show where to place your Foldable. Striped tabs indicate where to tape the Foldable.

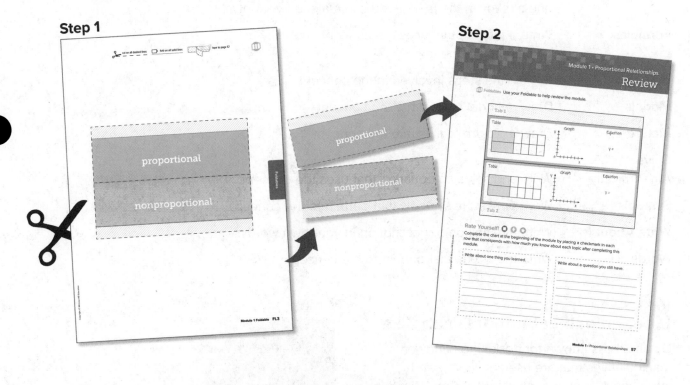

How Will I Know When to Use My Foldable?

You will be directed to work on your Foldable at the end of selected lessons. This lets you know that it is time to update it with concepts from that lesson. Once you've completed your Foldable, use it to study for the module test.

How Do I Complete My Foldable?

No two Foldables in your book will look alike. However, some will ask you to fill in similar information. Below are some of the instructions you'll see as you complete your Foldable. **HAVE FUN** learning math using Foldables!

Instructions and What They Mean

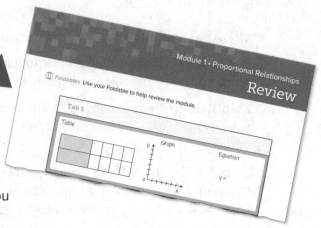

Best Used to...	Complete the sentence explaining when the concept should be used.
Definition	Write a definition in your own words.
Description	Describe the concept using words.
Equation	Write an equation that uses the concept. You may use one already in the text or you can make up your own.
Example	Write an example about the concept. You may use one already in the text or you can make up your own.
Formulas	Write a formula that uses the concept. You may use one already in the text.
How do I ...?	Explain the steps involved in the concept.
Models	Draw a model to illustrate the concept.
Picture	Draw a picture to illustrate the concept.
Solve Algebraically	Write and solve an equation that uses the concept.
Symbols	Write or use the symbols that pertain to the concept.
Write About It	Write a definition or description in your own words.
Words	Write the words that pertain to the concept.

Meet Foldables Author Dinah Zike

Dinah Zike is known for designing hands-on manipulatives that are used nationally and internationally by teachers and parents. Dinah is an explosion of energy and ideas. Her excitement and joy for learning inspires everyone she touches.

proportional

nonproportional

Foldables

Tab 1

Write About It

Write About It

Tab 2

cut on all dashed lines | fold on all solid lines | tape to page 121

Percents

percent of increase

percent of decrease

Foldables

Definition

Definition

Operations with Rational Numbers

+ or –
integers

× or ÷
integers

+ or –
rational numbers

× or ÷
rational numbers

Foldables

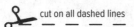 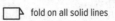

Examples

Examples

Examples

Examples

Tab 2

Tab 1

cut on all dashed lines

fold on all solid lines

tape to page 281

Laws of Exponents

Product of Powers

Quotient of Powers

Power of Powers

Power of Products

Foldables

Examples

Examples

Examples

Examples

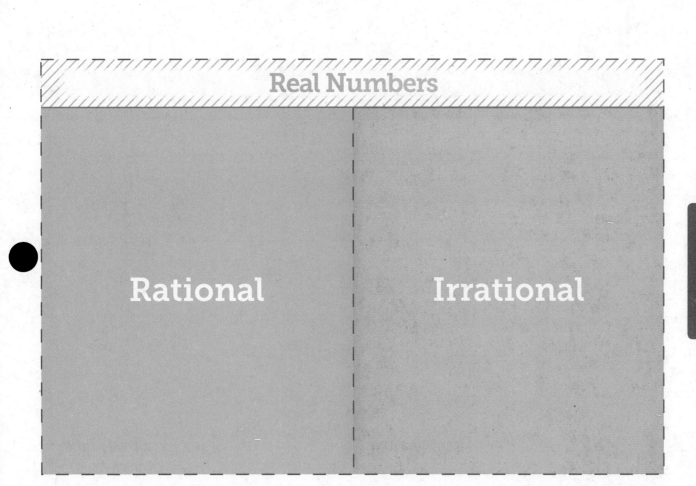

Real Numbers

Rational

Irrational

Foldables

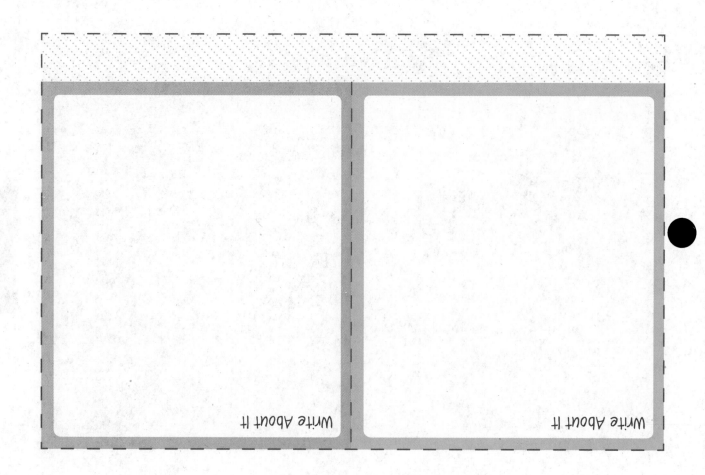

Write About It

Write About It

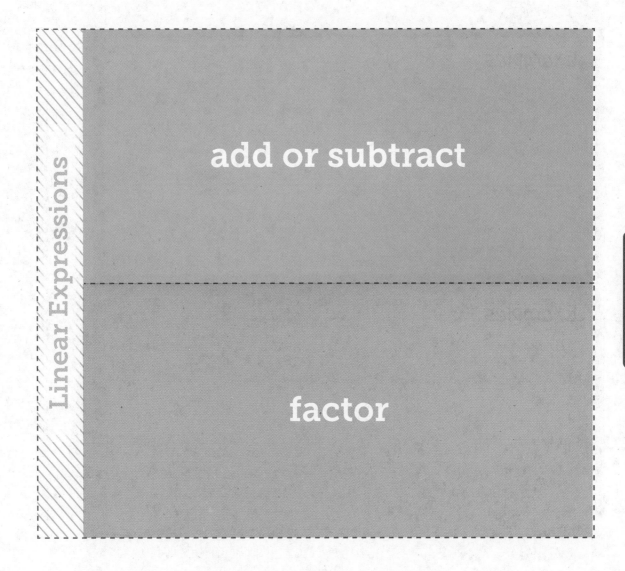

Linear Expressions

add or subtract

factor

Examples

Examples

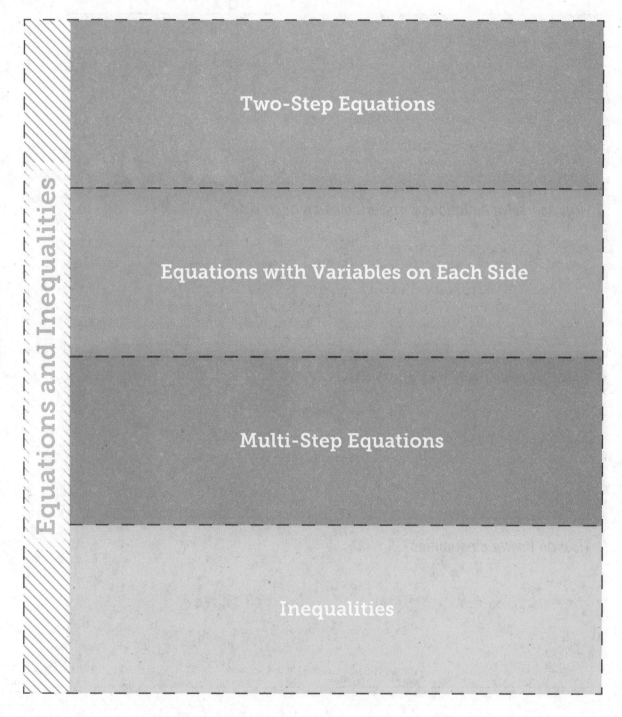

Equations and Inequalities

Two-Step Equations

Equations with Variables on Each Side

Multi-Step Equations

Inequalities

Foldables

How do I solve two-step equations?

How do I solve equations with variables on each side?

How do I solve multi-step equations?

How do I solve inequalities?

Glossary

The Multilingual eGlossary contains words and definitions in the following 14 languages:

Arabic	English	Hmong	Russian	Urdu
Bengali	French	Korean	Spanish	Vietnamese
Brazilian Portuguese	Haitian Creole	Mandarin	Tagalog	

English / Español

A

absolute value (Lesson 3-1)   The distance the number is from zero on a number line.

valor absoluto   Distancia a la que se encuentra un número de cero en la recta numérica.

acute angle (Lesson 11-1)   An angle with a measure greater than 0° and less than 90°.

ángulo agudo   Ángulo que mide más de 0° y menos de 90°.

acute triangle (Lesson 11-4)   A triangle having three acute angles.

triángulo acutángulo   Triángulo con tres ángulos agudos.

Addition Property of Equality (Lesson 7-2)   If you add the same number to each side of an equation, the two sides remain equal.

propiedad de adición de la igualdad   Si sumas el mismo número a ambos lados de una ecuación, los dos lados permanecen iguales.

Addition Property of Inequality (Lesson 7-6)   If you add the same number to each side of an inequality, the inequality remains true.

propiedad de desigualdad en la suma   Si se suma el mismo número a cada lado de una desigualdad, la desigualdad sigue siendo verdadera.

Additive Identity Property (Lesson 3-1)   The sum of any number and zero is the number.

propiedad de identidad de la suma   La suma de cualquier número y cero es el mismo número.

additive inverse (Lesson 3-1)   Two integers that are opposites. The sum of an integer and its additive inverse is zero.

inverso aditivo   Dos enteros opuestos. La suma de un entero y su inverso aditiva es cero.

Additive Inverse Property (Lesson 3-1)   The sum of any number and its additive inverse is zero.

propiedad inversa aditiva   La suma de cualquier número y su inversa aditiva es cero.

adjacent angles (Lesson 11-1)   Angles that have the same vertex, share a common side, and do not overlap.

ángulos adyacentes   Ángulos que comparten el mismo vértice y un común lado, pero no se sobreponen.

algebraic expression (Lesson 6-1)   A combination of variables, numbers, and at least one operation.

expresión algebraica   Combinación de variables, números y por lo menos una operación.

alternate exterior angles (**Lesson 11-3**)   Exterior angles that lie on opposite sides of the transversal.

ángulos alternos externos   Ángulos externos que se encuentran en lados opuestos de la transversal.

alternate interior angles (**Lesson 11-3**)   Interior angles that lie on opposite sides of the transversal.

ángulos alternos internos   Ángulos internos que se encuentran en lados opuestos de la transversal.

amount of error (**Lesson 2-4**)   The positive difference between the estimate and the actual amount.

cantidad de error   La diferencia positiva entre la estimación y la cantidad real.

angle (**Lesson 11-1**)   Two rays with a common endpoint form an angle. The rays and vertex are used to name the angle.

ángulo   Dos rayos con un extremo común forman un ángulo. Los rayos y el vértice se usan para nombrar el ángulo.

angle of rotation (**Lesson 13-3**)   The degree measure of the angle through which a figure is rotated.

ángulo de rotación   Medida en grados del ángulo sobre el cual se rota una figura.

area (**Lesson 12-2**)   The measure of the interior surface of a two-dimensional figure.

área   La medida de la superficie interior de una figura bidimensional.

asymmetric distribution (**Lesson 10-4**)   A distribution in which the shape of the graph on one side of the center is very different than the other side, or it has outliers that might affect the average.

distribución asimétrica   Una distribución en la que la forma del gráfico en un lado del centro es muy diferente del otro lado, o tiene valores atípicos que pueden afectar al promedio.

B

bar notation (**Lesson 3-6**)   In repeating decimals, the line or bar placed over the digits that repeat. For example, $2.\overline{63}$ indicates that the digits 63 repeat.

notación de barra   Línea o barra que se coloca sobre los dígitos que se repiten en decimales periódicos. Por ejemplo, $2.\overline{63}$ indica que los dígitos 63 se repiten.

base (**Lesson 4-1**)   In a power, the number that is the common factor. In 10^3, the base is 10. That is, $10^3 = 10 \cdot 10 \cdot 10$.

base   En una potencia, el número que es el factor común. En 10^3, la base es 10. Es decir, $10^3 = 10 \cdot 10 \cdot 10$.

base (**Lesson 11-7**)   One of the two parallel congruent faces of a prism.

base   Una de las dos caras paralelas congruentes de un prisma.

biased sample (**Lesson 10-1**)   A sample drawn in such a way that one or more parts of the population are favored over others.

muestra sesgada   Muestra en que se favorece una o más partes de una población.

box plot (**Lesson 10-4**)   A method of visually displaying a distribution of data values by using the median, quartiles, and extremes of the data set. A box shows the middle 50% of the data.

diagrama de caja   Un método de mostrar visualmente una distribución de valores usando la mediana, cuartiles y extremos del conjunto de datos. Una caja muestra el 50% del medio de los datos.

C

center (**Lesson 12-1**)   The point from which all points on a circle are the same distance.

centro   El punto desde el cual todos los puntos en una circunferencia están a la misma distancia.

center of dilation (Lesson 13-4)   The center point from which dilations are performed.

center of rotation (Lesson 13-3)   A fixed point around which shapes move in a circular motion to a new position.

circle (Lesson 12-1)   The set of all points in a plane that are the same distance from a given point called the center.

circumference (Lesson 12-1)   The distance around a circle.

coefficient (Lesson 6-1)   The numerical factor of a term that contains a variable.

commission (Lesson 2-5)   A payment equal to a percent of the amount of goods or services that an employee sells for the company.

common denominator (Lesson 3-6)   A common multiple of the denominators of two or more fractions. 24 is a common denominator for $\frac{1}{3}$, $\frac{5}{8}$, and $\frac{3}{4}$ because 24 is the LCM of 3, 8, and 4.

complementary angles (Lesson 11-2)   Two angles are complementary if the sum of their measures is 90°.

complementary events (Lesson 9-3)   Two events in which either one or the other must happen, but they cannot happen at the same time. The sum of the probability of an event and its complement is 1 or 100%.

complex fraction (Lesson 1-1)   A fraction $\frac{A}{B}$ where A and/or B are fractions and B does not equal zero.

composite figure (Lesson 12-3)   A figure that is made up of two or more figures.

composite solid (Lesson 12-9)   An object made up of more than one solid.

composition of transformations (Lesson 13-5) The resulting transformation when a transformation is applied to a figure and then another transformation is applied to its image.

compound event (Lesson 9-5)   An event consisting of two or more simple events.

centro de la homotecia   Punto fijo en torno al cual se realizan las homotecias.

centro de rotación   Punto fijo alrededor del cual se giran las figuras en movimiento circular alrededor de un punto fijo.

círculo   Conjunto de todos los puntos de un plano que están a la misma distancia de un punto dado denominado centro.

circunferencia   Distancia en torno a un círculo.

coeficiente   El factor numérico de un término que contiene una variable.

comisión   Un pago igual a un porcentaje de la cantidad de bienes o servicios que un empleado vende para la empresa.

común denominador   El múltiplo común de los denominadores de dos o más fracciones. 24 es un denominador común para $\frac{1}{3}$, $\frac{5}{8}$, y $\frac{3}{4}$ porque 24 es el mcm de 3, 8 y 4.

ángulos complementarios   Dos ángulos son complementarios si la suma de sus medidas es 90°.

eventos complementarios   Dos eventos en los cuales uno o el otro debe suceder, pero no pueden ocurrir al mismo tiempo. La suma de la probabilidad de un evento y su complemento es 1 o 100%.

fracción compleja   Una fracción $\frac{A}{B}$ en la cual A y/o B son fracciones y B no es igual a cero.

figura compuesta   Figura formada por dos o más figuras.

sólido complejo   Cuerpo compuesto de más de un sólido.

composición de transformaciones   Transformación que resulta cuando se aplica una transformación a una figura y luego se le aplica otra transformación a su imagen.

evento compuesto   Un evento que consiste en dos o más eventos simples.

cone (Lesson 11-7) A three-dimensional figure with one circular base connected by a curved surface to a single point.

cono Una figura tridimensional con una base circular conectada por una superficie curva para un solo punto.

congruent (Lesson 13-5) Having the same measure; if one image can be obtained from another by a sequence of rotations, reflections, or translations.

congruente Que tienen la misma medida; si una imagen puede obtenerse de otra por una secuencia de rotaciones, reflexiones o traslaciones.

congruent angles (Lesson 11-1) Angles that have the same measure.

ángulos congruentes Ángulos que tienen la misma medida.

congruent figures (Lesson 11-4) Figures that have the same size and same shape and corresponding sides and angles with equal measure.

figuras congruentes Figuras que tienen el mismo tamaño y la misma forma y los lados y los ángulos correspondientes tienen igual medida.

congruent segments (Lesson 11-4) Sides with the same length.

segmentos congruentes Lados con la misma longitud.

constant (Lesson 6-1) A term that does not contain a variable.

constante Término que no contiene ninguna variable.

constant of proportionality (Lesson 1-3) A constant ratio or unit rate of two variable quantities. It is also called the constant of variation.

constante de proporcionalidad Una razón constante o tasa por unidad de dos cantidades variables. También se llama constante de variación.

constant of variation (Lesson 1-3) The constant ratio in a direct variation. It is also called the constant of proportionality.

constante de variación Una razón constante o tasa por unidad de dos cantidades variables. También se llama constante de proporcionalidad.

constant rate of change (Lesson 1-3) The rate of change between any two points in a linear relationship is the same or *constant*.

tasa constante de cambio La tasa de cambio entre dos puntos cualesquiera en una relación lineal permanece igual o *constante*.

convenience sample (Lesson 10-1) A sample which consists of members of a population that are easily accessed.

muestra de conveniencia Muestra que incluye miembros de una población fácilmente accesibles.

corresponding angles (Lesson 11-3) Angles that are in the same position on two parallel lines in relation to a transversal.

ángulos correspondientes Ángulos que están en la misma posición sobre dos rectas paralelas en relación con la transversal.

corresponding parts (Lesson 8-3) Parts of congruent or similar figures that are in the same relative position.

partes correspondientes Partes de figuras congruentes o similares que están en la misma posición relativa.

counterexample (Lesson 5-3) A statement or example that shows a conjecture is false.

contraejemplo Ejemplo o enunciado que demuestra que una conjetura es falsa.

cross section (Lesson 11-7) The intersection of a solid and a plane.

sección transversal Intersección de un sólido con un plano.

cube root **(Lesson 4-1)**   One of three equal factors of a number. If $a^3 = b$, then a is the cube root of b. The cube root of 64 is 4 since $4^3 = 64$.

cylinder **(Lesson 11-7)**   A three-dimensional figure with two parallel congruent circular bases connected by a curved surface.

raíz cúbica   Uno de tres factores iguales de un número. Si $a^3 = b$, entonces a es la raíz cúbica de b. La raíz cúbica de 64 es 4, dado que $4^3 = 64$.

cilindro   Una figura tridimensional con dos paralelas congruentes circulares bases conectados por una superficie curva.

D

degrees **(Lesson 11-1)**   The most common unit of measure for angles. If a circle were divided into 360 equal-sized parts, each part would have an angle measure of 1 degree.

diameter **(Lesson 12-1)**   The distance across a circle through its center.

dilation **(Lesson 13-4)**   A transformation that enlarges or reduces a figure by a scale factor.

dimensional analysis **(Lesson 1-2)**   The process of including units of measurement when you compute.

direct variation **(Lesson 8-4)**   A relationship between two variable quantities with a constant ratio. A proportional linear relationship.

discount **(Lesson 2-8)**   The amount by which the regular price of an item is reduced.

distribution **(Lesson 10-4)**   The shape of a graph of data.

Distributive Property **(Lesson 3-3)**   To multiply a sum by a number, multiply each addend of the sum by the number outside the parentheses. For any numbers a, b, and c, $a(b + c) = ab + ac$ and $a(b - c) = ab - ac$.

Example: $2(5 + 3) = (2 \cdot 5) + (2 \cdot 3)$ and $2(5 - 3) = (2 \cdot 5) - (2 \cdot 3)$

Division Property of Equality **(Lesson 7-1)**   If you divide each side of an equation by the same nonzero number, the two sides remain equal.

Division Property of Inequality **(Lesson 7-7)**   When you divide each side of an inequality by a negative number, the inequality symbol must be reversed for the inequality to remain true.

grados   La unidad más común para medir ángulos. Si un círculo se divide en 360 partes iguales, cada parte tiene una medida angular de 1 grado.

diámetro   Segmento que pasa por el centro de un círculo y lo divide en dos partes iguales.

homotecia   Transformación que produce la ampliación o reducción de una imagen por un factor de escala.

análisis dimensional   Proceso que incluye las unidades de medida al hacer cálculos.

variación directa   Relación entre dos cantidades variables con una razón constante. Una relación lineal proporcional.

descuento   Cantidad que se le rebaja al precio regular de un artículo.

distribución   La forma de un gráfico de datos.

propiedad distributiva   Para multiplicar una suma por un número, multiplíquese cada sumando de la suma por el número que está fuera del paréntesis. Sean cuales fuere los números a, b, y c, $a(b + c) = ab + ac$ y $a(b - c) = ab - ac$.

Ejemplo: $2(5 + 3) = (2 \cdot 5) + (2 \cdot 3)$ y $2(5 - 3) = (2 \cdot 5) - (2 \cdot 3)$

propiedad de igualdad de la división   Si divides ambos lados de una ecuación entre el mismo número no nulo, los lados permanecen iguales.

propiedad de desigualdad en la división   Cuando se divide cada lado de una desigualdad entre un número negativo, el símbolo de desigualdad debe invertirse para que la desigualdad siga siendo verdadera.

double box plot (Lesson 10-4) Two box plots graphed on the same number line.

double dot plot (Lesson 10-4) A method of visually displaying a distribution of two sets of data values where each value is shown as a dot above a number line.

doble diagrama de caja Dos diagramas de caja sobre la misma recta numérica.

doble diagrama de puntos Un método de mostrar visualmente una distribución de dos conjuntos de valores donde cada valor se muestra como un punto arriba de una recta numérica.

E

edge (Lesson 11-7) The line segment where two faces of a polyhedron intersect.

borde El segmento de línea donde se cruzan dos caras de un poliedro.

enlargement (Lesson 11-6) An image larger than the original.

ampliación Imagen más grande que la original.

equiangular (Lesson 11-4) In a polygon, all of the angles are congruent.

equiangular En un polígono, todos los ángulos son congruentes.

equilateral (Lesson 11-4) In a polygon, all of the sides are congruent.

equilátero En un polígono, todos los lados son congruentes.

equilateral triangle (Lesson 11-4) A triangle having three congruent sides.

triángulo equilátero Triángulo con tres lados congruentes.

equivalent expressions (Lesson 6-1) Expressions that have the same value.

expresiones equivalentes Expresiones que tienen el mismo valor.

equivalent ratios (Lesson 1-2) Two ratios that have the same value.

razones equivalentes Dos razones que tienen el mismo valor.

evaluate (Lesson 4-1) To find the value of an expression.

evaluar Calcular el valor de una expresión.

event (Lesson 9-1) The desired outcome or set of outcomes in a probability experiment.

evento El resultado deseado o conjunto de resultados en un experimento de probabilidad.

experimental probability (Lesson 9-2) An estimated probability based on the relative frequency of positive outcomes occurring during an experiment. It is based on what *actually* occurred during such an experiment.

probabilidad experimental Probabilidad estimada que se basa en la frecuencia relativa de los resultados positivos que ocurren durante un experimento. Se basa en lo que en *realidad* ocurre durante dicho experimento.

exponent (Lesson 4-1) In a power, the number of times the base is used as a factor. In 10^3, the exponent is 3.

exponente En una potencia, el número de veces que la base se usa como factor. En 10^3, el exponente es 3.

exterior angle (Lesson 11-5) An angle between one side of a polygon and the extension of an adjacent side.

ángulo exterior Un ángulo entre un lado de un polígono y la extensión de un lado adyacente.

exterior angles (Lesson 11-3) The four outer angles formed by two lines cut by a transversal.

ángulo externo Los cuatro ángulos exteriores que se forman cuando una transversal corta dos rectas.

face **(Lesson 11-7)** A flat surface of a polyhedron.

cara Una superficie plana de un poliedro.

factor **(Lesson 6-4)** To write a number as a product of its factors.

factorizar Escribir un número como el producto de sus factores.

factored form **(Lesson 6-4)** An expression expressed as the product of its factors.

forma factorizada Una expresión expresada como el producto de sus factores.

factors **(Lesson 6-1)** Two or more numbers that are multiplied together to form a product.

factores Dos o más números que se multiplican entre sí para formar un producto.

fee **(Lesson 2-5)** A payment for a service. It can be a fixed amount, a percent of the charge, or both.

cuota Un pago por un servicio. Puede ser una cantidad fija, un porcentaje del cargo, o ambos.

gratuity **(Lesson 2-7)** Also known as a tip. It is a small amount of money in return for a service.

gratificación También conocida como propina. Es una cantidad pequeña de dinero en retribución por un servicio.

greatest common factor (GCF) of two monomials **(Lesson 6-4)** The greatest monomial that is a factor of both monomials. The greatest common factor also includes any variables that the monomials have in common.

mayor factor común (GCF) de dos monomios El monomio más grande que es un factor de ambos monomios. El factor común más grande también incluye las variables que los monomios tienen en común.

hemisphere **(Lesson 12-8)** One of two congruent halves of a sphere.

hemisferio Una de dos mitades congruentes de una esfera.

image **(Lesson 13-1)** The resulting figure after a transformation.

imagen Figura que resulta después de una transformación.

indirect measurement **(Lesson 13-7)** A technique using properties of similar polygons to find distances or lengths that are difficult to measure directly.

medición indirecta Técnica que usa las propiedades de polígonos semejantes para calcular distancias o longitudes difíciles de medir directamente.

inequality **(Lesson 7-6)** An open sentence that uses $<$, $>$, $\neq$, $\leq$, or $\geq$ to compare two quantities.

desigualdad Enunciado abierto que usa $<$, $>$, $\neq$, $\leq$, o $\geq$ para comparar dos cantidades.

inference **(Lesson 10-1)** A prediction made about a population.

inferencia Una predicción hecha sobre una población.

initial value **(Lesson 8-5)** The starting value in a real-world situation in which an equation can be written. The y-intercept of a linear function.

valor inicial El valor inicial en una situación real en la que se puede escribir una ecuación. La intersección y de una función lineal.

integer **(Lesson 3-1)** Any number from the set $\{..., -4, -3, -2, -1, 0, 1, 2, 3, 4, ...\}$, where ... means continues without end.

entero Cualquier número del conjunto $\{..., -4, -3, -2, -1, 0, 1, 2, 3, 4, ...\}$, donde ... significa que continúa sin fin.

interest **(Lesson 2-9)** The amount paid or earned for the use of the principal.

interés La cantidad pagada o ganada por el uso del principal.

interior angle **(Lesson 11-5)** An angle inside a polygon.

ángulo interno Ángulo dentro de un polígono.

interior angles **(Lesson 11-3)** The four inside angles formed by two lines cut by a transversal.

ángulo interno Los cuatro ángulos internos formados por dos rectas intersecadas por una transversal.

interquartile range (IQR) **(Lesson 10-4)** A measure of variation in a set of numerical data, the interquartile range is the distance between the first and third quartiles of the data set.

rango intercuartil (RIQ) El rango intercuartil, una medida de la variacion en un conjunto de datos numéricos, es la distancia entre el primer y el tercer cuartil del conjunto de datos

invalid inference **(Lesson 10-1)** An inference that is based on a biased sample or makes a conclusion not supported by the results of the sample.

inferencia inválida Una inferencia que se basa en una muestra sesgada o hace una conclusión no apoyada por los resultados de la muestra.

inverse operations **(Lesson 5-1)** Pairs of operations that undo each other. Addition and subtraction are inverse operations. Multiplication and division are inverse operations.

peraciones inversas Pares de operaciones que se anulan mutuamente. La adición y la sustracción son operaciones inversas. La multiplicación y la división son operaciones inversas.

irrational number **(Lesson 5-2)** A number that cannot be expressed as the ratio $\frac{a}{b}$, where a and b are integers and $b \neq 0$.

números irracionales Número que no se puede expresar como la proporción $\frac{a}{b}$, donde a y b son enteros y $b \neq 0$.

isosceles triangle **(Lesson 11-4)** A triangle having at least two congruent sides.

triángulo isósceles Triángulo que tiene por lo menos dos lados congruentes.

L

lateral face **(Lesson 12-5)** In a polyhedron, a face that is not a base.

cara lateral En un poliedro, las caras que no forman las bases.

lateral surface area **(Lesson 12-5)** The sum of the areas of all of the lateral faces of a solid.

área de superficie lateral Suma de las áreas de todas las caras de un sólido.

least common denominator (LCD) **(Lesson 3-6)** The least common multiple of the denominators of two or more fractions. You can use the LCD to compare fractions.

mínimo común denominador (mcd) El menor de los múltiplos de los denominadores de dos o más fracciones. Puedes usar el mínimo común denominador para comparar fracciones.

like terms (Lesson 6-1)   Terms that contain the same variable(s) raised to the same power. Example: 5x and 6x are like terms.

likelihood (Lesson 9-1)   The chance of an event occurring.

linear (Lesson 8-1)   To fall in a straight line.

linear equation (Lesson 8-1)   An equation with a graph that is a straight line.

linear expression (Lesson 6-2)   An algebraic expression in which the variable is raised to the first power, and variables are neither multiplied nor divided.

linear relationship (Lesson 1-4)   A relationship for which the graph is a straight line.

line of reflection (Lesson 13-2)   The line over which a figure is reflected in a transformation.

line segment (Lesson 11-5)   Part of a line containing two endpoints and all of the points between them.

términos semejante   Términos que contienen las mismas variable(s) elevadas a la misma potencia. Ejemplo: 5x y 6x son términos semejante.

probabilidad   La probabilidad de que ocurra un evento.

lineal   Que cae en una línea recta.

ecuación lineal   Ecuación cuya gráfica es una recta.

expresión lineal   Expresión algebraica en la cual la variable se eleva a la primera potencia.

relación lineal   Una relación para la cual la gráfica es una línea recta.

línea de reflexión   Línea a través de la cual se refleja una figura en una transformación.

segmento de línea   Parte de una línea que contiene dos extremos y todos los puntos entre ellos.

M

markdown (Lesson 2-8)   An amount by which the regular price of an item is reduced.

markup (Lesson 2-7)   The amount the price of an item is increased above the price the store paid for the item.

mean (Lesson 10-4)   The sum of the data divided by the number of items in the data set.

mean absolute deviation (MAD) (Lesson 10-4) A measure of variation in a set of numerical data, computed by adding the distances between each data value and the mean, then dividing by the number of data values.

measures of center (Lesson 10-4)   Numbers that are used to describe the center of a set of data. These measures include the mean, median, and mode.

measures of variation (Lesson 10-4)   A measure used to describe the distribution of data.

rebaja   Una cantidad por la cual el precio regular de un artículo se reduce.

margen de utilidad   Cantidad de aumento en el precio de un artículo por encima del precio que paga la tienda por dicho artículo.

media   La suma de los datos dividida entre el número total de artículos en el conjunto de datos.

desviación media absoluta   Una medida de variación en un conjunto de datos numéricos que se calcula sumando las distancias entre el valor de cada dato y la media, y luego dividiendo entre el número de valores.

medidas del centro   Números que se usan para describir el centro de un conjunto de datos. Estas medidas incluyen la media, la mediana y la moda.

medidas de variación   Medida usada para describir la distribución de los datos.

median (Lesson 10-4) A measure of center in a set of numerical data. The median of a list of values is the value appearing at the center of a sorted version of the list—or the mean of the two central values, if the list contains an even number of values.

monomial (Lesson 4-2) A number, variable, or product of a number and one or more variables.

Multiplication Property of Equality (Lesson 7-1) If you multiply each side of an equation by the same nonzero number, the two sides remain equal.

Multiplication Property of Inequality (Lesson 7-7) When you multiply each side of an inequality by a negative number, the inequality symbol must be reversed for the inequality to remain true.

Multiplicative Identity Property (Lesson 3-3) The product of any number and one is the number.

multiplicative inverse (Lesson 3-8) Two numbers with a product of 1. For example, the multiplicative inverse of $\frac{2}{3}$ is $\frac{3}{2}$.

Multiplicative Property of Zero (Lesson 3-3) The product of any number and zero is zero.

mediana Una medida del centro en un conjunto de dados númericos. La mediana de una lista de valores es el valor que aparece en el centro de una versíon ordenada de la lista, o la media de dos valores centrales si la lista contiene un número par de valores.

monomio Número, variable o producto de un número y una o más variables.

propiedad de multiplicación de la igualdad Si multiplicas ambos lados de una ecuación por el mismo número no nulo, lo lados permanecen iguales.

propiedad de desigualdad en la multiplicación Cuando se multiplica cada lado de una desigualdad por un número negativo, el símbolo de desigualdad debe invertirse para que la desigualdad siga siendo verdadera.

propiedad de identidad de la multiplicación El producto de cualquier número y uno es el mismo número.

inverso multiplicativo Dos números cuyo producto es 1. Por ejemplo, el inverso multiplicativo de $\frac{2}{3}$ es $\frac{3}{2}$.

propiedad del cero en la multiplicación El producto de cualquier número y cero es cero.

N

natural numbers (Lesson 5-1) The set of numbers used for counting.

negative exponent (Lesson 4-4) The result of repeated division used to represent very small numbers.

negative integer (Lesson 3-1) An integer that is less than zero. Negative integers are written with a − sign.

net (Lesson 12-5) A two-dimensional figure that can be used to build a three-dimensional figure.

números naturales El conjunto de números utilizado para el recuento.

exponente negativo El resultado de la división repetida se utiliza para representar números muy pequeños.

entero negativo Número menor que cero. Se escriben con el signo −.

red Figura bidimensional que sirve para hacer una figura tridimensional.

nonproportional (Lesson 1-3) The relationship between two ratios with a rate or ratio that is not constant.

no proporcional Relación entre dos razones cuya tasa o razón no es constante.

numerical expression (Lesson 6-1) A combination of numbers and operations.

expresión numérica Combinación de números y operaciones.

O

obtuse angle (Lesson 11-1) Any angle that measures greater than 90° but less than 180°.

ángulo obtuso Cualquier ángulo que mide más de 90° pero menos de 180°.

obtuse triangle (Lesson 11-4) A triangle having one obtuse angle.

triángulo obtusángulo Triángulo que tiene un ángulo obtuso.

opposites (Lesson 3-1) Two integers are opposites if they are represented on the number line by points that are the same distance from zero, but on opposite sides of zero. The sum of two opposites is zero.

opuestos Dos enteros son opuestos si, en la recta numérica, están representados por puntos que equidistan de cero, pero en direcciones opuestas. La suma de dos opuestos es cero.

order of operations (Lesson 3-9) The rules to follow when more than one operation is used in a numerical expression.

1. Evaluate the expressions inside grouping symbols.
2. Evaluate all powers.
3. Multiply and divide in order from left to right.
4. Add and subtract in order from left to right.

orden de las operaciones Reglas a seguir cuando se usa más de una operación en una expresión numérica.

1. Primero, evalúa las expresiones dentro de los símbolos de agrupación.
2. Evalúa todas las potencias.
3. Multiplica y divide en orden de izquierda a derecha.
4. Suma y resta en orden de izquierda a derecha.

ordered pair (Lesson 8-1) A pair of numbers used to locate a point in the coordinate plane. The ordered pair is written in this form: (*x*-coordinate, *y*-coordinate).

par ordenado Par de números que se utiliza para ubicar un punto en un plano de coordenadas. Se escribe de la siguiente forma: (coordenada *x*, coordenada *y*).

origin (Lesson 8-1) The point of intersection of the *x*-axis and *y*-axis in a coordinate plane.

origen Punto en que el eje *x* y el eje *y* se intersecan en un plano de coordenadas.

outcome (Lesson 9-1) Any one of the possible results of an action. For example, 4 is an outcome when a number cube is rolled.

resultado Cualquiera de los resultados posibles de una acción. Por ejemplo, 4 puede ser un resultado al lanzar un cubo numerado.

P

parallel lines (Lesson 11-3) Lines in the same plane that never intersect or cross. The symbol ‖ means parallel.

rectas paralelas Rectas que yacen en un mismo plano y que no se intersecan. El símbolo ‖ significa paralela a.

parallelogram (Lesson 12-3) A quadrilateral with opposite sides parallel and opposite sides congruent.

paralelogramo Cuadrilátero cuyos lados opuestos son paralelos y congruentes.

percent error (Lesson 2-4) A ratio that compares the inaccuracy of an estimate (amount of error) to the actual amount.

percent of change (Lesson 2-3) A ratio that compares the change in a quantity to the original amount.

$$\text{percent of change} = \frac{\text{amount of change}}{\text{original amount}} \cdot 100$$

percent of decrease (Lesson 2-3) A negative percent of change.

percent of increase (Lesson 2-3) A positive percent of change.

perfect cube (Lesson 5-1) A number whose cube root is an integer. 27 is a perfect cube because its cube root is 3.

perfect square (Lesson 5-1) A number whose square root is a whole number. 25 is a perfect square because its square root is 5.

perpendicular lines (Lesson 11-3) Two lines that intersect to form right angles.

pi (Lesson 12-1) The ratio of the circumference of a circle to its diameter. The Greek letter π represents this number. The value of pi is 3.1415926.... Approximations for pi are 3.14 and $\frac{22}{7}$.

plane (Lesson 11-7) A two-dimensional flat surface that extends in all directions.

polygon (Lesson 12-3) A simple closed figure formed by three or more straight line segments.

polyhedron (Lesson 11-7) A three-dimensional figure with faces that are polygons.

population (Lesson 10-1) The entire group of items or individuals from which the samples under consideration are taken.

positive integer (Lesson 3-1) An integer that is greater than zero. They are written with or without a + sign.

power (Lesson 4-1) A product of repeated factors using an exponent and a base. The power 7^3 is read *seven to the third power,* or *seven cubed.*

porcentaje de error Una razón que compara la inexactitud de una estimación (cantidad del error) con la cantidad real.

porcentaje de cambio Razón que compara el cambio en una cantidad a la cantidad original.

$$\text{porcentaje de cambio} = \frac{\text{cantidad del cambio}}{\text{cantidad original}} \cdot 100$$

porcentaje de disminución Porcentaje de cambio negativo.

porcentaje de aumento Porcentaje de cambio positivo.

cubo perfecto Número cuya raíz cúbica es un número entero. 27 es un cubo perfecto porque su raíz cúbica es 3.

cuadrados perfectos Número cuya raíz cuadrada es un número entero. 25 es un cuadrado perfecto porque su raíz cuadrada es 5.

rectas perpendiculares Dos rectas que se intersecan formando ángulos rectos.

pi Relación entre la circunferencia de un círculo y su diámetro. La letra griega π representa este número. El valor de pi es 3.1415926.... Las aproximaciones de pi son 3.14 y $\frac{22}{7}$.

plano Superficie bidimensional que se extiende en todas direcciones.

polígono Figura cerrada simple formada por tres o más segmentos de recta.

poliedro Una figura tridimensional con caras que son polígonos.

población El grupo total de individuos o de artículos del cual se toman las muestras bajo estudio.

entero positivo Entero que es mayor que cero; se escribe con o sin el signo +.

potencia Producto de factores repetidos con un exponente y una base. La potencia 7^3 se lee *siete a la tercera potencia* o *siete al cubo.*

Power of a Power Property (Lesson 4-3) A property that states to find the power of a power, multiply the exponents.

Power of a Product Property (Lesson 4-3) A property that states to find the power of a product, find the power of each factor and multiply.

preimage (Lesson 13-1) The original figure before a transformation.

principal (Lesson 2-9) The amount of money deposited or borrowed.

principal square root (Lesson 5-1) The positive square root of a number.

prism (Lesson 11-7) A polyhedron with two parallel congruent faces called bases.

probability (Lesson 9-2) The chance that some event will happen. It is the ratio of the number of favorable outcomes to the number of possible outcomes.

probability experiment (Lesson 9-2) When you perform an event to find the likelihood of an event.

probability model (Lesson 9-3) A model used to assign probabilities to outcomes of a chance process by examining the nature of the process.

Product of Powers Property (Lesson 4-2) A property that states to multiply powers with the same base, add their exponents.

properties (Lesson 3-1) Statements that are true for any number or variable.

proportion (Lesson 1-5) An equation stating that two ratios or rates are equivalent.

proportional (Lesson 1-3) The relationship between two ratios with a constant rate or ratio.

pyramid (Lesson 11-7) A polyhedron with one base that is a polygon and three or more triangular faces that meet at a common vertex.

potencia de una propiedad de potencia Una propiedad que declara encontrar el poder de un poder, multiplicar los exponentes.

potencia de una propiedad de producto Una propiedad que declara encontrar el poder de un producto, encuentra el poder de cada factor y se multiplica.

preimagen Figura original antes de una transformación.

capital Cantidad de dinero que se deposita o se toma prestada.

raíz cuadrada principal La raíz cuadrada positiva de un número.

prisma Un poliedro con dos caras congruentes paralelas llamadas bases.

probabilidad La posibilidad de que suceda un evento. Es la razón del número de resultados favorables al número de resultados posibles.

experimento de probabilidad Cuando realiza un evento para encontrar la probabilidad de un evento.

modelo de probabilidad Un modelo usado para asignar probabilidades a resultados de un proceso aleatorio examinando la naturaleza del proceso.

producto de la propiedad de los poderes Una propiedad que declara multiplicar poderes con la misma base, añade sus exponentes.

propiedades Enunciados que son verdaderos para cualquier número o variable.

proporción Ecuación que indica que dos razones o tasas son equivalentes.

proporcional Relación entre dos razones con una tasa o razón constante.

pirámide Un poliedro con una base que es un polígono y tres o más caras triangulares que se encuentran en un vértice común.

quadrants **(Lesson 8-1)** The four sections of the coordinate plane.

cuadrantes Las cuatro secciones del plano de coordenadas.

quadrilateral **(Lesson 12-3)** A closed figure having four sides and four angles.

cuadrilátero Figura cerrada que tiene cuatro lados y cuatro ángulos.

Quotient of Powers Property **(Lesson 4-2)** A property that states to divide powers with the same base, subtract their exponents.

propiedad del cociente de poderes Una propiedad que declara dividir poderes con la misma base, resta sus exponentes.

radical sign **(Lesson 5-1)** The symbol used to indicate a positive square root, $\sqrt{}$.

signo radical Símbolo que se usa para indicar una raíz cuadrada no positiva, $\sqrt{}$.

radius **(Lesson 12-1)** The distance from the center of a circle to any point on the circle.

radio Distancia desde el centro de un círculo hasta cualquiera de sus puntos.

random **(Lesson 9-2)** Outcomes occur at random if each outcome occurs by chance. For example, rolling a number on a number cube occurs at random.

azar Los resultados ocurren aleatoriamente si cada resultado ocurre por casualidad. Por ejemplo, sacar un número en un cubo numerado ocurre al azar.

rate **(Lesson 1-1)** A special kind of ratio in which the units are different.

tasa Un tipo especial de relación en el que las unidades son diferentes.

rate of change **(Lesson 8-1)** A rate that describes how one quantity changes in relation to another quantity.

tasa de cambio Una tasa que describe cómo cambia una cantidad en relación con otra cantidad.

ratio **(Lesson 1-1)** A comparison between two quantities, in which for every a units of one quantity, there are b units of another quantity.

razón Una comparación entre dos cantidades, en la que por cada a unidades de una cantidad, hay unidades b de otra cantidad.

rational numbers **(Lesson 3-6)** The set of numbers that can be written in the form $\frac{a}{b}$, where a and b are integers and $b \neq 0$.

números racionales Conjunto de números que puede escribirse en la forma $\frac{a}{b}$ donde a y b son números enteros y $b \neq 0$.

Examples: $1 = \frac{1}{1}, \frac{2}{9}, -2.3 = \frac{-23}{10}$

Ejemplos: $1 = \frac{1}{1}, \frac{2}{9}, -2.3 = \frac{-23}{10}$

real numbers **(Lesson 5-2)** The set of rational numbers together with the set of irrational numbers.

números reales El conjunto de números racionales junto con el conjunto de números irracionales.

reciprocal **(Lesson 3-8)** The multiplicative inverse of a number.

recíproco El inverso multiplicativo de un número.

rectangular prism **(Lesson 12-4)** A prism that has two parallel congruent bases that are rectangles.

prisma rectangular Un prisma con dos bases paralelas congruentes que son rectángulos.

reduction (Lesson 11-6) An image smaller than the original.

reflection (Lesson 13-2) A transformation where a figure is flipped over a line. Also called a flip.

regular polygon (Lesson 12-3) A polygon that has all sides congruent and all angles congruent.

regular pyramid (Lesson 12-5) A pyramid whose base is a regular polygon and in which the segment from the vertex to the center of the base is the altitude.

relative frequency (Lesson 9-2) A ratio that compares the frequency of each category to the total.

relative frequency graph (Lesson 9-2) A graph used to organize occurrences compared to a total.

relative frequency table (Lesson 9-2) A table used to organize occurrences compared to a total.

remote interior angles (Lesson 11-5) The angles of a triangle that are not adjacent to a given exterior angle.

repeating decimal (Lesson 3-6) A decimal in which 1 or more digits repeat.

rhombus (Lesson 12-3) A parallelogram having four congruent sides.

right angle (Lesson 11-1) An angle that measures exactly 90°.

right triangle (Lesson 11-4) A triangle having one right angle.

rise (Lesson 8-2) The vertical change between any two points on a line.

rotation (Lesson 13-3) A transformation in which a figure is turned about a fixed point.

run (Lesson 8-2) The horizontal change between any two points on a line.

reducción Imagen más pequeña que la original.

reflexión Transformación en la cual una figura se voltea sobre una recta. También se conoce como simetría de espejo.

polígono regular Polígono con todos los lados y todos los ángulos congruentes.

pirámide regular Pirámide cuya base es un polígono regular y en la cual el segmento desde el vértice hasta el centro de la base es la altura.

frecuencia relativa Razón que compara la frecuencia de cada categoría al total.

gráfico de frecuencia relativa Gráfico utilizado para organizar las ocurrencias en comparación con un total.

tabla de frecuencia relativa Una tabla utilizada para organizar las ocurrencias en comparación con un total.

ángulos internos no adyacentes Ángulos de un triángulo que no son adya centes a un ángulo exterior dado.

decimal periódico Un decimal en el que se repiten 1 o más dígitos.

rombo Paralelogramo que tiene cuatro lados congruentes.

ángulo recto Ángulo que mide exactamente 90°.

triángulo rectángulo Triángulo que tiene un ángulo recto.

elevación El cambio vertical entre cualquier par de puntos en una recta.

rotación Transformación en la cual una figura se gira alrededor de un punto fijo.

carrera El cambio horizontal entre cualquier par de puntos en una recta.

sales tax (Lesson 2-6) An additional amount of money charged on items that people buy.

impuesto sobre las ventas Cantidad de dinero adicional que se cobra por los artículos que se compran.

sample (Lesson 10-1) A randomly selected group chosen for the purpose of collecting data.

muestra Grupo escogido al azar o aleatoriamente que se usa con el propósito de recoger datos.

sample space (Lesson 9-3) The set of all possible outcomes of a probability experiment.

espacio muestral Conjunto de todos los resultados posibles de un experimento probabilístico.

scale (Lesson 11-6) The scale that gives the ratio that compares the measurements of a drawing or model to the measurements of the real object.

escala Razón que compara las medidas de un dibujo o modelo a las medidas del objeto real.

scale drawing (Lesson 11-6) A drawing that is used to represent objects that are too large or too small to be drawn at actual size.

dibujo a escala Dibujo que se usa para representar objetos que son demasiado grandes o demasiado pequeños como para dibujarlos de tamaño natural.

scale factor (Lesson 11-6) A scale written as a ratio without units in simplest form.

factor de escala Escala escrita como una razón sin unidades en forma simplificada.

scale factor (Lesson 13-4) The ratio of the lengths of two corresponding sides of two similar polygons.

factor de escala La razón de las longitudes de dos lados correspondientes de dos polígonos semejantes.

scale model (Lesson 11-6) A model used to represent objects that are too large or too small to be built at actual size.

modelo a escala Réplica de un objeto real, el cual es demasiado grande o demasiado pequeño como para construirlo de tamaño natural.

scalene triangle (Lesson 11-4) A triangle having no congruent sides.

triángulo escaleno Triángulo sin lados congruentes.

scientific notation (Lesson 4-5) A compact way of writing numbers with absolute values that are very large or very small. In scientific notation, 5,500 is 5.5×10^3.

notación científica Manera abreviada de escribir números con valores absolutos que son muy grandes o muy pequeños. En notación científica, 5,500 es 5.5×10^3.

selling price (Lesson 2-7) The amount the customer pays for an item.

precio de venta Cantidad de dinero que paga un consumidor por un artículo.

semicircle (Lesson 12-2) Half of a circle. The formula for the area of a semicircle is $A = \frac{1}{2}\pi r^2$.

semicírculo Medio círculo. La fórmula para el área de un semicírculo es $A = \frac{1}{2}\pi r^2$.

similar (Lesson 13-6) If one image can be obtained from another by a sequence of transformations and dilations.

similar Si una imagen puede obtenerse de otra mediante una secuencia de transformaciones y dilataciones.

similar figures (Lesson 8-3) Figures that have the same shape but not necessarily the same size.

figuras semejantes Figuras que tienen la misma forma, pero no necesariamente el mismo tamaño.

simple event (Lesson 9-2) One outcome or a collection of outcomes.

eventos simples Un resultado o una colección de resultados.

simple interest (Lesson 2-9) The amount paid or earned for the use of money. The formula for simple interest is $I = prt$.

interés simple Cantidad que se paga o que se gana por el uso del dinero. La fórmula para calcular el interés simple es $I = prt$.

simple random sample (Lesson 10-1) An unbiased sample where each item or person in the population is as likely to be chosen as any other.

muestra aleatoria simple Muestra de una población que tiene la misma probabilidad de escogerse que cualquier otra.

simplest form (Lesson 6-1) An expression is in simplest form when it is replaced by an equivalent expression having no like terms or parentheses.

expresión mínima Expresión en su forma más simple cuando es reemplazada por una expresión equivalente que no tiene términos similares ni paréntesis.

simplify (Lesson 6-5) Write an expression in simplest form.

simplificar Escribir una expresión en su forma más simple.

simulation (Lesson 9-6) An experiment that is designed to model the action in a given situation.

simulación Un experimento diseñado para modelar la acción en una situación dada.

slant height (Lesson 12-5) The height of each lateral face.

altura oblicua Altura de cada cara lateral.

slope (Lesson 8-1) The rate of change between any two points on a line. The ratio of the rise, or vertical change, to the run, or horizontal change.

pendiente Razón de cambio entre cualquier par de puntos en una recta. La razón de la altura, o cambio vertical, a la carrera, o cambio horizontal.

slope-intercept form (Lesson 8-5) An equation written in the form $y = mx + b$, where m is the slope and b is the y-intercept.

forma pendiente intersección Ecuación de la forma $y = mx + b$, donde m es la pendiente y b es la intersección y.

slope triangles (Lesson 8-3) Right triangles that fall on the same line on the coordinate plane.

triángulos de pendiente Triángulos rectos que caen en la misma línea en el plano de coordenadas.

solid (Lesson 12-6) A three-dimensional figure formed by intersecting planes.

sólido Figura tridimensional formada por planos que se intersecan.

sphere (Lesson 12-8) The set of all points in space that are a given distance from a given point called the center.

esfera Conjunto de todos los puntos en el espacio que están a una distancia dada de un punto dado llamado centro.

square root (Lesson 5-1) One of the two equal factors of a number. If $a^2 = b$, then a is the square root of b. A square root of 144 is 12 since $12^2 = 144$.

raíz cuadrada Uno de dos factores iguales de un número. Si $a^2 = b$, la a es la raíz cuadrada de b. Una raíz cuadrada de 144 es 12 porque $12^2 = 144$.

standard form (Lesson 4-5) Numbers written without exponents.

forma estándar Números escritos sin exponentes.

statistics (Lesson 10-1) The study of collecting, organizing, and interpreting data.

estadística Estudio que consiste en recopilar, organizar e interpretar datos.

straight angle (**Lesson 11-1**) An angle that measures exactly 180°.

ángulo llano Ángulo que mide exactamente 180°.

stratified random sample (**Lesson 10-1**) A sample in which the population is divided into groups with similar traits that do not overlap. A simple random sample is then selected from each group.

muestra aleatoria estratificada Una muestra en la que la población se divide en grupos con rasgos similares que no se superponen. A continuación, se selecciona una muestra aleatoria simple de cada grupo.

Subtraction Property of Equality (**Lesson 7-1**) If you subtract the same number from each side of an equation, the two sides remain equal.

propiedad de sustracción de la igualdad Si restas el mismo número de ambos lados de una ecuación, los dos lados permanecen iguales.

Subtraction Property of Inequality (**Lesson 7-6**) If you subtract the same number from each side of an inequality, the inequality remains true.

propiedad de desigualdad en la resta Si se resta el mismo número a cada lado de una desigualdad, la desigualdad sigue siendo verdadera.

supplementary angles (**Lesson 11-2**) Two angles are supplementary if the sum of their measures is 180°.

ángulos suplementarios Dos ángulos son suplementarios si la suma de sus medidas es 180°.

surface area (**Lesson 12-5**) The sum of the areas of all the surfaces (faces) of a three-dimensional figure.

área de superficie La suma de las áreas de todas las superficies (caras) de una figura tridimensional.

survey (**Lesson 10-1**) A question or set of questions designed to collect data about a specific group of people, or population.

encuesta Pregunta o conjunto de preguntas diseñadas para recoger datos sobre un grupo específico de personas o población.

symmetric distribution (**Lesson 10-4**) A distribution in which the shape of the graph on each side of the center is similar.

distribución simétrica Distribución en la que la forma de la gráfica en cada lado del centro es similar.

systematic random sample (**Lesson 10-1**) A sample where the items or people are selected according to a specific time or item interval.

muestra aleatoria sistemática Muestra en que los elementos o personas se eligen según un intervalo de tiempo o elemento específico.

T

term (**Lesson 4-2**) Each part of an algebraic expression separated by an addition or subtraction sign.

término Cada parte de un expresión algebraica separada por un signo adición o un signo sustracción.

terminating decimal (**Lesson 3-6**) A decimal with a repeating digit of 0.

decimal finito Un decimal que tiene un dígito que se repite que es 0.

theoretical probability (**Lesson 9-3**) The ratio of the number of ways an event can occur to the number of possible outcomes in the sample space. It is based on what *should* happen when conducting a probability experiment.

probabilidad teórica Razón del número de maneras en que puede ocurrir un evento al número de resultados posibles en el espacio muestral. Se basa en lo que *debería* pasar cuando se conduce un experiment probabilístico.

theoretical probability of a compound event (Lesson 9-5)   The ratio of the number of ways an event can occur to the number of possible outcomes in the sample space. It is based on what *should* happen when conducting a probability experiment.

probabilidad teórica de un evento compuesto   Razón del número de maneras en que puede ocurrir un evento al número de resultados posibles en el espacio muestral. Se basa en lo que *debería* pasar cuando se conduce un experimento probabilístico.

three-dimensional figure (Lesson 11-7)   A figure with length, width, and height.

figura tridimensional   Figura que tiene largo, ancho y alto.

tip (Lesson 2-7)   Also known as a gratuity, it is a small amount of money in return for a service.

propina   También conocida como gratificación; es una cantidad pequeña de dinero en recompensa por un servicio.

transformation (Lesson 13-1)   An operation that maps a geometric figure, preimage, onto a new figure, image.

transformación   Operación que convierte una figura geométrica, la pre-imagen, en una figura nueva, la imagen.

translation (Lesson 13-1)   A transformation that slides a figure from one position to another without turning.

traslación   Transformación en la cual una figura se desliza de una posición a otra sin hacerla girar.

transversal (Lesson 11-3)   A line that intersects two or more other lines.

transversal   Recta que interseca dos o más rectas.

trapezoid (Lesson 12-3)   A quadrilateral with one pair of parallel sides.

trapecio   Cuadrilátero con un único par de lados paralelos.

tree diagram (Lesson 9-5)   A diagram used to show the sample space.

diagrama de árbol   Diagrama que se usa para mostrar el espacio muestral.

triangle (Lesson 11-4)   A figure with three sides and three angles.

triángulo   Figura con tres lados y tres ángulos.

triangular prism (Lesson 12-4)   A prism that has two parallel congruent bases that are triangles.

prisma triangular   Un prisma que tiene dos bases congruentes paralelas que triángulos.

truncating (Lesson 5-3)   A process of approximating a decimal number by eliminating all decimal places past a certain point without rounding.

truncando   Proceso de aproximación de un número decimal eliminando todos los decimales más allá de un cierto punto sin redondear.

two-step equation (Lesson 7-1)   An equation having two different operations.

ecuación de dos pasos   Ecuación que contiene dos operaciones distintas.

two-step inequality (Lesson 7-8)   An inequality that contains two operations.

desigualdad de dos pasos   Desigualdad que contiene dos operaciones.

U

unbiased sample (Lesson 10-1)   A sample representative of the entire population.

muestra no sesgada   Muestra que se selecciona de modo que se representativa de la población entera.

uniform probability model (Lesson 9-3)
A probability model which assigns equal probability to all outcomes.

unit rate (Lesson 1-1) A rate in which the first quantity is compared to 1 unit of the second quantity.

unit ratio (Lesson 1-2) A ratio in which the first quantity is compared to every 1 unit of the second quantity.

modelo de probabilidad uniforme Un modelo de probabilidad que asigna igual probabilidad a todos los resultados.

tasa unitaria Una tasa en la que la primera candidad se compara con 1 unidad de la segunda candidad.

razón unitaria Una relación en la que la primera cantidad se compara con cada 1 unidad de la segunda cantidad.

valid inference (Lesson 10-1) A prediction, made about a population, based on an unbiased sample that is representative of the population.

valid sampling method (Lesson 10-1) A sampling method that is: representative of the population selected at random, where each member has an equal chance of being selected, and large enough to provide accurate data.

variability (Lesson 10-3) A measure that describes the amount of diversity in values within a sample or samples.

vertex (Lesson 11-1) A vertex of an angle is the common endpoint of the rays forming the angle.

vertex (Lesson 11-7) The point where three or more faces of a polyhedron intersect.

vertical angles (Lesson 11-1) Opposite angles formed by the intersection of two lines. Vertical angles are congruent.

vertices (Lesson 11-7) Plural of vertex.

visual overlap (Lesson 10-4) A visual demonstration that compares the centers of two distributions with their variation, or spread.

volume (Lesson 12-4) The measure of the space occupied by a solid. Standard units of measure are cubic units such as in^3 or ft^3.

voluntary response sample (Lesson 10-1) A sample which involves only those who want to participate in the sampling.

inferencia válida Una predicción, hecha sobre una población, basada en una muestra imparcial que es representativa de la población.

método de muestreo válido Un método de muestreo que es: representativo de la población seleccionada al azar, donde cada miembro tiene la misma oportunidad de ser seleccionado y suficientemente grande para proporcionar datos precisos.

variabilidad Medida que describe la cantidad de diversidad en valores dentro de una muestra o muestras.

vértice El vértice de un ángulo es el extremo común de los rayos que lo forman.

vértice El punto donde tres o más caras de un poliedro se cruzan.

ángulos opuestos por el vértice Ángulos opuestos formados por la intersección de dos rectas. Los ángulos opuestos por el vértice son congruentes.

vértices Plural de verticé.

superposición visual Una demostración visual que compara los centros de dos distribuciones con su variación, o magnitud.

volumen Medida del espacio que ocupa un sólido. Unidades de medida estándar son unidades cúbicas tales como pulg^3 o pies^3.

muestra de respuesta voluntaria Muestra que involucra sólo aquellos que quieren participar en el muestreo.

W

wholesale cost **(Lesson 2-7)**   The amount the store pays for an item.

coste al por mayor   La cantidad que la tienda paga por un artículo.

X

x-axis **(Lesson 8-1)**   The horizontal number line that helps to form the coordinate plane.

eje x   La recta numérica horizontal que ayuda a formar el plano de coordenadas.

x-coordinate **(Lesson 8-1)**   The first number of an ordered pair.

coordenada x   El primer número de un par ordenado.

x-intercept **(Lesson 8-1)**   The x-coordinate of the point where the line crosses the x-axis.

intersección x   La coordenada x del punto donde cruza la gráfica el eje x.

Y

y-axis **(Lesson 8-1)**   The vertical number line that helps to form the coordinate plane.

eje y   La recta numérica vertical que ayuda a formar el plano de coordenadas.

y-coordinate **(Lesson 8-1)**   The second number of an ordered pair.

coordenada y   El segundo número de un par ordenado.

y-intercept **(Lesson 8-5)**   The y-coordinate of the point where the line crosses the y-axis.

intersección y   La coordenada y del punto donde cruza la gráfica el eje y.

Z

zero angle **(Lesson 11-1)**   An angle that measures exactly 0 degrees.

ángulo cero   Un ángulo que mide exactamente 0 grados.

Zero Exponent Rule **(Lesson 4-4)**   A rule that states that any nonzero number to the zero power is equivalent to 1.

regla de exponente cero   Una regla que establece que cualquier número diferente de cero a la potencia cero es equivalente a 1.

zero pair **(Lesson 3-1)**   The result when one positive counter is paired with one negative counter. The value of a zero pair is 0.

par nulo   Resultado de hacer coordinar una ficha positiva con una negativa. El valor de un par nulo es 0.

Index

Double box plot, 657

Double dot plot, 657

Drawing, triangles, 712–718

Lesson 1-1 Unit Rates Involving Ratios of Fractions, Practice Pages 11–12

1. 64 miles in one hour **3.** 124.8 miles in one hour **5.** $\frac{5}{9}$ feet in one month **7.** $\frac{3}{5}$ mile in one minute **9.** penny; about 56.4 kilometers per hour faster **11.** Sample answer: The student who runs $\frac{3}{4}$ mile in 6 minutes will run 1 mile in 8 minutes if the rate is constant. The student who runs $\frac{1}{2}$ mile in 5 minutes will run 1 mile in 10 minutes if the rate is constant. **13.** Sample answer: She set up her rate as $\frac{\frac{3}{4} \text{ hour}}{9 \text{ greeting cards}}$. Her unit rate of $\frac{1}{12}$ is hour per greeting card, not card per hour.

Lesson 1-2 Understand Proportional Relationships, Practice Pages 19–20

1. yes; Both have a ratio of 3 : 1. **3.** yes; Both have a ratio of 7 : 200. **5.** no; One ratio is 4 cm to 1 year and the other is 3 cm to 1 year. **7.** 5 ft³ minerals, 5 ft³ peat moss, 10 ft³ compost; 12 ft³ minerals, 12 ft³ peat moss, 24 ft³ compost; 20 ft³ minerals, 20 ft³ peat moss, 40 ft³ compost; 100 ft³ minerals, 100 ft³ peat moss, 200 ft³ compost **9.** $\frac{1}{2}$ cup **11.** Sample answer: In the first cleaning solution, the ratio of vinegar to water is 1 : 2. The second solution, however, has a ratio of 2 : 3. The ratios are not equivalent. **13.** Sample answer: The unit rate for each ratio of a situation is the quantity compared to 1 unit of another quantity. If a relationship is proportional, then each ratio would have the same unit rate.

Lesson 1-3 Tables of Proportional Relationships, Practice Pages 29–30

1.

Lunches Bought	1	2	3	4
Total Cost ($)	2.50	5.00	7.50	10.00

yes; The ratios between the quantities are all equal and have a unit rate of $2.50 per lunch.

3.

Hours	1	2	3	4
Cost ($)	55	75	95	115

no; The ratios between the quantities are not equal.

5. 0.50 **7.** 3.1 **9.** yes; $9.00 **11.** yes; Sample answer: There is a proportional relationship between the number of fluid ounces and the number of cups. So, the number of cups would increase by the same factor as the number of fluid ounces. **13.** Sample answer: The number of laps increases at the same rate and the time increases at the same rate, but the ratios of laps to time are not equal.

Lesson 1-4 Graphs of Proportional Relationships, Practice Pages 39–40

1.

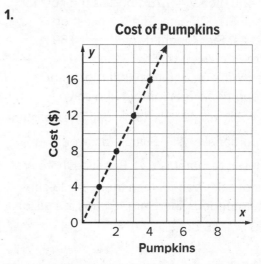

Cost of Pumpkins

The cost is proportional to the number of pumpkins bought because the graph is a straight line through the origin.

3. 15

5.

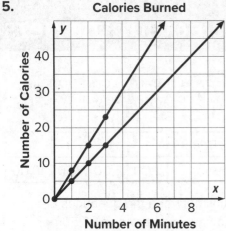

The point (1, 5) represents the unit rate.

7. Sample answer: A line can have a constant rate and not be proportional because it does not pass through the origin. For example, renting a boat where there is an initial fee and then a cost per hour would have a constant rate but not pass through the origin.
9. Sample answer: The graph shows the number of feet per yard. The constant of proportionality is 3, so, for every yard, there are 3 feet.

Lesson 1-5 Equations of Proportional Relationships, Practice Pages 47–48

1. 4.75; Liv earns $4.75 for each bracelet she sells. **3.** $y = 1.5x$ **5.** $y = 1\frac{3}{4}x$; $10\frac{1}{2}$ c
7. The equation $y = 7.75x$ models the cost for tickets at Star Cinema. The total cost of 9 tickets at Star Cinema would cost $69.75.
9. 91 free throws **11.** $12.75; Sample answer: Write an equation that represents the proportional relationship and solve for x. $38.25 = 3x$; $x = 12.75$ **13.** $1\frac{1}{8}$ gal; The constant of proportionality is $\frac{1}{3}$. Sample answer: The ratio of salt to water is $\frac{1}{3}$: 1. Because there are 6 cups of salt needed, then there are 18 cups of water needed. There are 16 cups in 1 gallon. $18 ÷ 16 = 1.125$ or $1\frac{1}{8}$.

Lesson 1-6 Solve Problems Involving Proportional Relationships, Practice Pages 55–56

1. 177 girls **3.** 75 pieces **5.** 19.2 min
7. 315 adults **9.** 0.52 gal; Sample answer: The area of the fence is 26(7) or 182 square feet. Using the equation $y = 350x$ to represent the proportional relationship, she will need $182 ÷ 350$ or 0.52 gal. **11.** Sample answer: Enrique goes through a 12-ounce bottle of shampoo in 16 weeks. How long would you expect an 18-ounce bottle of the same brand to last him? 24 weeks

Module 1 Review Pages 59–60

1. D **3.** 8 cups flour, 2 cups salt, 4 cups water; 6 cups flour, $1\frac{1}{2}$ cups salt, 3 cups water
5. no; The ratios $\frac{2}{1}$, $\frac{5}{4}$, $\frac{7}{6}$, and $\frac{10}{8}$ are not equivalent ratios. **7.** The relationship is proportional. The point (9, 18) satisfies this relationship. The constant of proportionality is 2. **9A.** $y = 22.5x$ **9B.** $292.50

Lesson 2-1 Percent of Change, Practice Pages 71–72

1. 25%; increase **3.** 37.5%; decrease
5. 48.5% **7.** 10.7% **9.** 56.6% **11.** area; 600%
13. Sample answer: Last month the amount of snowfall was 0.5 inch. This month it was 0.75 inch. What is the percent of increase in snowfall?; 50% **15.** true; Sample answer: A decrease indicates a lesser amount. So, the original amount must be more than the decreased amount.

Lesson 2-2 Tax, Practice Pages 81–82

1. $19.26   **3.** $52.45   **5.** $79.82
7. $359.13   **9.** D   **11.** $34.54   **13.** Sample answer: $1.06 \cdot a$; $a + 0.06a$; Both expressions multiply the cost by 100% and by the sales tax rate to give the total cost.   **15.** greater than; Sample answer: 10% of $160 is $16 so 5% of $160 is $8. The tax rate is greater than 5% so the tax will be more than $8.

Lesson 2-3 Tips and Markups, Practice Pages 89–90

1. $22.00   **3.** $194.25   **5.** $47.40   **7.** $30
9. 55%   **11.** $342.37   **13.** yes; Sample answer: The total would be $77 which is less than $80.
15. Sample answer: $1.18x$; $x + 0.18x$; Both expressions multiply the cost by 100% and by the tip to give the total cost.

Lesson 2-4 Discounts, Practice Pages 97–98

1. $126.00   **3.** $276.25   **5.** $24.70
7. $375.15   **9.** $47.89   **11.** Skye Bouncer; $3.76   **13.** $11\frac{1}{9}\%$; Sample answer: Suppose the cost of the item is $50. So, 10% of $50 is $5 and $50 − $5 = $45. Then set up a proportion to find what percent of $45 is $5. Solve for r: $\frac{5}{45} = \frac{r}{100}$. $r = 11\frac{1}{9}\%$ **15.** $34.30; about 14%

Lesson 2-5 Interest, Practice Pages 105–106

1. $31.80   **3.** $157.50   **5.** $112.50   **7.** $288.75
9. $157.50   **11.** $247.50   **13.** 3.75%; Use the simple interest formula and solve for r.
$328.13 = 2,500 \cdot 3.5 \cdot r$; $r = 3.75\%$ **15.** Sample answer: If the rate is doubled, then the interest is doubled to $80. If the time is doubled, then the interest is also doubled to $80.

Lesson 2-6 Commission and Fees, Practice Pages 113–114

1. $391   **3.** $27,273   **5.** $20,000   **7.** $12
9. Job 2; $108.75   **11.** Sample answer: The student wrote the percent as 65 out of 100. It should be 6.5 out of 100. The commission should be $34.13.   **13.** true; Sample answer: You could earn 0.5% of a sale of $5,000 worth of jewelry. 0.5% of $5,000 is $25.

Lesson 2-7 Percent Error, Practice Pages 119–120

1. 30%   **3.** 36.9%   **5.** 1.01%   **7.** 34.1%   **9.** 4%
11. The student forgot to write the amount of error as a percent. $0.12 = 12\%$ **13.** The denominator would be 0 in the calculation. A denominator of 0 is undefined so the percent of error would be undefined.

Module 2 Review Pages 123–124

1. 48%   **3.** B
5.

	yes	no
subtotal: $129.50 tip: $23.27		X
subtotal: $98.40 tip: $17.71	X	
subtotal: $142.17 tip: $22.75		X

7. 47.19   **9.** There are actually 118 elements. The amount of error in Clark's estimate is 10. To the nearest tenth percent, the percent error in Clark's estimate is 8.5%.   **11.** $64,815; I solved the proportion $\frac{3,500}{x} = \frac{5.4}{100}$.

Lesson 3-1 Add Integers, Practice Pages 137–138

1. −11 **3.** −26 **5.** −17 **7.** −40 **9.** −20
11. −1 **13.** 76 m higher **15.** Tasha; $1
17a. −1 **17b.** 3 **17c.** 0 **19.** Sample answer:
2 and −2 are additive inverses and the sum
of any number and its additive inverse is zero.
The integer 3 is positive, so the sum will be
positive.

Lesson 3-2 Subtract Integers, Practice Pages 147–148

1. 11 **3.** 76 **5.** −10 **7.** −6 **9.** 22 **11.** −37
13. 19 units **15.** 440 feet **17.** Utah; Nevada
19. The student incorrectly wrote 4 − 2 instead
of 4 + 2. The correct solution is 6.
21. sometimes; Sample answer: For example,
−10 − (−40) = 30 and −28 − (−13) = −15.

Lesson 3-3 Multiply Integers, Practice Pages 157–158

1. −28 **3.** −108 **5.** 100 **7.** −140 **9.** −192
11. 70 **13.** 756 **15.** −63 points **17.** yes; At
the end of January, he had $1,300 −$1,250 or
$50 left in his account. $50 · 12 = $600 and
$600 > $500. **19a.** Multiplicative Identity
Property **19b.** Commutative Property of
Multiplication **21.** −8 and 8

Lesson 3-4 Divide Integers, Practice Pages 165–166

1. −11 **3.** −25 **5.** 3 **7.** −16 **9.** −4
11. 3 **13.** −60 **15.** 16 weeks **17.** no; Sample
answer: The Associative Property is not true for
the division of integers because the way the
integers are grouped affects the solution.
[12 ÷ (−6)] ÷ 2 = −1; 12 ÷ [(−6 ÷ 2)] = −4
19. Sample answer: Lucy borrowed $50 from her
mother over 5 days in equal amounts. What was
the change in the amount she borrowed from
her mom each day? −$10

Lesson 3-5 Apply Integer Operations, Practice Pages 169–170

1. −39 **3.** 22 **5.** −105 **7.** −6 **9.** −7 **11.** 10
13. −$200 **15.** Chicago; Oklahoma City; The
difference between the extremes for Chicago is
40°C − (−33°C) or 77°C. The difference between the extremes for Nashville is
42°C − (−27°C) or 69°C. The difference
between the extremes for Oklahoma City is
43°C − (−22°C) or 65°C. 77 > 69 > 65
17. The student subtracted 9 instead of adding
its additive inverse. The correct answer should
be −63. **19.** no; According to the order of
operations, multiplication should be performed
from left to right before subtraction.

Lesson 3-6 Rational Numbers, Practice Pages 179–180

1. 0.625; terminating **3.** $0.\overline{2}$; non-terminating
5. −0.8; terminating **7.** $\frac{8}{9}$ **9.** $-1\frac{5}{9}$
11A. 0.3; terminating **11B.** $0.0\overline{3}$; non-
terminating **13.** Cho, Kevin, Sydney
15. no; Sample answer: $0.\overline{5} = \frac{5}{9}$ **17.** $0.\overline{2}$, $0.\overline{50}$,
and $0.\overline{98}$; Sample answer: When the
denominator of the fraction is 9 or 99, the
numerator of the fraction is the repeating part
of the decimal.

Lesson 3-7 Add and Subtract Rational Numbers, Practice Pages 193–194

1. $\frac{1}{2}$ **3.** $-\frac{9}{10}$ **5.** $37.20; Sample answer:
Quinn gave $37.20 to his brother for his
mother's birthday gift. **7.** $1\frac{17}{24}$ or $1.708\overline{3}$
9. $4\frac{13}{30}$ or $4.4\overline{3}$ **11.** 1.45 or $1\frac{9}{20}$ **13.** $11\frac{1}{6}$ or
$11.1\overline{6}$ **15.** 20.8 in. or $20\frac{4}{5}$ in. **17.** 55 pounds
19. Sample answer: $3\frac{1}{2} + 1\frac{1}{16}$; $4\frac{9}{16}$ **21.** Sample
answer: The student found a common
denominator but not the least common
denominator. The least common denominator
of 9, 3, and 6 is 18.

Lesson 3-8 Multiply and Divide Rational Numbers, Practice Pages 207–208

1. $\frac{2}{5}$ **3.** $-7\frac{9}{16}$ **5.** -4 or $-\frac{4}{1}$ **7.** $-2\frac{11}{17}$
9. -6 or $-\frac{6}{1}$ **11.** -1 or $-\frac{1}{1}$ **13.** -8 or $-\frac{8}{1}$
15. -3.99 or $-3\frac{99}{100}$ **17.** $135.30   **19.** false;
Sample answer: $\frac{1}{2}\left(2\frac{1}{4}\right) = 1\frac{1}{8}$ and $1\frac{1}{8} < 2\frac{1}{4}$
21. $20 \div \frac{3}{4}$; Multiplying 20 by a number less than 1 will result in a number less than 20. Dividing 20 by a number less than 1 will result in a number greater than 20.

Lesson 3-9 Apply Rational Numbers Operations, Practice Pages 213–214

1. -2.55 or $-2\frac{11}{20}$ **3.** -1.825 or $-1\frac{33}{40}$
5. 0.5 or $\frac{1}{2}$ **7.** -1.85 or $-1\frac{17}{20}$ **9.** 0.26 or $\frac{13}{50}$
11. 0.28 or $\frac{7}{25}$ **13.** 33.6 **15.** $\frac{3}{4}$ cup oats;
$1\frac{7}{8}$ cups of blueberries
17. $\left(\frac{1}{2} \times 120 \times 0.25\right) + \left(\frac{1}{5} \times 120 \times 0.75\right) +$
$\left(\frac{3}{10} \times 120 \times 1.50\right)$; $87   **19.** Sample answer: A homeowner is enclosing his rectangular property with fencing. The rectangular property is $25\frac{1}{3}$ yards wide and $30\frac{2}{3}$ yards long. Fencing is sold in 8-foot sections and costs $45.50 per section. How much will it cost to fence in the rectangular property? $1,911

Module 3 Review Pages 217–218

1A. Commutative Property; Sample answer: One of the methods that can be used to add three or more integers is to group like signs together, then add.   **1B.** 35   **3.** -72
5. Freezing point is 0°C. Boiling point is 100°C.
7. 102.875 or $107\frac{7}{8}$ pounds   **9.** A   **11.** $47.25; Sample answer: I divided 31.5 by the thickness of the DVD case, 0.6 to get 52.5. So, each shelf can hold at most 52 DVDs. Then I divided 132 by 52 to get approximately 2.5. Therefore, I know that I need 3 shelves. Then I multiplied 3 by the cost per shelf, $15.75.

Lesson 4-1 Powers and Exponents, Practice Pages 229–230

1. $(-7)^2 \cdot 5^4$ **3.** 65   **5.** $7\frac{7}{16}$ **7.** $>$ **9.** $\left(-\frac{4}{5}\right)^3$;
$3^5 - 10^4$; $(9.8)^2 - 10^2$ **11.** 8,200 lakes
13. Sample answer: He substituted -4 for x, but should have taken -4 to the second power, not just 4. Placing parentheses around -4 would have helped him to take the correct value to the second power. The correct value of the expression is 729.   **15.** It is incorrect. Because $4^2 = 2^4$, the first set of expressions are equal. This is not the case for $\left(\frac{1}{3}\right)^2$ and $\left(\frac{1}{2}\right)^3$ because $\left(\frac{1}{3}\right)^2 = \frac{1}{9}$ and $\left(\frac{1}{2}\right)^3 = \frac{1}{8}$.

Lesson 4-2 Multiply and Divide Monomials, Practice Pages 241–242

1. 3^9 or 19,683   **3.** $24m^4n^5$ **5.** b^7 **7.** 10^3 or 1,000   **9.** a^2c^5 **11.** Dish B   **13.** 4^{16}; Sample answer: *four times* 4^{15} translates to $4 \cdot 4^{15}$ which simplifies to 4^{16}.   **15.** $n = 5$

Lesson 4-3 Powers of Monomials, Practice Pages 249–250

1. 7^6 or 117,649   **3.** d^{42} **5.** $64m^{30}$
7. $-243w^{15}z^{40}$ **9.** $(6^2)^3$; $(6^2)^3 = 46,656$ and $46,656 > 1,000$ **11.** $49x^6y^{10}$ tiles
13. Sample answer: Using the Power of a Power law of exponents, both expressions simplify to 4^8. Multiplication is commutative, so $2 \cdot 4$ is the same as $4 \cdot 2$.   **15.** $[(-4)^4]^5$; Sample answer: Using the Power of a Power property, $[(-4)^4]^5$ simplifies to $(-4)^{20}$, a positive number. The expression $-[(4^{12})^3]$ simplifies to $-(4^{36})$, a negative number.

Lesson 4-4 Zero and Negative Exponents, Practice Pages 259–260

1. 1 **3.** $\frac{1}{8^4}$ **5.** d^{-6} **7.** $\frac{1}{81}$ **9.** x^4 **11.** 1
13. 10^{12} times larger **15.** 5^{-7}, 5^0, 5^4; Sample answer: Written with a positive exponent, 5^{-7} is $\frac{1}{5^7}$ and is less than 1, 5^0 equals 1, and 5^4 is greater than 1. **17.** They are equivalent; Sample answer: Both expressions simplify to 1.

Lesson 4-5 Scientific Notation, Practice Pages 271–272

1. 1,600 **3.** 0.0000083 **5.** 2.204×10^9
7. 5 kilometers; Sample answer: The number 5×10^6 millimeters is unnecessarily large and would be difficult to visualize. Choosing the larger unit of measure is more appropriate.
9. about 2×10^{-3} inch **11.** 1,089,822 insects **13.** 1.5×10^5 **15.** Sample answer: Katrina did not take into account place value and the powers of ten. The number 3.5×10^4, in standard form, is 35,000. The number 2.1×10^6 is 2,100,000. Therefore, 2.1×10^6 is greater.

Lesson 4-6 Compute with Scientific Notation, Practice Pages 279–280

1. 3×10^{22} **3.** 1.2913×10^5 **5.** about 2.72×10^8 seconds **7.** 9.6×10^{-4} gram
9. Sample answer: Each number has a factor that is a power of 10. Since the bases are the same, these properties can be applied to multiply or divide the powers of 10.
11. Sample answer: He found $5.78 \div 2$ as 2.89, but incorrectly found the power of ten. The numerator is 10^5 and the denominator is 10^{-6}. When dividing, the resulting power of ten is 10^{11}. The correct answer should be 2.89×10^{11}.

Module 4 Review Pages 283–284

1. B **3.** Harrisburg, Liberty Crossing, Glenview
5.

	true	false
$(d^4)^3 = d^{12}$	X	
$(n^5)^{10} = n^{15}$		X
$(k^7)^3 = k^{21}$	X	

7. 6^{-4} **9.** A **11.** 0.0072 **13.** 7.75×10^7

Lesson 5-1 Roots, Practice Pages 297–298

1. 19 **3.** $-\frac{3}{4}$ **5.** ± 0.2 **7.** -8 **9.** 14 plants
11. 24 feet **13.** Sample answer: The rational number, 2, when cubed, results in 8. However, there is not a rational number that, when multiplied by itself, results in 8. **15.** 9,261

Lesson 5-2 Real Numbers, Practice Pages 307–308

1. irrational **3.** rational **5.** rational
7. irrational **9.** A, C, D, E **11.** A, C **13.** true; Sample answer: All integers can be expressed as a ratio $\frac{a}{b}$, where a and b are integers and $b \neq 0$, which is the definition of a rational number. Therefore, all integers are rational numbers. **15.** Sample answer: I would use a calculator to find $\sqrt{8}$. The calculator shows 2.82842712474619. Then I would multiply that answer by itself, without using the x^2 button. If the solution is 8, then it is a terminating decimal. If the solution is not 8, then I know that the original solution was rounded by the calculator, and it is not a terminating decimal. **17.** false; The definition of a rational number is a number expressed as a ratio $\frac{a}{b}$, where a and b are integers and $b \neq 0$. In the ratio $\frac{\sqrt{2}}{1}$, $\sqrt{2}$ is not an integer, therefore, it does not satisfy the definition of a rational number.

Lesson 5-3 Estimate Irrational Numbers, Practice Pages 317–318

1. 11 **3.** 4 **5.** 17.2 **7.** 3.3 **9.** 21.2 miles per hour **11.** 22.1 **13.** about 12 feet **15.** Sample answer: To write the exact value for the square root of a non-perfect square, such as $\sqrt{13}$, I would leave it written as $\sqrt{13}$. The decimal form would have to be rounded, no matter how many decimal places I wrote. **17.** Sample answer: In the same way I can estimate square roots and cube roots, I could find the nearest fourth root of 20. Since 16 < 20 < 81, the fourth root is between 2 and 3. Since 20 is closer to 16, the fourth root of 20 is about 2.

Lesson 5-4 Compare and Order Real Numbers, Practice Pages 329–330

1. <

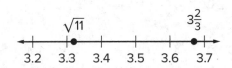

3. <

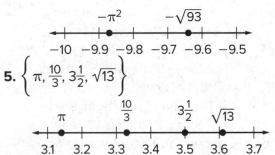

5. $\left\{ \pi, \dfrac{10}{3}, 3\dfrac{1}{2}, \sqrt{13} \right\}$

7. greater than **9.** 0.3 second **11.** π; Sample answer: 3.14 can be extended to 3.14000... . The number π written as a decimal is 3.141... and 3.141 > 3.140. **13.** Sample answer: 1.7 and $\sqrt{3}$; 1.7 < $\sqrt{3}$

Module 5 Review Pages 333–334

1. −1.3; 1.3 **3.** $-\dfrac{9}{11}$; −0.8181... **5A.** false
5B. Sample answer: −1

7.

	7	8	9
$\sqrt{70}$		X	
$\sqrt{79}$			X
$\sqrt{88}$			X
$\sqrt{52}$	X		
$\sqrt{60}$		X	
$\sqrt{47}$	X		
$\sqrt{65}$		X	

9. 6.92 **11A.** >

11B.

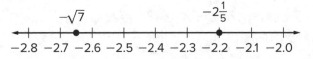

Lesson 6-1 Simplify Algebraic Expressions, Practice Pages 345–346

1. 1.25c **3.** −17y − 16z + 4 **5.** −7c − 2d − 6
7. $\dfrac{3}{10}y + 1\dfrac{1}{8}$ **9.** −6x + 10 **11.** 15y − 10z
13. −7x + 14

15.

Triangle Side	Length (units)
1	2(a + 3)
2	3a − 1
3	a − 2

17. The perimeter of Quadrilateral 1 is 13x + 6. The perimeter of Quadrilateral 2 is 14x − 5. Quadrilateral 1 has the greater perimeter if x = 3. Yes, if x = 4, the perimeter of Quadrilateral 1 is 58 units and the perimeter of Quadrilateral 2 is 51 units. **19.** Sample answer: The student multiplied 4 by 3 instead of −3. 5x − 3(x + 4) = 2x − 12

Lesson 6-2 Add Linear Expressions, Practice Pages 353–354

1. $2x + 7$ **3.** $-3x - 4$ **5.** $-14x + 1$
7. $\frac{7}{16}x + 2$ **9.** $x + \frac{1}{3}$ **11.** $-\frac{5}{12}x - 8$
13. $10x + 4$ units or $2(5x + 2)$ units   **15.** $258
17. $5x - 3$ **19.** The sum will be zero when the coefficients of the x–terms are opposites.

Lesson 6-3 Subtract Linear Expressions, Practice Pages 361–362

1. $-3x + 6$ **3.** $4x + 8$ **5.** $2x + 1$ **7.** $-8x - 9$
9. $\frac{4}{15}x + \frac{7}{8}$ **11.** $-\frac{1}{2}x - \frac{1}{3}$ **13.** $(3x + 4)$ points
15. $374   **17.** $6x + 3$ **19.** Sample answer: The student only added the additive inverse of $-x$ and not 5. $(6x - 2) - (-x + 5) = 7x - 7$

Lesson 6-4 Factor Linear Expressions, Practice Pages 369–370

1. $4y$ **3.** $8m$ **5.** $14ab$ **7.** $5(x + 7)$
9. cannot be factored   **11.** $18x(4 - y)$
13. cannot be factored   **15.** $\frac{1}{2}(x + 1)$
17. $9   **19.** Sample answer: $3x$, $9x$
21. c. $7x + 3$; All the other expressions can be factored but $7x + 3$ cannot be factored.

Lesson 6-5 Combine Operations with Linear Expressions, Practice Pages 375–376

1. $4(2x + 3)$ **3.** $15(x + 2)$ **5.** $\frac{13}{24}x - 3$
7. $\frac{3}{4}x + 5$ **9.** $2(7x + 11)$ **11.** $12(x - 1)$
13. $2(9x + 4)$ **15.** $5(3x + 13)$ **17.** Sample answer: The student forgot to distribute -3 to both terms inside the parentheses. The student only distributed it to the first term. The correct answer is $3(x - 2)$.   **19a.** Sample answer: $\frac{1}{2}(x + 12)$ **19b.** Sample answer: $\frac{3}{4}(x - 24)$

Module 6 Review Pages 379–380

1. $1.15c$ **3A.** Triangle 1: $-6x + 105$; Triangle 2: $20x - 30$ **3B.** Triangle 1   **5.** C   **7.** $-\frac{1}{2}x - 2$; Sample answer: I used the Distributive Property for the subtraction sign through the second parentheses. Then I used the Commutative Property to group like terms. Then I combined like terms.   **9.** C   **11.** A   **13.** $2x - 3$

Lesson 7-1 Write and Solve Two-Step Equations: $px + q = r$, Practice Pages 393–394

1. $x = 3$ **3.** $x = 3$ **5.** $x = -36$ **7.** $29.50p + 15 = 133$; 4 people   **9.** $100\frac{1}{5} - 10\frac{4}{5}m = 57$; 4 min   **11.** 5 fruit baskets   **13.** $237.50
15. 16 years old   **17.** The student added -5 to both sides instead of adding $+5$ to both sides. The solution is $x = 10$.

Lesson 7-2 Write and Solve Two-Step Equations: $p(x + q) = r$, Practice Pages 405–406

1. $x = -13$ **3.** $x = -64$ **5.** $x = 14$
7. $6\left(1\frac{1}{4} + b\right) = 10$; $\frac{5}{12}$ yd   **9.** $3(b - 8) = 21$; 15 balloons   **11.** $5(p - 1.50) = 48.75$; $11.25
13. 15.88 m   **15.** Let Keith's age $= x$. Trina's age is $x - 5$. The sum of their ages is $(x) + (x - 5)$. $2(2x - 5) = 62$; Keith is 18 years old.
17. Sample answer: You and 11 friends go bowling. Shoe rental costs $2.50. The total cost of one game and one shoe rental for everyone is $78. What is the cost of one game of bowling for one person?; $4

Lesson 7-3 Write and Solve Equations with Variables on Each Side, Practice Pages 417–418

1. $x = -3$ **3.** $x = 14$ **5.** Let $m =$ the number of months; $45 + 4m = 61 + 2m$; 8 months
7. $15 - 0.75g = 13 - 0.50g$; 8 games   **9.** $780
11. Let $m =$ the number of hours worked on Monday; $m + (m + 3) + (2m + 1) = 5m + 2$; 2 hours

Lesson 7-4 Write and Solve Multi-Step Equations, Practice Pages 429–430

1. $x = -2$ **3.** $x = 4$ **5.** Let $\ell =$ the length; $86\frac{1}{2} = 2\ell + 2\left(2\ell - 40\frac{3}{4}\right)$; 28 feet
7. −5 **9.** $3x + 4x + 5x = 4(x + 2)$; $x = 1$; 3 units
11. $(2x + 3) + 2(2x + 3) + 2[(2x + 3) + 2(2x + 3)] = 477$; $53

Lesson 7-5 Determine the Number of Solutions, Practice Pages 439–440

1. no solution **3.** infinitely many solutions
5. $-5x - 70$ **7.** Sample constant: $-12x + 6$
9. D **11.** Fatima **13.** Sample answer: The solution $x = 0$ means that 0 is the solution to the equation and the equation has one solution, 0. **15.** Sample answer: one solution: $-4x + 7 = -5x + 9$; no solution: $-4x + 7 = -4x + 8$; infinite solutions: $-4x + 7 = -4x + 7$

Lesson 7-6 Write and Solve One-Step Addition and Subtraction Inequalities, Practice Pages 451–452

1. $x \geq 6$

-3 -2 -1 0 1 2 3 4 5 6 7 8 9 10

3. $x \geq 2\frac{2}{3}$

-3 -2 -1 0 1 2 3 4 5 6 7

5. $x \leq -5$

-8 -7 -6 -5 -4 -3 -2 -1 0 1 2

7. $-50 + x > -35$; $x > 15$; The dolphin ascended more than 15 feet.
9. $x - (-2) < 27$; $x < 25$; The monthly high temperature was less than 25° Fahrenheit.
11. Sample answer: one popcorn ball and one candy stick **13.** Sample answer: A school bus can hold at most 40 students and there are currently 30 students on the bus. How many more students can board the bus? **15.** Sample answer: Petra must write a report with more than 1,000 words for her history class. So far, she has written 684 words. Write and solve an inequality to find how many more words Petra needs to write for her report. $684 + x > 1,000$; $x > 316$ words.

Lesson 7-7 Write and Solve One-Step Multiplication and Division Inequalities, Practice Pages 465–466

1. $x \leq -2$

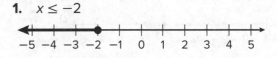

-5 -4 -3 -2 -1 0 1 2 3 4 5

3. $x > -8$

-10 -9 -8 -7 -6 -5 -4 -3 -2 -1 0

5. $x \geq 3$

-5 -4 -3 -2 -1 0 1 2 3 4 5

7. $6x \geq 168$; $x \geq 28$; Hermes needs to babysit 28 or more hours. **9.** $\frac{x}{4.25} \leq 12$; $x \leq 51$; Chase should buy at most 51 inches of fabric.
11. at most $367.50 **13.** Sample answer: Kail earns $4 for every dog he walks. He needs to earn at least $100 for new ski boots. Write and solve an inequality to determine the number of dogs he must walk to earn enough for the ski boots. **15.** $\frac{x}{-3} \geq -2$; This inequality's solution is $x \leq 6$. The other three inequalities' solutions are $x \geq 6$.

Lesson 7-8 Write and Solve
Two-Step Inequalities, Practice
Pages 477–478

1. $x < -5$

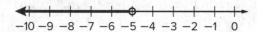

3. $x \le 3$

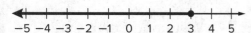

5. $x < 1\frac{3}{4}$

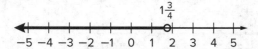

7. $15 + 4x \le 27$, $x \le 3$; Margie can rent the bicycle for up to 3 hours.   **9.** $20 - 3.5x \ge 7.75$, $x \le 3.5$; Douglas can race the go–karts no more than 3 times.   **11.** a minimum score of 16 points
13. $x > -4$

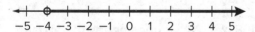

15. Sample answer: Add $2x$ to each side of the inequality before solving.

Module 7 Review Pages 481–482

1. $t = -3$ **3.** -7 **5.** $5m + 20 = 7m$; $m = 10$
7. A   **9.** Let $x =$ cost of T-shirt; $59 + 2x \le 75$; $x \le 8$; Sample answer: Based upon the solution, each T-shirt can cost $8 or less.
11. < 3 **13.** $z \le -1.25$; Sample answer: The direction of the inequality symbol is not reversed because I did not have to multiply or divide each side of the inequality by a negative number in order to solve for z.